THE ASIATIC MODE OF
PRODUCTION

Ruth Hamilton

THE ASIATIC MODE OF PRODUCTION
Science and Politics

Edited by
Anne M. Bailey

and
Josep R. Llobera

ROUTLEDGE & KEGAN PAUL
London, Boston and Henley

*First published in 1981
by Routledge & Kegan Paul Ltd
39 Store Street,
London WC 1E 7DD,
9 Park Street,
Boston, Mass. 02108, USA and
Broadway House,
Newtown Road,
Henley-on-Thames,
Oxon RG9 1EN
Printed in Great Britain by
Billing & Sons Limited
Guildford, London, Oxford and Worcester
Selection, translations and editorial matter
© Anne M. Bailey and J.R. Llobera 1981*

British Library Cataloguing in Publication Data

*The Asiatic mode of production.
1. Asiatic mode of production - Addresses,
essays, lectures
I. Bailey, Anne M.   II. Llobera, Josep
338'. 001   HB 97. 5*

ISBN 0-7100-0738-8

CONTENTS

Contents   vii

PREFACE

Over the past few years there has been a growing interest in Marxist approaches to pre-capitalist societies. One of the central concepts in the contemporary discussions is that of the Asiatic mode of production (AMP). The English reading public has had limited access to both these recent debates (carried out in France, Italy, Eastern Europe, etc.) and to the earlier controversies surrounding the AMP (at the turn of the century and in the 1920s and early 1930s). The present work aims to document this highly significant and controversial chapter in the history of the social sciences and Marxism. The articles and excerpts published here represent contributions by scholars from a variety of disciplines. Many are published here in English for the first time. In order to keep the book to a reasonable length, we have been forced to select only a few representative texts out of what is now a vast literature.

In the General Introduction we outline the basic epistemological and historiographic assumptions that have informed the structuring of this reader. Part I provides a résumé of the recent literature on the historical, political, and epistemological parameters of the genealogy of the concept of the AMP. Limitations of space prevented the inclusion of selections from early authors who developed the ideas of oriental society and despotism. On the other hand, the writings of Marx and Engels on the AMP are generally available and we thought it unnecessary to reprint them here.

Part II traces the fate of Marx's ideas on the AMP from the time of his death until Stalin's consolidation of power. It includes texts from Plekhanov and Lenin, and excerpts from L. Mad'iar and the Leningrad discussions of 1931. Although we have concentrated on Russian Marxism in our selection, the brief introduction to Part II provides a guide-line to the views on the AMP held by other Marxist thinkers of this period.

Part III focuses on the political and theoretical impact of Karl Wittfogel's work on Marxists and non-Marxists from various disciplines. To begin this section, we have reproduced Wittfogel's earlier and less polemical formulation of the concepts of oriental society and despotism. We also present a cross-section of the reviews of his 'Oriental Despotism' (1957). The repercussions of Wittfogel's theory among anthropologists and archeologists is highlighted in the articles by J. Steward, E. Leach, and B. Price.

Finally, Part IV brings together a number of contributions to the renewed Marxist discussion of the AMP which began in the late

1950s. The articles by F. Tökei, M. Godelier, and I. Banu address
the subject at a more theoretical level, while those by E. Welskopf,
Le Thanh Khoi, J. Golte, and H. Islamoğlu and C. Keyder illustrate the range of societies to which the concept of the AMP has
been applied. The excerpt from L. Krader's 'The Asiatic Mode of
Production' (1975) serves as a conclusion - or rather as a starting
point - for further research. Parts II, III, and IV are preceded
by brief introductions intended to situate the selections within the
intellectual and political contexts within which they first appeared.
An extensive bibliography, containing both the works referred to
in the general and short introductions to each part, and additional
sources on the AMP, completes the present volume.

While we take full responsibility for the final content of this
book, we would like to thank a number of persons who have helped
in its preparation. The pioneering works of M. Godelier and L.
Krader have had a great influence on our general approach to the
AMP as is apparent in Part I of this book. Both have encouraged
our endeavour and offered constructive suggestions. We have
been fortunate to draw upon P. Skalnik's extensive knowledge of
the recent Eastern European literature on the AMP and G.L.
Ulmen's expertise on Wittfogel's work. We have also benefited
from informal discussions with T. Asad, J. Friedman, P. Glavanis,
J. Kahn, and C. Levitt.

Robert Croskey selected and translated the texts by Mad'iar
and from the Leningrad discussions of 1931. The translations from
the German of the excerpts by Plekhanov, Lewin, and Welskopf
were made by J. Gordon-Kerr, and Levada's review has been translated by Peter Skalnik from the Russian. The editors have translated
the articles by Tökei, Banu and Le Thanh Khoi from the French
and that of Golte from the Spanish.

We should like to thank Adam Kuper, who expressed an interest
in this project from its inception and has fostered its publication.
Our thanks also to Eileen Lee and Teresa Weatherston who helped
in typing parts of the manuscript. This book was researched,
compiled, and written while we were associated with the University
of Hull and we would like to express our gratitude to that institution for the research facilities extended to us.

Anne M. Bailey
Josep R. Llobera

ACKNOWLEDGMENTS

The editors and the publishers are grateful for permission to reproduce the following copyright material: Lawrence & Wishart Ltd
for Plekhanov, 'Selected Philosophical Works', 1961, vol. 1, pp.
271-5; Professor Karl A. Wittfogel for 'The Foundations and Stages
of Chinese Economic History'; 'Readings in Anthropology', vol. II
(Cultural Anthropology), 2nd edition by Morton H. Fried. Copyright © 1968 by Harper & Row, Publishers, Inc.; the American
Political Science Association for Toynbee's review and Wittfogel's
reply; the American Anthropological Association for Murdock's
review; the author and the American Sociological Association for
Eberhard's review; the University of Illinois Press for Steward,
'Wittfogel's Irrigation Hypothesis'; the author and the Past and
Present Society, Corpus Christi College, Oxford, for 'Hydraulic
Society in Ceylon'; Barbara Price and 'Latin American Research
Review', vol. 6, no. 3, copyright 1971, for 'Prehispanic Irrigation
Agriculture in Nuclear America'; Akademie-Verlag, Berlin, for
Welskopf, 'Die Produktionsverhältnisse im Alten Orient und in der
griechisch-römischen Antike'; the author and Akadémiai Kiadó for
'Sur le mode de production asiatique'; Frank Cass & Co. Ltd for
Godelier in D. Seddon (ed.), 'Relations of Production', 1978;
Editions Sociales, Paris, for Le Thanh Khoi, 'Contribution à
l'étude . . .', 'La Pensée', no. 171, 1973; Jürgen Golte for 'The
Economy of the Inca State . . .'; The Fernand Braudel Center,
State University of New York at Binghamton, for Islamoğlu and
Keyder, 'Agenda for Ottoman History'; Van Gorcum & Comp.
B.V. for Krader, 'The Asiatic Mode of Production', 1975.

GENERAL INTRODUCTION
Anne M. Bailey and
Josep R. Llobera

PRELUDE

The Asiatic mode of production (hereafter AMP) is one of the most
controversial concepts in the history of Marxism. The term was
originally coined by Marx to account for a type of society outside
the mainline of Western development (the term 'Asiatic' was not
restricted to the geographical area of Asia). The theoretical
status of the concept of the AMP has never been too secure for
three main reasons. First, the formulation of the concept was
precarious in the work of Marx. Second, within the Marxist
scheme of evolution, the AMP was an anomaly and, as such, has
been and still is considered anathema for those nostalgic for
orthodoxy and eager to embrace a unilineal and mechanical conception of history. Third, the tremendous political potential of
the concept has triggered off all sorts of ideological manipula-
tions destined to suit short- or long-term national and /or party
interests; this is especially clear in the characterization of certain
societies as 'Asiatic' in different historical moments. From an
'orthodox' Marxist perspective, a society defined as 'Asiatic'
(or 'feudal' for that matter) can not be transformed into a
'socialist' one before going through the purgatory of a 'capitalist'
period.
   As we have stated in the Preface, we view the present volume
as a contribution to the history of the social sciences and Marxism. Such history was long seen as unproblematic; it was assumed
that no special skills were required to approach the subject: that
it was sufficient to be a social scientist or a Marxist to have the
necessary insights. We believe that this picture corresponds by
and large to a situation that lasted until the mid-1960s. This is
not to suggest that prior to this time there had not been scholarly work on the history of the social sciences and Marxism; the
'Journal of the History of Ideas' is proof to the contrary. However, the impact of such work on mainstream social sciences and
Marxism was negligible. After 1965 a noticeable change occurred
as a result of the following interrelated factors:
1   The prodigious development of the history of the sciences,
which had begun after the Second World War, began to have its
effects on the social sciences. The Kuhnian revolution, in particular, proved to be very appealing for social scientists, though
often for the wrong reasons. By challenging the objectivity of
the natural sciences, it added respectability to the social sciences.
2   The crisis that the social sciences underwent in the late 1960s

and early 1970s subverted the most cherished beliefs of its practitioners. At the theoretical level, it challenged the foundations
of the dominant paradigm: structural-functionalism; at the epistemological level, it shook the foundations of the most popular
philosophy of science: positivism; and at the political level, it
destroyed many illusions of value-neutrality, and pressed for
relevance.
3   Marxism irrupted on to the horizon of the social sciences and
became an actual and promising alternative, partly as a result of
the factors mentioned above, partly because of its own dynamics
(the end of Stalinism). However, it would be wrong to assume
that the term 'Marxism' corresponds to a clearly defined and unequivocally agreed-upon meaning. It is in the nature of the
recent interest in Marx that it rethinks his contribution to the
social sciences from different standpoints and consequently
arrives at different conclusions. The Althusserian approach is
one of many, though in terms of research elicited, it was and
probably still is one of the most influential.

Althusser's suggestion that an epistemological break took place
in the work of Marx, which made the constitution of a science of
history possible, was undoubtedly a momentous idea. It may well
be that it created more problems than it solved, but it had a
tremendous impact on a significant number of Marxist intellectuals
who were impressed by the forceful arguments, the impeccable
logic, and the concrete intimations of a promised land of scientific
bliss. In this sense, it triggered off a fair amount of research,
both in France and elsewhere. In retrospect, it is easy to say
that the emperor was naked, but the truth is that Althusser's
work appeared to many of us as one of the few serious attempts
to show that Marx's contribution to the foundation of a science
of history was paramount. Althusser's achievement was, on the
whole, the result of rethinking Marx in the light of contemporary
(French) epistemology and structuralism.

It is now fashionable to treat Althusser as a 'dead dog', but
those who proceed in such a way are often the same persons who
some years ago blindly accepted his ideas. In this introduction
there are a number of implicit and explicit criticisms of Althusser;
however, we would like to point out that many of our ideas (particularly in the area of epistemology and history of the sciences)
have been forged in an intensive, though far from exclusive,
dialogue with his work.

We shall first of all set out the broad outlines of the fundamental
epistemological and theoretical assumptions belying our general
approach to the history of the social sciences and Marxism. In
particular we will explore the relevance of the problematic of the
history of the sciences to the history of the social sciences and
Marxism. Second, in Part I, we examine the sources, formation,
and development of the concept of the AMP in the works of Marx
and Engels. This epistemological history of the AMP is taken up
further in the sectional introductions to Parts II, III, and IV.

THE PROBLEMATIC OF THE HISTORY OF THE SCIENCES

The new historiography of the sciences emerged in the inter-
stitial space between history, epistemology, and the sciences. It
is by no means uniform in its theoretical approach, but I think
that there is an underlying problematic, in the sense used by
Althusser, of a total and autonomous intellectual structure which
can be expressed in two pairs of oppositions: internalism versus
externalism and continuism versus discontinuism.

The opposition between internalism and externalism refers to
the *focus* of the research. In order to explain scientific development, the internalist concentrates almost exclusively on the
scientific works (theoretical and experimental problems as defined
by a scientific community) while the externalist also considers
other influences, such as technological, socio-economic, institutional, political, and ideological factors. For the former, the
interaction of scientific ideas (or in a wider sense, the intellectual interaction) suffices to explain the dynamics of science,
while for the latter other conditions - external to science - are
required.

The controversy between internalists and externalists centred
upon the problem of the so-called scientific revolution of the sixteenth and seventeenth centuries. For the internalists the
scientific revolution was fundamentally an intellectual revolution;
for the externalists its origins had to be sought elsewhere in the
development of capitalism. This opposition was subsequently
generalized to include not only the appearance of modern science,
but also to explain its steady growth in the following centuries
and the developments up to the present day. In its extreme forms,
the internalist places science outside of society, while the externalists challenge its objectivity independent of an hic et nunc
society.

The controversy between internalists and externalists is far
from resolved in spite of statements from both sides that they have
won the war; it will probably rage for a long time. Unfortunately
the rules of the game are not yet clearly defined. There is neither
firm agreement on the meaning of such terms as 'pure science'
and 'technology' (though most internalists tend to equate science
with 'pure science'), nor is there a clear delineation of the range
and scope of external factors. What are these factors supposed to
influence? The cognitive aspect of science, its origins, its
developments, its shifts? And is their influence equal for all
times and cultures or does it vary from one society to another?

The opposition between continuism and discontinuism refers
to the vexed question of whether there is a continuous development
of knowledge from common sense to scientific knowledge. The
continuist states that progress and historical change take place
step by step, gradually, and that scientists are greatly indebted
to their predecessors. The discontinuists see knowledge as being
subverted, as changing from one period to another. For the continuist, science has always existed, albeit in rudimentary forms.

Ideally, the discontinuist sees science (and sometimes each science in particular) as an epistemological irruption that emerges in a particular historical period.

Some continuists conceive of scientific progress as a process of indiscriminate accumulation, others prefer to look at it as a process of selective accumulation. All use the idea of precursors: this is based on the principle that to every thinker one can find a list of forerunners who show different degrees of intellectual kinship. As a technique, this idea facilitates the transition from one period to another, thus contributing to the idea that there are no revolutions or abrupt changes in science.

The discontinuists think that scientific progress takes place by abrupt and sudden leaps forward which subvert the ancient order. They talk about scientific revolutions. For example, they believe that modern science, that originating in the sixteenth and seventeenth centuries, represents a radical change, a discontinuity, with respect to medieval science. Discontinuists often concern themselves with the problem of the beginnings of science, be it the beginnings of science in general which they place in a particular historical period (Mesopotamia, Greece, Western Europe) or with the beginnings of each specific science.

Obviously the discontinuists reject the idea of precursors, but they accept that scientific ideas which were born in one individual can be developed and brought to completion by another. Their main complaint is against the assumption that theories, concepts, experiments and the like can belong to different periods and can be easily transferred from one intellectual space to another. This activity – which may be condoned in the practising scientist who, for pedagogical reasons or with the purpose of enlisting some scientific authority of the past, is trying to trace the filiation of his theories to a forerunner – is inexcusable for the proper historian in so far as his attitude represents a distortion of the past in a way that makes it practically unintelligible.

As a result of the polemic between continuists and discontinuists a number of areas of research have been delineated. Three of them seem to us of the greatest interest: the emergence of two different types of history, the question of the beginnings of science, and the relations between history of science and epistemology.

The first effect of the dispute between continuists and discontinuists has made it possible to distinguish clearly between two types of histories. On the one hand, there is the type of history in which the theories of the past are classified as correct or incorrect according to whether they do or do not conform with the current practice of science. This attitude has been referred to by Butterfield as the 'Whig interpretation of history'. It informs the typical history of scientists-historians, inductivists, and positivists. Its only concern is to show the triumphant progress of science from the beginnings up to the present, always looking at past achievements from the standpoint of the scientific attitudes of today.

There is another kind of history that looks at the past in a
different way. It does not look only at the concepts that are
confirmed by the scientific practice of today, but also at those
that have been abandoned. It works under the assumption that
what now has been abandoned and excluded was once held to be
true and might have been considered indissociable from what we
still consider to be true today. In other words, it tries to understand the work of the scientist of the past as a whole and not
as a mixture of scientific and non-scientific theories. This kind
of history 'is an effort to investigate and try to understand to
what extent superseded notions, attitudes, or methods were in
their time an advance, and consequently how the superseded past
remains the past of an activity for which we should retain the
name of scientific' (Canguilhem, 1968, p. 14).

These two ways of writing history have been labelled differently.
For example, in the history of anthropology, G. W. Stocking
(1968) talks about 'presentism' and 'historicism' respectively,
but we would prefer to avoid the term 'historicism' altogether
since it has so many different connotations. A similar, but no
means equal, classification is the one put forward by G. Bachelard (1951). He distinguishes between sanctioned history (histoire
sanctionée) and outdated history (histoire perimée); the former
is a history of thoughts that have been confirmed by contemporary
science, the latter is a history of thoughts that do not make
sense from the present state of scientific rationality.

There is a crucial difference between Whiggism and sanctioned
history, in that although both use the concept of recurrence,
Bachelard does not believe in linear progress, but instead a
dialectical one - progress that takes place by sudden, abrupt
mutations. Of course, the proper historian has to pay attention
to both histories.

The second area of crucial interest that has developed around
the controversies between continuists and discontinuists concerns the question of the beginnings of science or of each specific
science. This is, of course, related to the wider and more complex question of what science is. This problem does not seem to
bother the natural scientist too much, but it is time-consuming
for philosophers of science and social scientists.

A number of answers have been given to these two questions.
As a matter of fact, answers can only be provided - short of
adopting the diffuse attitude of the extreme continuists - by
accepting an epistemological intervention of one kind or another.
The whole issue can thus only be tackled if one is aware of the
different epistemological issues that exist. For a positivist, for
example, the crucial element that defines the beginnings of
science is methodological; the imposition of certain standards of
observation and experience; the empiricist sees the primordial
feature in the collection of facts; others prefer to talk about
the delineation and definition of an object; finally, the rational
materialists insist that the definition of a science is its history -
the history of the real conditions of production of its concepts

(the formation of the concepts and theories) - and see a science as constituting itself by departing from a previous ideology.

The third area of interest refers to the relation between history of science and epistemology. It is clear from what has been said up to now that it is neither possible nor desirable to keep history of science and epistemology separate. We can agree with Lakatos that epistemology without history is empty and that history without epistemology is blind, but otherwise they should not be confused. There is always the danger, in a pure sanctioned type of history, of reducing history to epistemology, while the other extreme would be to reduce history to a pure narration of events without any evaluation.

As to epistemology it can either mean the universalizing project of the philosophy of the sciences or a certain normative stand which claims to draw its judgments from a continuous and up-to-date contact with each science, and in consequence is bound to be provisional and changing.

SOME CRUCIAL ISSUES IN THE HISTORY OF THE SOCIAL SCIENCES AND MARXISM

The general scheme that we have presented delineates in its very broad lines the quintessence of the current problematic of the history of the sciences. It is our intention now to examine the suitability of this scheme for the history of the social sciences and Marxism.

The first thing to be noticed, as T. S. Kuhn has aptly re-marked, is that 'the new historiography [of science] has not yet touched the history of the social sciences' (1968, p. 77). Kuhn does not mention the question of whether he thinks that the tools developed to deal with the fully fledged sciences can also be applied to the social sciences, the scientificity of which is sus-pect - or uncertain to say the least. It is true that in the last chapter of Kuhn's famous 'The Structure of Scientific Revolutions' (1962), a reference is made to the pre-paradigmatic stage in which the social sciences are found.

In what follows we will examine two issues which we consider crucial and preliminary to any attempt to write a proper history of the social sciences and Marxism. We believe that until now these issues have received either inappropriate attention, or no attention at all. We will refer to them respectively as the problem of the epistemic status of the social sciences and Marxism (i.e. the reliability of their knowledge) and the problem of theoretical externalism.

1 *The problem of the epistemic status of the social sciences and Marxism*

Unless it is willing to start on the wrong footing, a history of the social sciences and Marxism can not avoid a consideration of their epistemic status. This, of course, requires the use of

certain criteria to separate science from non-science, unless
one is prepared to accept that there is nothing specific and
differential that can be called scientific knowledge. But if
scientific knowledge, as we endorse, is different from religious,
magical, and other forms of knowledge, it is crucial that we
provide the criteria for this distinction.

A number of criteria are available (verifiability, falsifiability,
logical consistency, etc.) which have been put forward by
philosophers of science. As is well known, most of these criteria,
if applied rigorously, come up with the result that the social
sciences do not stand up to the standards of scientificity.

But against the imperialist philosophies of science that impose
abstract criteria resulting from experience at best limited to the
natural sciences, it is always possible to demand specific criteria
for the social sciences. It is unfortunate that most of the epistemological studies on the social sciences fail to recognize this
specificity and rely heavily upon the natural scientific model.

It is true that in the social sciences there has always been a
significant, though minoritarian, trend (exemplified today by
the phenomenological and Frankfurt schools among others)
against blindly following the natural scientific mode, but this
trend has often taken a completely anti-scientific bias, thus
throwing out the baby with the bathwater.

One of the things that social scientists of one denomination or
another have accepted uncritically is their own object of knowledge. That there should be, for example, a science of man is
seen as unproblematic, but Michel Foucault has been able to
show when and how this specific object of science appeared in
the Western world, and why this event should be considered an
irruption in the realm of knowledge. He also considers the
possible disappearance of this object (Foucault, 1970). Whether
'The Order of Things' is for the human sciences what Kant's
'Critique of Pure Reason' was for the natural sciences - as
Canguilhem (1967, p. 618) has suggested - is to be seen, but
there is no doubt that Foucault's book, and his work in general,
requires closer attention from social scientists than it has received until now.

The answer we give to the question of the epistemic status of
the social sciences and Marxism has immediate consequences for
the methodology of research into their history. If we accept
their cognitive autonomy, in other words if we assert their
scientific character, there will be certain limits to what an external history can explain. Some extreme externalists and vulgar
Marxists may find this statement unacceptable but we tend to
agree with D. Lecourt (1975, p. 14) that

the effects of the external determinations (social, economic,
ideological and political determinations) are subject to the internal conditions (the norm of the true) of scientific practice.
Here is a principle which rules out from the start all epistemological economism, sociologism, and psychologism: it is indeed
impossible to achieve a *genesis* of scientific concepts on the

basis of what are known as the social, economic, psychological
(or even biological) conditions of scientific practice.
On the contrary, if we take the social sciences and Marxism to
be non-scientific it is apparent that the external determinisms
(what Mannheim used to call Seinaverbundenheit) will be fully
operative and that the criteria of the true cannot be found at
the internal level. Of course, this is not to deny the relative
autonomy of each ideological formation.
We are aware of the fact that any solution to this question can
only be very fragile. That does not mean that one can avoid
taking an implicit of explicit epistemological stand. Indeed, most
of our historians of the social sciences and Marxism are still
practising within the traditional framework and consequently
the dilemma would only affect them in a rather superficial way.
A provisional way out of the question of the epistemic status
of the social sciences is to consider them as sciences in formation, the foundations of which were laid down by the Enlightenment, but which never consolidated as sciences because of a
number of epistemological and ideological obstacles that emerged
around the discipline in the nineteenth century and that have
persisted into the twentieth century. These epistemological
obstacles are the result of uncritically accepting the natural
scientific model; they include inductivism, empiricism, mechanical
materialism, and fixist evolutionism. They are the result, con-
secrated by Comte and Stuart Mill among others, of a misreading
of the natural scientific practice and a denial of the specificity
of the social sciences. We will return to this topic later when
dealing with the problems of theoretical externalism. The ideological obstacles are the result of the class character of the
society that produced and developed the social sciences. The
effect of these obstacles is to drag the social sciences from the
purely cognitive level to the more practico-social one, where it
can be used, directly or indirectly, by the ideology and interests of the dominant class. Though born as a critical science
of men and society, Marxism is not immune to such epistemological
and ideological obstacles.

2   *The problem of theoretical externalism*
There is one peculiar characteristic of the social sciences that
we would like to emphasize. We have already referred to it as
'theoretical externalism' and by this term we purport to describe
the fact that since its beginnings, the social sciences have
adopted a mimetic attitude with respect to the natural sciences.
This situation has not been without serious negative theoretical
effects, the implications of which - although adumbrated - have
not been fully spelt out. One of the consequences is that a
history of the social sciences cannot be written independently
of a general history of scientific thought. In other words, we
maintain that in each historical period the problematic of the
social sciences has tended to depend on an imported model of
scientific practice. If we add that this borrowing has often been

mechanical and has ignored the specificity of the social sciences, we can explain why what were intended as applications of the 'scientific method' or of successful theories in other sciences resulted in epistemological obstacles for the social sciences.

It is our contention that a number of epistemological obstacles developed around the social sciences. Over time this has resulted in the creation of a resistant web of obstacles which make scientific progress difficult. Without wanting to be exhaustive we would like to mention a few of those which we think particularly relevant. We want to insist that these obstacles are the result of what we could call certain scientistic fixations deeply embedded in the social sciences.

One of the early obstacles could be referred to as mechanicism. It is the result of applying the idea that phenomena can be explained totally on mechanical principles to the social sciences. It can be said to derive from the theories of Galileo, Descartes, and Newton. Underlying this conception is an idea of causality which assumes an immediate relation of cause and effect; on the other hand, it only allows for the existence of efficient but not final causes. A related obstacle can be designated as vulgar materialism. This conception suggests that the techno-economic elements of society determine all the others; that the latter are pure epiphenomena. In a sense it is suggested that the existence of the techno-economic sphere is necessary and sufficient cause for the existence of the other spheres. These two obstacles, along with a lineal and fixist interpretation of evolution, have often been present in Marxism and evolutionism. As a result, the social whole has been usually conceived of, not as an articulation of different spheres with a causality which has to be deciphered, but as a whole in which the social, political, and ideological elements are seen as a mere emanation from the techno-economic structure.

Another obstacle that derives from a certain misunderstanding of Newton can be referred to as inductivism. The basic point of inductivism is that what is scientific must be provable by reference to facts. This assumes that theories must be deduced from facts and consequently denies the need for a free conceptual construction in the development of theories. In the social sciences this obstacle has often been identified with the prescription against conjectural history. In general, its effects have been to bring theoretical work to a near standstill.

Empiricism is another of the obstacles that has pervaded the social sciences from the beginnings, especially in the Anglo-Saxon world. There are two dimensions that we would like briefly to refer to. Firstly, empiricism equates science with collection of facts; secondly, and more importantly, it contains an ontological assumption according to which universals or laws are to be found at the empirical (behavioural) level. No distinction is made between nature as sensed and nature as perceived by science.

With the concept of theoretical externalism, we have tried to point out broadly the situation of scientific dependency in which the social sciences have found and continue to find themselves today. This situation is not without important consequences, for if it is true, social scientists have systematically misrepresented scientific practice and have not usually taken into account the specificity of the social sciences. This holds true for Marxists as well.

PART I
THE AMP: SOURCES AND FORMATION OF THE CONCEPT

Anne M. Bailey and
Josep R. Llobera

PRELIMINARIES

The history of the concept of the AMP is particularly appropriate
to see the effects of ideological and epistemological obstacles. At
the ideological level, the concept has often been a weapon used
to justify political positions. On the other hand, the fact that
the concept has been apprehended within a materialist and
evolutionary theoretical perspective, throughout most of its
career, makes it particularly adequate for a case study.

Before embarking on a genealogy of the concept of the AMP,
the parameters of such a task should be made explicit. First of
all, there is the question of whether or not there is a continuity
between Marx's formulation of an AMP and earlier writers' con-
ceptualizations of a specific geographical area, Asia, or of a
specific social totality. The first step, of course, is to establish
the specificity of Marx's concept of the AMP. (1) In recent years,
attempts to isolate Marx's concept of the AMP have adopted two
opposing viewpoints with regard to Marx's writings on Asia or,
more generally, on non-European societies. One position is to
affirm that all of Marx's writings on these societies must be given
equal weight in reconstructing his concept of the AMP. Another
is to adopt the view that only certain of his writings can be considered pertinent to the formation of the concept in Marx. The
position adopted here rejects an extreme continuism, but considers Marx's journalistic articles of the 1850s, taken as a whole,
as a foundation upon which the concept of the AMP was elaborated.

Second, bearing in mind our thesis of the greater weight of
external factors on sciences in formation like the social sciences,
there is the problem of continuity/discontinuity among earlier
thinkers, particularly between political theorists of the sixteenth
and seventeenth centuries like Bodin, Machiavelli, or Hobbes
and the Enlightenment; and between both of these and the early
nineteenth century, particularly German classical philosophy and
English political economy. A thorough-going analysis of this
problem is beyond the scope of this introduction, but an attempt
will be made to present certain lines of inquiry which could
prove fruitful to this endeavour.

Finally, and following from the previous points, a word should
be said on the criteria employed by contemporary historiography
on the sources of influence on Marx's concept of the AMP. For
the most part these criteria are not grounded in a critical epistemological stance. The net effect is that arguments as to the

13

continuity or discontinuity between Marx and earlier writers
are highly ideological. Such epistemologically uninformed criteria
include 1   the intentionality of the author, and 2   a stress on
the socially conditioned nature of the knowledge produced. The
highly subjective results of employing solely externalist criteria
can be illustrated by comparing K. Wittfogel (1957) and P. Anderson (1974), who have both opted for a continuist position between
Marx and earlier writers.

In his introduction to 'Oriental Despotism', Wittfogel briefly
summarizes the 'extraordinary insights . . . [of] keen-minded
Western travellers and scholars' from the sixteenth to the early
nineteenth century (1957, p. 1). The discovery of oriental
despotism by these early writers was the result of their critical
attitudes towards tyrannical governments of their day. But,
Wittfogel questions why these insights, the starting point for
a systematic and comparative study of total power, were not
further developed until he took up the problem. He finds the
answer in the political developments of Western Europe (1957, p.1):

> Absolutism prevailed in Europe when Bernier described his
> experiences in the Near East and Mogul India and when Montes-
> quieu wrote 'The Spirit of the Laws'. But by the middle of the
> nineteenth century representative governments were established
> in almost all industrially advanced countries. It was then that
> social science turned to what seemed to be more pressing
> problems.

His own revival of the concept is likewise attributed to the present
state of the world (ibid., p. 2). For Wittfogel, the politically
conditioned nature of the knowledge produced from the sixteenth
to eighteenth centuries or his own, for that matter, does not
appear to affect the validity of the concepts produced (although
he is quick to dismiss certain parts of Marx's writings on the
AMP on the basis of political intentions which he attributes to
him) (ibid., pp. 380-2).

Anderson, who suggests that Marx and Engels reproduced the
traditional European discourse with minimal modifications (1974,
p. 492), argues that the fact that these previous knowledges
were politically conditioned seriously affects their reliability for
the explanation of Asian history. Unfortunately, for the sake
of his own argument, Anderson is left without concepts which
he would deem scientific for his analyses of the Ottoman state,
and Islamic and Chinese civilizations (ibid., pp. 365 and 548). Of
course, behind Wittfogel's acceptance and Anderson's rejection
of these knowledges lies an empiricist conception of their adequacy
to the facts of specific non-European societies.

It is perhaps one of the most outstanding merits of Krader's
scholarly work on the sources and the theory of the AMP to try
to advance beyond this sort of epistemologically blind historiography. While Krader takes into consideration external factors
in the formation of ideas about non-European civilizations, he
also attempts to place those ideas in the context of development
of the social sciences.

On the political and social conditions affecting the production
of concepts by earlier writers, Krader writes (1975, pp. 6-7):
> The prehistory of the theory of the Asiatic mode of production
> is found in the early capitalist period, as a part of the attempt
> of the writers and thinkers of that time to grasp their own
> history and comprehend their society. The initial ideas and
> observations that contributed to the formation of the theory
> came from the travellers, merchants, sailors, diplomats, who
> went to the East in the seventeenth century seeking careers
> or private gain or commercial advantage for their respective
> countries. Their writings were reflected on and digested or
> caricatured by the philosophers, historians, political economists
> of the eighteenth century, few of whom were concerned with
> the study of Asia for its own sake, but were rather formulators
> and agents of policies of their own lands. This judgment is
> intentionally negative, and should be modified in the case of
> Adam Smith who, by the breadth and acumen of his insights,
> stood above the others of his time in his studies of the Orient,
> at least as compared to those who were not themselves Oriental-
> ists. Among the latter Abraham-Hyacinthe Anquetil-Duperron
> and Sir William Jones stand out above average. Generally the
> works of that time were written with particular policies at
> home in view and abstractions concerning the distant lands.
> The speculations about oriental despotism or tyranny, the
> forms of land ownership in Asia, the oriental society, supported
> at one time by the advocates of free trade, the East India Company,
> at another the utilitarians, the liberal interests, and the
> colonialists throughout.

But in addition to these external factors, Krader gives at least
equal weight to what might be called an internalist explanation
(1975, p. 1):
> The social sciences have passed through a repeated cycle of
> development at different times. Those who professed these
> studies at first drank heavy drafts of politics before turning
> to the sober concerns of economy and society. At the begin-
> ning of the modern capitalist period of history, Machiavelli,
> Jean Bodin and certain Jesuits, Calvinists, Lutherans of the
> time were political writers in the first place; if they thought
> of the society they did not think of it as a whole. That politics
> is a part of the social whole, indeed a subordinate part, and is
> only apparently its leading principle, was not an early thought,
> but a late one. The study of the Orient does not differ from
> this general rule. The European writers who dealt with Asia
> in the seventeenth century at first took up the study of the
> forms of political authority, which they characterized with
> the catchwords, despotism, tyranny. *Only in the nineteenth
> century was the society in the countries of Asia acknowledged
> as a subject of study unto itself.* At first the study was taken
> up as an abstraction, as though there were but one oriental
> society later as the many societies in Asia existing in its
> singularity, concretely. *There is no secret about this transition*

*from political to social concerns, nor about the transition from
society as an abstraction to society as a concretion: the idea
of society as we now understand it is the product of the
nineteenth-century thought and so is the oriental society, both
abstractly and concretely* (emphasis added).

THE INTELLECTUAL SOURCES OF THE CONCEPT

The sixteenth century
The late fifteenth and early sixteenth centuries have been hailed
as marking the advent of the 'modern world system', not a world-
empire or political unit, but a 'world economy' encompassing
empires, city-states, and the emerging 'nation-states' (Wallerstein, 1974, p. 15). While this new European economy included
lands beyond the continent of Europe, it excluded Muscovite
Russia and the Ottoman Empire and Asia (ibid., p. 68) which
were not incorporated as the periphery to this system until a
century or more later. The sixteenth century saw the rise of
the absolutist state in Western Europe; centralized monarchies
in France, England, and Spain introduced standing armies,
national taxation, codified law and a permanent bureaucracy.(2)
Furthermore, this period marked the end of a unified European
Christian church and fundamental re-examination of doctrine.
    Among the witnesses to these dramatic changes who attempted
to understand and shape the policy of the emerging 'nation-state'
were Niccolò Machiavelli (1469-1527) and Jean Bodin (1530-96).
Their principal sources were the Bible and the classics, but in
illustrating their studies of comparative government they drew
on the contemporary examples of the Ottoman and Muscovite
Empires. For Machiavelli, 'The entire Turkish empire is ruled by
one master, and all other men are his servants' (Anderson, 1974,
p. 397). The author of 'The Prince',(3) for whom the main bases
of government are 'good laws' and 'good arms', the latter entrain-
ing the former, writes of the military might of the Turkish prince:
'He controls a permanent army of 12,000 infantry and 15,000
cavalry, on which the security of his realm rests; the supreme
principle of his power is to safeguard its loyalty' (ibid., p. 398).
    Machiavelli has been cited as the 'first theorist to use the
Ottoman State as the antithesis of a European monarchy' (Anderson, 1974, p. 397). Like Bacon after him, he noted the lack of
hereditary nobility as a distinctive feature of Europe's neighbour.
The effects of these two distinct forms of government, despotism
and that by a monarch and his councils, result in great empires
with a servile population in the first case, and the creation of a
multitude of states, favouring the development of its inhabitants'
creativity in the latter case (Stelling-Michaud, 1960-1, p. 332).
    Loys le Roy's 1568 translation of Aristotle's 'Politics'(4) was to
provide the immediate source of inspiration for Jean Bodin's
'Six Livres de la Republique' (1579) (Koebner, 1951, pp. 284-5).
Bodin distinguished between three forms of monarchy-seigneurial,
royal, and tyrannical:

In the royal monarchy, the subjects obey the laws of the
monarch, the laws of nature, natural liberty and ownership of
goods remaining with the subjects. The seigneurial monarchy
is that in which the prince is made seigneur of the goods and
persons by right of arms and war, governing his subjects as
the father of the family governs his slaves. The tyrannical
monarchy is that in which the monarch, despising the laws of
nature,(5) abuses free persons as slaves (Krader, 1975, p. 22).
For Bodin, seigneurial monarchy was the first form of monarchy
among men, but was still to be found among the Turks and
Muscovites of his day.(6) Bodin's view that in a seigneurial
monarchy, in contrast to a royal monarchy, the sovereign was
the owner of the goods and persons within his realm and obeyed
no other law than his own, was a theme that influenced, and was
corroborated by, seventeenth-century travellers – particularly
Bernier.

As we shall see later, there are certain parallels between the
Turkophile-Turkophobe(7) preoccupations of the sixteenth century
and the Sinophobe-Sinophile debates of the French Enlightenment.
However, future externalist explanations must no doubt go beyond lumping these parallels under the umbrella of 'political
arguments before the triumph of capitalism' to borrow Hirschman's
phrase (1977). Such explanations must await, perhaps, a more
sophisticated periodization of the expansion of the world system
and its differential incorporation economically, politically, and
ideologically of coexisting world economies or empires.

In early seventeenth-century England, Francis Bacon was to
emphasize that the fundamental distinction between the Ottoman
Empire and European forms of government was the absence of a
hereditary aristocracy, the net result of which was tyranny
(Krader, 1975, pp. 27 and 314; Anderson, 1974, p. 398).
Thomas Hobbes incorporated Bodin's concept of 'seigneurial
monarchy' into his 'Elements of Law, Natural and Politic' (1650),
rechristening the form in its Greek original as 'kingdom despotical'.
Such kingdoms were confined neither to barbarians nor peoples
of the Orient, but were the form of rule found in 'conquest
states'. Furthermore, despotic monarchies implied 'no harsher
rule than any other government in which the principles of social
order [were] consistently carried through' (Koebner, 1951,
p. 290).

There is no doubt that the ideas of Machiavelli, Bodin, Bacon,
and Hobbes concerning forms of government reflected the tumul-
tuous time in which they lived, when neither the Bible nor the
classics could any longer provide a viable guide to order. Their
discourse looms large in the history of Western political thought.
However, whereas we are immediately struck by them as political
writers, who held certain opinions on forms of government beyond
Europe, it is perhaps misleading to reduce their concern with the
'other' to the political (Krader, 1975, p. 1). They were also
equally concerned with the manners and mores of non-Europeans.(8

Seventeenth-century travellers and polemics on absolutism
The seventeenth century brought an expansion of European
trade, with the Dutch, French, and English taking over the
formerly Portuguese-Asian trade. Reports from merchants and
ambassadors to the courts of Persia and Mogul India began to
speculate on the sources of despotic power there. Within Europe
there was a mounting reaction against absolutism.

In contrast to Marco Polo's tales of the great wealth of the
Orient, seventeenth-century travellers(9) compared the riches
of the courts to the poverty of the general populace, stressing
the lack of any intermediate social orders, in a manner reminis-
cent of Machiavelli and Bodin. In their roles as ambassadors and
merchants, however, they were more interested in the life of the
courts, particularly in the source of the 'oriental despot's' power.
With few exceptions,(10) there was agreement that the ruler was
the sole proprietor of the lands of his realm. However, there was
some dispute about whether his ownership was the source of his
power (the thesis promoted by Bernier and later taken up by
the physiocrats, Adam Smith and Marx) or whether he 'derived
the power over property and persons from the absolute power
over the sovereignty and not the other way around' (Krader,
1975, pp. 23, and 68-9). Not until the eighteenth century would
the idea that the rulers of Persia and Mogul India owned all the
land in their territories, be seriously challenged.

Throughout the seventeenth century, the notion of 'despotic'
monarchy and its relevance to the French monarchy was bandied
about. It was employed by Fronde pamphleteers to describe
Mazarin's regency, but after Louis XIV became king there was
an attempt to censure the word completely, replacing it with
the term 'arbitrary' which was opposed to 'absolute rule' (Koebner,
1951, pp. 293-6). 'Les Soupirs de la France esclave, qui aspire
après la liberté' (1689 or 1690), published in Holland by contested
authorship, questioned whether there was any essential difference
between French and Turkish domination, a theme to recur in
the debates between Voltaire and Montesquieu (Stelling-Michaud,
1960-1, pp. 335-7).

The French and Scottish Enlightenments
Montesquieu is often considered the codifier of the concept of
'despotism', one of the three forms of government he described
in 'L'Esprit des lois' (1748). It has been suggested (Cassirer,
1951, p. 210) that Montesquieu was more concerned with constructing an 'ideal type' than accurately describing particular
societies. On the existence of despotism he stated in his
'Considérations' (1734, Ch. 22) that 'C'est une erreur de croire
qu'il ait dans le monde un autorité à tous les égards despotique;
il n'y en a jamais eu et il n'y aura jamais.' However, while he
drew examples to illustrate his concept of despotism from Africa,
the Roman Empire, and England of Henry VIII's reign, there
seems little doubt that Montesquieu believed that the governments
of Turkey, Persia, India, and China best approximated his concept.

Montesquieu's principal distinction was between monarchy and despotism. In the latter there were no intervening social orders between the despot and the people, among whom there was extreme equality: they were equal in being 'nothing'. Indeed, Montesquieu saw despotic society as a society that lacks fundamental political laws, commerce,(11) and perhaps is totally without a 'social structure' (Althusser, 1958, pp. 597-8). Criticized for his acritical attitude towards his sources by Voltaire and Gibbon, Montesquieu followed Bernier in his characterization of the worst type of despotism as one where the ruler was the sole proprietor of the land.

Voltaire, who had joined the ranks of eighteenth-century Sinophiles, interpreted China as the land of philosopher kings.(12) He viewed Montesquieu's 'L'Esprit des lois' as an attack on absolutism – or 'enlightened despotism' as it came to be called. For Voltaire, the rule of reason, freed of the constraints of Catholic doctrine, was not to be confused with despotic rule and its recently acquired derogatory connotations, which depicted it as government by 'a ferocious madman following nothing but his personal whims; by a barbarian making his courtiers prostrate themselves before him or one who takes pleasure in ordering his retainers to go hither and thither strangling and empaling' (Koebner, 1951, p. 275).

Support for Voltaire's attack on Montesquieu's concept of despotism and his characterization of China was given by the leading physiocrat, François Quesnay, who put forth the distinction between arbitrary and legal despotism.(13) The latter, exemplified by China, was considered a model of physiocratic political proposals. The physiocrats had

> formulated in the laissez-faire principle the other, better known harmony-of-interests doctrine according to which the public good is the outcome of the free pursuit by everyone of his own self-interest. Being located at the intersection of these two *Harmonielehren*, the physiocrats oddly advocate both freedom from governmental interference and the enforcement of this freedom by an all-powerful ruler whose self-interest was tied up with the 'right' economic system. The latter arrangement is referred to by them as 'legal despotism', which they oppose to . . .'arbitrary despotism' (Hirschman, 1977, pp. 97-8).

For Quesnay, the 'right' economic system lay in limiting taxation to a direct tax on agricultural production with the government thus being 'co-proprietor' of the net agricultural product (Sawer, 1974, p. 1509). Linguet, critic of both Montesquieu and the physiocrats, 'carried this manner of reasoning to its ultimate conclusion . . . he . . . came out in favor of *total* ownership of all national wealth by the ruler. With great consistency he praises "oriental" or "Asian despotism" ' (Hirschamn, 1977, p. 99).

However, not all the physiocrats saw in Asia the ideal representation of this model. Bandeau wrote with great disdain of the 'usurpatory and destructive policy adopted by the great arbitrary despots of Asia' (Krader, 1975, p. 33). While less concerned with

the realities of China or Asia in general, the physiocrats appear
as the first to formulate a systematic economic model(14) for the
notion of 'oriental despotism' (Sawer, 1974, p. 1508).

Whereas Montesquieu had looked to the influence of climate,
manners, and morals for the origins of oriental despotism,
Boulanger's 'Recherches sur l'origine du despotisme orientale'
(1761) sought its basis in religion. Boulanger's treatise was to
be attacked by the French orientalist Anquetil-Duperron, today
a controversial figure in the historiography of oriental despotism.(15) Anquetil, drawing on his own experiences and those of
travellers to the East, disputed the 'despotic' nature of India,
according to Montesquieu's usage, as a government without laws
or property rights. His 'Legislation orientale' (1778), 'Recherches
historiques et géographiques sur l'Inde' (1786), and 'Description
historique et chronologique de l'Inde' (1786-91) provide a detailed account of the land tenure system of the Coromandel Coast
and the Mogul Empire, dispelling the prevailing notion of the
sovereign as sole proprietor. At the same time, he attacked
Alexander Dalrymple ('A Short Account of the Gentoo Mode of
Collecting the Revenues on the Coast of Choromandel' (1783))
who had argued that land was owned collectively by the villages.(16)
Forms of land tenure in India were to become subject to extended
debate by British colonial officials at the beginning of the nineteenth century.

Among the members of the Encyclopedia, Turgot and Condorcet
considered the empires of Asia and China, in particular, notably
low on their evolutionary scheme of the progress of human
reason. While noting the technological achievements of the
Chinese and Egyptians, these peoples had lost their initial
advantage, dominated as they were by superstition and false
dogmas (Krader, 1975, pp. 42-3).

In the second half of the eighteenth century, attempts were
made to incorporate contemporary views on Asia with the so-
called 'four-stage theory'. Herder ('Ideen zur Philosophie der
Geschichte der Menschheit' (1784-91)) associated agriculture
with despotism (Krader, 1975, p. 31; Meek, 1976, p. 196). In
Scotland, Steuart and Ferguson, pursuing certain aspects of
Montesquieu's theory, put forth the view that despotism, where-
ever found, is the product of the low state of commercial arts.
Adam Smith in his 'Wealth of Nations' sought the reasons for the
pre-commercial stage of the Chinese and Indian economies: agriculture rather than manufacture was held in higher esteem and
better remunerated in Asia; and wealth in the form of gold and
silver was concentrated in the hands of 'grandees' who would
not invest or allow others to do so. Furthermore, the state or
the ruler, as proprietor of the land,(17) was actively interested
in promoting agriculture and maintained roads and canals for
purposes of communication and irrigation.(18) In India, the
caste system was likened to the European guilds in their effects
on the market.

*The nineteenth century: German philosophy and British
political economy*
Hegel was familiar with the works of the French philosophers
and Adam Smith.(19) In both his 'Philosophie des Rechts' and
'Vorlesungen über die Philosophie der Weltgeschichte', he contrasted the Orient with the West, and China with India. Whereas in Europe there had been a progressive unfolding of inner
self-consciousness, China, Persia and Turkey had developed
abstract, external moral consciousness. China was a patriarchal
empire where the feeling of self was extremely limited; in contrast, India totally lacked the feeling of self in the individual
and moral consciousness. The Chinese had a written history,
a chronicle of events, but no philosophy of history; India,
dominated by an extremely parochial vision, derived from the
caste and village systems, was completely lacking in history
(Hegel, 1830a, pp. 161-2). While both China and India had remained stationary for millennia, China had developed a theocratic,
patriarchal state where right and morality were indistinguishable
(ibid., pp. 160-1):

> In the Chinese State the moral will of the Emperor is the law,
> but so that subjective, inward freedom is thereby repressed,
> and the Law of Freedom governing individuals only as from
> without. In India, the primary aspect of subjectivity - viz.
> that of the imagination - presents a union of the Natural and
> the Spiritual, in which Nature on the one Hand, does not
> present itself as a world embodying Reason, nor the Spiritual
> on the other hand, as consciousness in contrast with Nature.
> Here the antithesis in the (above-stated) principle is wanting.
> Freedom both as abstract will and as subjective freedom is
> absent. The proper basis of the State, the principle of free-
> dom is altogether absent, there cannot therefore be any State
> in the true sense of the term.

Politics, whether in the form of palace revolutions or invasions,
was a matter of indifference to the people bound within villages.

The village community and its economic self-sufficiency as
the basis of the stationary state of India prior to British rule,
is taken up by James Mill in his 'History of British India' (1821).
Like Bernier, the elder Mill considered that oriental despotism
in the subcontinent stemmed from the ruler's ownership of the
land.(20)

John Stuart Mill is attributed with coining the term 'oriental
society' for the stage of agriculture in the East ('Principles of
Political Economy' (1848)). The path from hunting and gathering,
through pastoralism, culminated in the Orient in sedentary agriculture which relied on irrigation organized by the government
or the villages themselves. The younger Mill employed India in
the construction of his model of oriental society. The state was
considered the great land-owner, although peasants retained
usufruct as long as they paid customary rents. Agriculture and
handicrafts were combined within the villages, the city as a site
of manufacture was absent. J.S. Mill laid stress on the stationary

character of India, its widespread poverty, and inefficient
economic practices, seeing British penetration 'if not improving
the actual condition of the population . . . [creating] in them
new wants, increased ambition, and greater thought for the
future' (Sawer, 1974, p. 1515).

Richard Jones, lecturer on political economy in East India
College of Haileybury, supported Bernier's view that the
sovereign's exclusive ownership of the soil(21) was the foundation of 'the unbroken despotism of the Eastern world, as it is
of the revenue of the sovereign, and of the form which society
assumes beneath their feet' ('An Essay on the Distribution of
Wealth and Sources of Taxation' (1831, p. 8) cited in Anderson,
1974, p. 470). Jones proceeded to analyse the rents paid by
Indian peasants (ryot rents) as one of the four categories of
rent (i.e. rent in labour, rent in kind, ryot rent - i.e. rent
as tax- and cottier rents), developing Adam Smith's ideas on
the 'rent/tax' combination. The ryot rents of India (peasant
rents in China, Persia, and central Asia were thought to be
similar) doomed the country to stagnation, preventing independent accumulation and perpetuating despotism. Jones believed,
following Bernier, that Indian craftsmen were dependent on
revenues distributed by the sovereign.

*The nineteenth century: the village community: evolutionary
and revolutionary perspectives*
The idea that Asia in general (and India in particular) had remained unchanged for millennia, coupled with the epistemological
revolution in linguistics(22) and the substantive conclusions
on the Sanskritic origins of the major European languages,
provided the impetus for a major re-examination of European
and, by extension, world history. The Indian village community
served as the illustrative example of original conditions. In contrast to previous preoccupations with government, considered
from the vantage point of the rulers' powers and possessions,
the focus was now on the village unit and the development of
particular institutions within it. Whereas supposed isolation and
self-sufficiency had previously given a partial explanation of
the 'stagnation' of the over-all society, these ideas now provided
the theoretical justification for treating the community as the
unit of analysis. The idea of the village communities as isolated
political units - as little republics and as groups of 'co-owners
of the soil' - among nineteenth-century British colonialists and
historians of India(23) has been described by Louis Dumont
(1966). Sir Henry Maine, drawing upon this material on Indian,
Slavic,(24) Germanic, village communities(25) and Celtic traditions, set out to trace the evolution of legal practices and institutions within a supposed historical Indo-European unity. Maine's
work(26) provided the working hypothesis for further comparative research into the Indian, Russian, and Slavic village landholding practices.(27)

The academic controversy provoked by the theory of agrarian

communism, a controversy still with us, preoccupied the principal social thinkers of the latter half of the nineteenth and early
twentieth centuries. This controversy over social evolution goes
beyond academic debate, occurring as it did and does within a
conflict of socialist and capitalist ideologies, and ultimately reduced to opposing views of human nature.(28)

While in Western Europe, the existence, historicity, and import
of 'agrarian communism' implicitly resorted to ideological assumptions. Within tsarist Russia the controversy was directly linked
to questions of revolutionary strategy; a correct understanding
of the Russian mir in its historical development and present
state was necessary in order that its role in a future socialist
society could be appreciated. Chernyshevsky, in his 'Criticism
of Philosophical Prejudices against the Communal Ownership of
the Land' (1859), operated with a dialectical view of progress,
'thus, primitive communal collectivism is similar in form to the
developed collectivism of a socialist society and can make easier
a direct transition to it' (Walicki, 1969, p. 18). By the 1870s,
however, political work among the Russian peasantry(29) and
the reception of Marx's 'Capital'(30) by the populists led them
to question both the historical concomitants of the Russian mir,
its communist nature, and the possibilities of it serving as the
foundation for a future socialist society.

MARX AND ENGELS ON ORIENTAL SOCIETIES AND THE ASIATIC MODE OF PRODUCTION

Marx's development of the concept of a specific social totality,
the Asiatic mode of production, spanned a period of thirty years,
beginning with his newspaper articles of the 1850s, extending
through his critiques of political economy, and culminating in
his correspondence and ethnological research of the last years
of his life. In certain writings, particular elements of this
totality - property, the division of labour, surplus appropriation,
exchange, and commodity production - are treated in detail. However, Marx never achieved a systematic exposition of his theory
of the AMP.

With the exception of his ethnological researches of 1879-81,
Marx's delineation of the elements of the AMP is ultimately related to his critique of political economy and his analysis of the
capitalist mode of production. Starting from the perspective of
the impact of capitalism (a provisional conceptualization, centring
on British industrial capital) on the AMP (a provisional conceptualization based on India), Marx went on to elaborate further
the specific nature of these totalities and their elements, and to
speculate on their genesis. The process by which significant
areas of the world, particularly Asia, were becoming incorporated
within capitalism raised a series of theoretical and ideological
problems concerning the relationship between capitalism and
'non-capitalism', and genetically, between 'pre-capitalist forms'

and capitalism, as well as between capitalism and socialism:
What did this process of incorporation indicate about the nature
of capitalism itself? The manner in which British capital had
'destroyed' or 'revolutionized' these non-capitalist societies
invited the analysis of the particularity and history of the latter.
Were the characteristics of these non-capitalist societies peculiar
to Asia or pre-Columbian America or could they be conceived of
as pre-capitalist in the sense that such characteristics had
existed in the history or prehistory of Europe? In terms of
political and ideological practice, what was the historic role of
capitalism in the development of socialism: were the remnants
of these non-capitalist/pre-capitalist elements a hindrance or a
foundation for the development of a future socialist society?

The works of Marx and Engels on oriental societies and the
AMP will be considered chronologically. Marx's development of
the concept of the AMP moves from a focus on the concrete to
a more abstract, logico-formal level and back to the concrete
in his ethnological researches.(31)

Correspondence and articles on India
As the London correspondent of the 'New York Daily Tribune',
Marx wrote a series of articles on India and China during the
1850s when he covered the parliamentary debates on the renewal
of the East India Company Charter, the Taiping rebellions, and
the mutiny of the Indian Army. Looking into the history of
British rule in India, Marx and Engels, in their correspondence,
and Marx, in his articles, attempted to set out the distinctive
features of oriental societies which included India, the East
Indies, Persia and, to some extent, China.(32)

Writing to Engels on 2 June 1853, Marx quotes extensively from
F. Bernier's 'Voyages contenant la description des états du grand
Mogul, de l'Indoustan, du Royaume de Cachemire, etc.' (1670)
on the military system and the likeness of cities to military camps,
concluding that 'Bernier rightly considered the basis of all
phenomena in the East - he refers to Turkey, Persia, Hindostan -
to be the *absence of private property* in land. This is the real
key even to the oriental heaven' (Marx and Engels, 1965, p. 81).

In this letter Marx rephrased Hegel's question of the lack of
history in the East, asking why the history of the East appeared
as the history of religions (ibid., p. 80). He found the solution
in the absence of private property in land. In the 'Formen',
Marx would consider the concept of property in greater detail
and the specific form of property/possession in the Asiatic mode.
However, at this point Marx is less concerned with the origins
of this absence of private property, which he considered to have
been established as a general principle throughout Asia by the
Muhammadans(33) (Marx to Engels, 14 June 1853, ibid., p. 86).

Engels, on the other hand, suggested to Marx(34) that the
origins of this absence of private property lay in climatic conditions, and Marx, in his article of 25 June 1853, modifies Engels's
explanation to focus on the role played by the government in
oriental agriculture:(35)

There have been in Asia, generally, from immemorial times,
but three departments of Government: that of Finance, or the
plunder of the interior; that of War, or the plunder of the
exterior; and finally, the department of Public Works. Climate
and territorial conditions, especially the vast tracts of desert,
extending from the Sahara, through Arabia, Persia, India and
Tartary, to the most elevated Asiatic highlands, constituted
artificial irrigation by canals and waterworks, the basis of
Oriental agriculture. As in Egypt and India, inundations are
used for fertilizing the soil in Mesopotamia, Persia, etc.;
advantage is taken of a high level for feeding irrigative canals.
This prime necessity of an economical and common use of water,
which, in the Occident, drove private enterprise to voluntary
association, as in Flanders and Italy, necessitated in the Orient
where civilization was too low and the territorial extent too
vast to call into life voluntary association, the interference of
the centralizing power of Government. Hence an economical
function devolved upon all Asiatic Governments, the function
of providing public works (Avineri, 1969, p. 90).(36)
In addition to singling out the absence of private property in
land and the role of government in public works as distinctive
features of oriental societies, Marx breaches the stationary
character of such societies, focusing on the division of labour
of the society as a whole and within the village communities.
This stationary character of this part of Asia (India) – despite
all the aimless movements on the political surface – is fully
explained by two circumstances which supplement each other:
1) the public works were the business of the central government; 2) besides this the whole empire, not counting the few
larger towns, was divided into *villages*, each of which possessed
a completely separate organization and formed a little world in
itself (Marx and Engels, 1965, p. 85).
Marx proceeds to copy out a list of twelve village professions (37)
or occupations, thought to be immemorial. Land in some of these
communities is cultivated in common, but in the majority the
villages till their own plots, although there is common pasture
land. Within the villages there are both slavery and caste distinctions.
In this same letter to Engels (14 June 1853) Marx outlined his
attack on the American economist H.C. Carey who was advocating
a union of industry and agriculture in opposition to the growing
centralization of industry. Perhaps spurred on by his critique
of Carey, Marx accounts for the lack of progress in oriental
societies by the central control of public works and the decentralized nature of production in the village system, prior to the
arrival of the British:
These small stereotype forms of social organism have to the
greater part dissolved, and are disappearing, not so much
through the brutal interference of the British tax-gatherer
and the British soldier, as to the working of English steam
and English free trade. Those family-communities were based

on domestic industry, in that peculiar combination of hand-weaving, hand-spinning and hand-tilling agriculture which gave them self-supporting power (Avineri, 1969, p. 93). Far from being idyllic, these communities combining handicraft and agriculture provided the foundation of oriental despotism and stunted the individual, restraining 'the human mind within the smallest possible compass, making it the unresisting tool of superstition, enslaving it beneath traditional rules, depriving it of all grandeur and historical energies' (ibid., p. 94).(38) In this article, which Marx himself acknowledges as polemical,(39) the impact of capitalism has promoted the only social revolution in the history of Asia and is seen as a progressive force.

In his article of 8 August 1853, a forecast of the results of British rule in India, Marx places rule in a broader historical perspective. Although the last in the succession of conquerors of India, the British were the first invaders who were 'superior', who rather than having been absorbed as had the Arabs, Turks, Tartars, and Moguls before them, were themselves breaking up the foundation, i.e. village communities, and absorbing India into their own civilization. While the destruction of the union of agriculture and handicrafts had been achieved through the introduction of cheap cotton and the building of the railways, the introduction of industry would hasten this process, dissolving the hereditary division of labour based on caste. This destruction of the old Asiatic society was a necessary one, historically necessary to capital and to the development of socialism. India, once the fabulous country of travellers' tales, would be annexed to the West with the development of commerce and technology. What appears as destruction now, is in fact the basis for the further evolution of mankind.

In these writings of 1853 Marx contrasts specific features of British industrial capitalism and oriental society (with particular reference to India) - the division of labour, the nature of cities, the role of the state versus voluntary associations in public works. Simultaneously, he is focusing on the destruction of the pre-existing system. While emphasizing the 'superiority' of British industrial capitalism over the pre-colonial society, he does not relate the two in a genetic way - oriental society is non-capitalist, not pre-capitalist.

The 'Grundrisse'
In 1857 Marx set about putting his studies of economics in order and elaborating his distinctive method of political economy. The results of this feverish activity (McLellan, 1973, p. 290), 'Die Grundrisse der Kritik der Politischen Oekonomie', were published only in 1939-41 - and then only in a limited edition.(40)

In the 'Introduction' Marx tackles the concepts of production, consumption, distribution and exchange, and their interrelationship. Criticizing Smith, Ricardo, and Rousseau for beginning their analysis from the perspective of individual producers, Marx

proposes to replace this point of departure with 'socially determined individual production' (Marx, 1857-8a, p. 83). Smith and
Ricardo's approach to production or Petty and Ricardo's concept
of value (Marx to Engels, 2 April 1958, in Marx and Engels,
1965, pp. 104-8) are historical abstractions which could only be
adopted on the basis of a particular economic development of
society. Against the eternalization of concepts, in this case of
production, Marx proposed (1857-8a, p. 85):

> Whenever we speak of production, then, what is meant is always
> production at a definite state of social development - production
> by social individuals. . . . *Production in general* is an abstrac-
> tion, but a rational abstraction in so far as it really brings out
> and fixes the common element and thus saves us repetition.
> Still, this *general* category, this common element sifted out
> by comparison, is itself segmented many times over and splits
> into different determinations. Some determinations belong to all
> epochs, others only to a few. [Some] determinations will be
> shared by the most modern epoch and the most ancient.

Another historical abstraction that receives Marx's attention is
that of property, conceived of as distinct from production and
equated with private property:

> All production is appropriation of nature on the part of an
> individual within and through a specific form of society. In
> this sense it is a tautology to say that property (appropriation)
> is a precondition of production. But it is altogether ridiculous
> to leap from that to a specific form of property, e.g. private
> property. . . . History rather shows common property (e.g. in
> India, among the Slavs, the early Celts, etc.) to be the more
> original form, a form which long continues to play a significant
> role in the shape of communal property (ibid., pp. 87-8).

Thus, Marx takes the example of common property in India(41)
to illustrate the historical nature of the concept, asserting a
similitude between property forms in European and Indian history.

Marx issues a caveat against an oversimplified periodization,
based on particular elements, neglecting the totality:

> Thus in this respect it may be said that the simpler category can
> express the dominant relations of a less developed whole, or else
> those subordinate relations of a more developed whole which
> already had a historic existence before the whole developed in
> the direction expressed by a more concrete category. . . . It
> may be said on the other hand that there are very developed
> but nevertheless historically less mature forms of society, in
> which the highest forms of economy, e.g. co-operation, a
> developed division of labour, etc., are found, even though
> there is no kind of money, e.g. Peru. Among the Slav com-
> munities also, money and the exchange which determines it
> play little or no role within the individual communities, but only
> on their boundaries, in traffic with others; it is simply wrong
> to place exchange at the centre of communal society as the
> original constituent element. It originally appears, rather in
> the connection of the different communities with one another,

not in the relations between the different members of a single
community (ibid., pp. 102-3).

In absolute terms, then, Marx conceives of original appropriation as common or communal and of exchange occurring first
between communities rather than within them. This periodization
of categories should be kept in mind in the consideration of his
abstract analysis of the forms preceding capitalist production.

Following his decision not to present economic categories in
the historical sequence of their appearance, but rather in their
relation to one another in modern bourgeois society, Marx begins
his study of the forms that precede capitalist production with
two basic presuppositions of that mode of production:

1   free labour and the exchange of this free labour for money(42)
2   the separation of free labour from the means of labour and
    the material for labour, particularly land.

His aim is to conceptualize both the ways in which labour has
directly confronted the object and means of labour, subjectively
and objectively, and the processes by which these links or ties
of 'unfreedom' are dissolved.

The separation of labour arises from the 'dissolution of small,
free landed property as well as of communal land-ownership resting on the oriental commune' (Marx, 1857-8a, p. 471). How then
were these labourers tied to the land originally and how did this
dissolution come about?

First of all, both of these forms of property involving settled
agriculturalists had developed from a more original, primeval
condition: Marx postulates that early man was organized into
clans (Stamm) or combinations of clans which led a migratory
existence, commonly based on pastoralism. In settling, both the
particular features of this kin-based organization and external,
ecological conditions effected modifications of the original community:

This naturally arisen clan community, or, if one will, pastoral
society, is the first presupposition - the communality (Gemein-
schaftlichkeit) of blood, language, customs - for the appropriation of the objective conditions of their life, and their life's
reproducing and objectifying activity (activity as herdsmen,
hunters, tillers, etc.). The earth is the great workshop, the
arsenal which furnishes both means and material of labour, as
well as the seat, the base of the community (Marx, 1857-8a,
p. 472).

Marx then addresses the manner in which the individual producer relates to the objective conditions of labour in one variety
of these settled agricultural communities: the Asiatic:

Each individual conducts himself only as a link, as a member of
this community as proprietor or possessor. The real appropriation through labour happens under these presuppositions,
which are not themselves the product of labour, but appear as
its natural or divine presuppositions. This form, with the same
land-relation as its foundation, can realize itself in very
different ways. E.g. it is not in the least contradiction to it

that, as in most of the *Asiatic* land forms, the *comprehensive unity* standing above all these little communities appears as the higher *proprietor* or as the *sole proprietor*; the real communities hence only as hereditary possessor (ibid., pp. 472-3).

Here Marx attempts to go beyond the appearances that informed the analysis of oriental society, following Bernier's interpretation of the ruler as the only landlord.(43) While this appearance of a sole landlord had been interpreted juridically by Western observers, for members of these communes the notion could have either a religious or political interpretation:

Because the *unity* is the real proprietor and the real presupposition of communal property, it follows that this unity can appear as a *particular* entity above the many real particular communities, where the individual is then in fact propertyless, or property - i.e. the relation of the individual to the *natural* conditions of labour and of reproduction as belonging to him, as the objective, nature-given inorganic body of his subjectivity - appears mediated for him through a cession by the higher unity - a unity realized in the form of a despot, the father of many communities - to the individual, through the mediation of the particular commune. . . . A part of [the commune's] surplus labour belongs to the higher community, which exists ultimately as a *person*, and this surplus labour takes the form of tribute etc., as well as of common labour for the exaltation of the unity, partly of the real despot, partly of the imagined clan-being, the god (Marx, 1857-8, p. 473).

Thus, despite the juridical appearances of despotism and the absence of property, the material foundation of this form is clan or communal property, in which both agriculture and handicrafts are combined. These communities are self-sufficient, containing 'all the conditions of reproduction and surplus production' within them (ibid.).

The realization of agricultural labour within these communes may vary from families working individual plots assigned to them or actually working the land in common. Within the community, its unity may be embodied in a chief or a council ('the relation of the patriarchs among one another' (ibid.)), in a more despotic or more democratic form, respectively.

Although these communes contain all the conditions of surplus production and reproduction within them, the public works carried out by them appear as 'the work of the higher unity of the despotic regime hovering over these communes' (ibid.). The accumulation of surplus product by this higher unity takes the form of revenue which is spent as a labour fund, i.e. in the remuneration of artisans and in public works.

Marx conceives of these Asiatic forms of communal property as particularly resistant to dissolution because of their self-sustaining nature (ibid., p. 486). Whereas in some communal forms slavery and serfdom modify these forms, in the Asiatic the individual 'never becomes a proprietor, but only a possessor, he is at bottom himself the property of him in whom the unity of the commune exists' (ibid., p. 493).

In the 'Formen' Marx has moved from the concrete analysis of the impact of British capitalism on India to an abstract analysis of the presuppositions or historical preconditions for the development of capitalism. One of the principal distinctions between capitalist enterprise and earlier forms of production is that between free and unfree labour, leading Marx to conceptualize the subjective and objective bases of unfree labour, i.e. the manner in which the producers relate to their conditions of production and reproduction and the material conditions of their production and reproduction:

The real appropriation takes place not in the mental but in the real active relations to these conditions - in their real positing as the conditions of his subjective activity. . . . Not only do the objective conditions change in the act of reproduction, e.g. the village becomes a town, the wilderness a cleared field etc. . . . but the producers change, too. . . . The older and more traditional the mode of production itself - and this lasts a long time in agriculture; even more in the oriental supplementation of agriculture with manufacture - i.e. the longer the *real process* of appropriation remains constant, the more constant will be the old forms of property and hence the community generally (ibid., pp. 493-4).

Among the earlier forms, the Asiatic communities are considered more original and more persistent because of their combination of agriculture and handicrafts. The dissolution of these earlier forms is related to the process of accumulation of monetary wealth, (44) but the accumulation of monetary wealth does not automatically lead to capitalism. Marx points to the case of ancient Rome and Byzantium where the development of monetary wealth through trade was bound up with the dissolution of old property relations, but instead of leading to industry 'led in fact to the supremacy of the countryside over the city' (ibid., p. 506).

In the 'Formen' and later in 'Capital', Marx's concern with precapitalist forms of production is to relate specific categories - property, labour, surplus appropriation - to capitalism in an abstract formal manner, rather than account for a specific or universal evolutionary process. (45) As has been noted, he was aware of the problems of an oversimplified periodization based on singular economic categories, although in his 'Preface' to 'A Contribution to the Critique of Political Economy' (1859), he seems to have moved towards a periodization of modes of production.

'A Contribution to the Critique of Political Economy' (1859)
In his succinct 'Preface', Marx introduces the notion of a succession of modes of production in a historical-chronological sequence, a sequence measured by 'progress'. 'In broad outline, the Asiatic ancient, feudal and modern bourgeois modes of production may be designated as epochs marking progress in the economic development of society' (Marx, 1859, p. 21).

The AMP is seen as the first 'progressive epoch'. (46) Further-

more, communal property which provides the foundation of this
mode is not geographically restricted:

At present an absurdly biased view is widely held, namely that
primitive communal property is a specifically Slavonic, or even
an exclusively Russian, phenomenon. It is an early form which
can be found among the Romans, Teutons and Celts, and of
which a whole collection of diverse patterns (though sometimes
only remnants survive) is still in existence in India. A careful
study of Asiatic, particularly Indian, forms of communal
property would indicate the disintegration of different forms
of primitive communal ownership gives rise to diverse forms
of property. For instance, various prototypes of Roman and
Germanic private property can be traced back to certain forms
of Indian communal property (Marx, 1859, p. 33).

India (the source of Indo-European languages) is seen as a
laboratory for the study of dissolution of communal property,
although given the impact of British colonialism there reconstruc
tion is difficult. Furthermore, the idea that the varieties of communal property which had existed in India had given rise to
distinct forms of property gives additional credence to the notion
that Marx envisioned evolution in 'multilinear' terms. Not only is
communal property now not specifically confined to one geographical area, but the variations of this form of property within
a specific geographical area preclude the assumption of beginning
with a political unit, e.g. India, in conceptualizing the mode of
production.

Turning from the forms of property and labour, Marx gives
attention to the form of accumulation in Asia, where 'hoarding'
or the accumulation of wealth in precious metals plays a sub
ordinate part 'in the total mechanism of production . . . [and]
is still considered the final goal' (ibid., pp. 34-5) in contrast
to capitalism.

With regard to the production of commodities in these communities,
Marx reconsiders his previous notion that exchange originates
between, rather than within, these communities, this time from
the appearance of the individual commodity producer:

In this case the social character of labour is evidently not
effected by the labour of the individual assuming the abstract
form of universal labour or his product assuming the form of
a universal equivalent. The communal system on which this
mode of production is based prevents the labour of the individual from becoming private labour and his product the private
product of a separate individual; it causes individual labour to
appear rather as the direct function of a member of the social
organization (ibid., pp. 33-4).

'Capital' and 'Theories of Surplus Value'
In the course of the three volumes of 'Capital', Marx makes
scattered references to the AMP, India, or Peru in trying to
differentiate historically the validity of particular abstractions.
The point of reference is again the capitalist mode of production.

For instance, on the logical relationship between the division
of labour and the production of commodities, Marx uses the
Indian community as a counter-case (1867, pp. 49-50):

> To all the different varieties of values in use there correspond
> as many different kinds of useful labour, classified according
> to the order, genus, species, and variety to which they belong
> in the social division of labour. This division of labour is a
> necessary condition for the production of commodities, but it
> does not follow conversely, that the production of commodities
> is a necessary condition for the division of labour. In the
> primitive Indian community, there is a social division of labour
> without production of commodities.

He distinguishes between the division of labour within the self-
sufficient communities where production is governed by use-
value and the division of labour within Indian society as a whole,
where only a portion of the surplus appropriated by the state
becomes a commodity (ibid., p. 337).

Marx contrasts the developed commodity production and circulation of industrial capitalism with that of earlier modes of production. In the Asiatic, part of the surplus is converted into commodity exchange between villages, another portion of surplus in
the form of state revenue is exchanged for products of artisans
retained by the state, or enters into long-distance trade (ibid.,
pp. 83 and 337). The development of commodity production is
taken as an indice of the dissolution of production based on the
combination of handicrafts and agriculture within village communes.

The impressive results of simple co-operation in the monumental
architecture of ancient Asia, Egypt, and the Etruscans is distinguished from the co-operation of the capitalist joint-stock
company by the opposition between free and unfree labour (ibid.,
p. 316):

> Co-operation, such as we find at the dawn of human develop-
> ment, among races who live by chase, or say, in the agriculture
> of Indian communities, is based, on the one hand, on owner-
> ship in common of the means of production, and on the other
> hand, that in those cases, each individual has no more torn
> himself off from the navelstring of his tribe or community,
> than each bee has freed himself from connexion with the hive.

In Volume III of 'Capital', Marx examines the question of how
surplus is to be defined and how it is appropriated in the Asiatic.
Presupposing some form of social production (Marx cites as
examples the primitive Indian community or 'the ingeniously
developed communism of the Peruvians'), one can distinguish
between the part of the product consumed by the direct producers, that which is consumed in production, and that which
constitutes surplus labour which 'serves constantly to satisfy
the general social needs, no matter how this surplus product
may be divided, and no matter who may function as representative
of these social needs' (Marx, 1894, p. 877).

Marx furthermore postulates a periodization of the forms which
the appropriation of surplus labour in agriculture can take: rent

in labour to rent in kind. The coexistence of corvée-labour with rent in kind, whether the landlord is a private person or the state, is an indication of the survival of an earlier form at a 'higher stage of civilization for the direct producer, i.e. a higher level of development of his labour and of society in general' (ibid., p. 794).

In conceptualizing the appropriation of surplus labour by the state in Asia, Marx follows on from Adam Smith's idea of coincidence of rent and tax. The extortion of surplus labour, not only in the Asiatic, but in other modes in which the direct producer still controls the means of production, proceeds by extra-economic means. Here Marx drops his idea of 'property' as real appropriate of the 'Grundrisse', returning to Bernier's notion of the state as the landlord:'Sovereignty here consists in the ownership of land concentrated on a national scale. But, on the other hand, no private ownership of land exists, although there is private and common possession and use of land' (ibid., p. 791).

In 'Capital' and in the 'Theories of Surplus Value', Marx recapitulates many of his ideas on the nature and conditions of unfree labour within the Asiatic commune; however, in contrast to the 'Grundrisse' and his newspaper articles, he directs his attention to the significance of the appropriation of surplus in the Asiatic for an appreciation of the appropriation of surplus value in the capitalist mode.

Following on from Adam Smith's definition of productive labour as that which enters into direct exchange with capital, Marx draws attention to the implications for the definition of unproductive labour from the perspective of the capitalist mode (1905, p. 157):

It is labour which is not exchanged with capital, but *directly* with revenue, that is, with wages or profit. . . . Where all labour in part still pays itself (like for example the agricultural labour of the serfs) and in part is directly exchanged for revenue (like the manufacturing labour in the cities of Asia), no capital and no wage labour exists in the sense of bourgeois political economy. These characteristics are therefore not derived from the material characteristics of labour, neither from the particular character of the labour as concrete labour, but from the definite social form, the social relations of production within which labour is realized.

Manufacturers are few in number in Asiatic countries and able to sell their commodities at monopoly prices, appropriating part of the revenue of the state. Such manufacturers profit not only from selling 'unpaid labour', but by selling commodities at over the quantity of labour contained in them (ibid., p. 277).

Drawing on Richard Jones's theory of accumulation in India Marx identifies the sources of accumulation in the Asiatic mode as wages and rent (Marx, 1894, p. 420).

In these texts Marx moves back and forth from a more originary view of the Asiatic mode, focusing on the self-sufficient communities and then on the relationship among these communities

and between them and the state, between the state and manufacturers' external trade. In this way the Asiatic mode is
conceived of as having an historical dimension, with a specific
dynamic. The AMP is not considered stagnant, in the sense that
'stagnation' is defined in an a priori fashion; it is pictured as an
historically differentiated whole which never spontaneously
evolved into capitalism.

*Marx's ethnological research and his correspondence on the
Russian mir*(47)
In the last years of his life Marx returned to his interest in pre-
capitalist societies, this time devoting himself to a more concrete
study of these societies rather than attempting to relate them to
the categories of the capitalist mode of production. In reading
Morgan and Kovalevsky he began to see communal systems as a
set of variants. In his critical appraisal of Maine and Phear he
returned to India as the laboratory for the study of the dissolution of communal forms.

In his critical notes on Maine, Marx reproaches Maine's critique
of John Austin's theory of sovereignty.(48) Maine had failed to
distinguish between government, society, and the state. The
state, in all its forms, appears at a certain stage of social evolution, once there has been a process of individuation from the
communal bonds of the group. These 'individualities' or interests
are themselves class interests: 'hence this individuality is itself
class, etc., individuality, and these interests all have, in the
last analysis, economic conditions at the basis. On these bases
the state is built and presupposes them' (Krader, 1972, p. 329).

Thus the development of the state and the various forms it
has taken cannot be understood as a sequence of juridicial forms
but must be related to the relations of production. In the development of the AMP, as seen in the Indian 'laboratory', it is not the
direct producer of the village community who is freed from the
communal bonds, rather:

The individuality that is torn free is that of the consumers of
the surplus product, these are members of the ruling class,
the sovereign, clients, his retainers, the courtiers, the wealthy
in the rural life, and money-lenders, usurers, zamindars
(Krader, 1975, p. 224).

Despite Marx's opposition to Maine's general theory of the
development of the state out of the patriarchal family, Marx
found in Maine confirmation for his own view of the self-sustaining
characteristics of village communities.

Marx criticized both Phear and Kovalevsky for the suggestion
that feudal relations of production were to be found in India.
Kovalevsky had based his argument on the existence of the ikta
in India, land grants for military service rendered. Marx pointed
out that such benefices were not uniquely feudal, but had existed
in ancient Rome as well. Furthermore, the form of social labour in
India could not be characterized as serfdom; in India land could
only be alienated through the consent of the village community;

and finally, in contrast to Europe, the ruler or superior lord had no claim over the administration of justice on the domain of his vassal, whereas in India the sovereign had direct control over his tax-collecting/policing agents (Levitt, 1978). In the midst of his ethnological research, Marx was called upon to relate his work on the village commune to the prospects of achieving socialism on the basis of pre-capitalist relations of production. In November 1877, he had written a letter to the editorial board of 'Otechestvenniye Zapiski', contesting the idea that his theory of the development of the capitalist mode of production in Western Europe was 'an historico-philosophic theory of the general path that every people is fated to tread' (Marx and Engels, 1965, p. 313). From his own researches Marx felt that Russia since the mid-nineteenth century was tending towards becoming a capitalist nation.

In 1881, Marx was called upon to predict the fate of Russia. Was the complete development of capitalism in Russia a necessary stage to the ultimate creation of socialism or did the remaining institutions of the Russian commune provide a basis for socialism? In his reply to U. Zazulich, Marx affirms that such a possibility of the development of socialism existed, not just because of the survival of these pre-capitalist relations, but because their survival was contemporaneous with developed capitalism in Western Europe, by which Russia had not been enslaved as is the case in the East Indies. This path would entail eliminating the private property which had developed within the mir and incorporating the positive developments (technology) of capitalism. The Russian commune, in which arable land was privately owned and pastures, forests, etc. were held in common, is seen by Marx as the most recent form of the agricultural commune. Whereas in the more archaic forms the individual was bound through kinship to the commune, the more evolved mir was an association of individuals.

Marx's ethnological researches and his correspondence on the Russian commune today constitute evidence of Marx's interest in the development of non-Western societies per se, his critical appreciation of leading ethnologists of his day. These writings reaffirm his method expounded in the 'Grundrisse', particularly his wariness of periodization focusing on a specific element abstracted from the totality.

Engels's 'Anti-Dühring' (1878) and articles on Russia (1875, 1894)

Engels's 'Herrn Eugen Dührings Umwälzung der Wissenschaft' is a polemical work aimed at refuting Dühring's force theory of history, wherein major changes in the history of human society were attributed to warfare and conquest. Oriental societies, with a history of numerous conquests but a relatively stable base, constituted a good counter-example to Dühring's theory. Engels noted that the age-old village communities of India had evolved from tribal communal property to the parcellation of land and

differences of wealth among the population; these changes had
come about through intercourse with the outside: exchange had
resulted in unequal distribution within the communities which in
turn had led to the parcellation of land and the creation of small
and large property holders. Such changes had not been effected
by the conquerors; conquerors and members of these village
communities had had little contact. Economic processes rather
than political events were at the basis of the change that had
occurred within these communities.

In arguing specifically against the force theory of the development of the state, Engels postulated a two-stage development of
the state; it first arose out of functions necessary to, and serving,
the common interest (generally protection, and in the Orient the
necessity of irrigation-works). Once in existence, the state
developed into a repressive force serving the interests of the
ruling class. For Engels, oriental despotism was the most primitive form of the state since it rested on the most elementary form
of rent, rent in labour.

In Engels's 'The Origin of the Family, Private Property and the
State' (1884), composed on the basis of Marx's notes on Morgan
and Engels's own reading on communal institutions in Europe,
oriental despotism is not considered.

Whereas in 'Anti-Dühring' Engels had focused on the historical-
civilized period, he now projected his analysis back to the prehistoric condition of man. His method is more chronological or
historical, setting out a linear typology of forms, from simple to
complex. The basis of the typology is a common end point, which
is defined by a particular element, a strong centralized government. The state and its origins in Asia do not enter into Engels's
evolutionary scheme which is centered on the development of the
Mediterranean and Europe using Morgan's ideas on American
Indians in his recreation of early developments.

Engels also entered the debate on the Russian commune. The
Russian commune was a more evolved form of communal property
than that found in India; within the commune there were already
significant differences of wealth. From all indications it was
leading towards complete dissolution into private property.
Furthermore, the isolation of these communities and the narrow-
ness of the members' outlook hardly provided the foundation for
a transition to socialism:

> Such a complete isolation of the individual communities from
> one another, which creates throughout the country similar,
> but the very opposite of common, interests, is the natural
> basis for *oriental despotism*, and from India to Russia this
> form of society, wherever it prevailed, has always produced it
> and always found its complement in it (Marx and Engels, 1969,
> p. 394).

Whatever vestiges of communal labour that remained in Russia
could provide a partial basis for the construction of socialism in
Russia, but:

> A victory of the West-European proletariat over the bourgeoisie

and the consequent substitution of a socially managed economy
for capitalist production - there is the necessary precondition
for the raising of the Russian community to the same stage of
development (ibid., p. 402).

NOTES

1   The two most systematic historiographical works on Marx's
    concept of the AMP (Godelier, 1970, and Krader, 1975) agree
    on the ideological nature of two notions frequently attributed
    to both Asiatic social formations and to Marx's concept of the
    AMP, i.e. the notions of despotism, or oriental despotism,
    and stagnation. However, Krader has been careful to delimit
    and periodize the concepts of oriental despotism, oriental
    society and the Asiatic mode of production in Marx (1975,
    pp. 1-2):
        Karl Marx brought out the theory of the Asiatic mode of
        production in several stages. Before he developed that
        theory he swiftly recapitulated the steps which European
        social thought had taken: in his youth Marx made a fleeting
        reference to the oriental despotism, but this characterization
        was a mere rhetorical flourish ('Rheinische Zeitung', no. 139,
        1842). In the second stage, or the first actual one, he
        addressed himself to the concept of the oriental society as
        a whole. In it he brought out the economic factors at work,
        but paid the most attention to the political and social
        characterization of the society, and it remained at a dis-
        tance from the subject matter. This stage is represented in
        his articles written in 1853 for the 'New York Daily Tribune',
        and in his correspondence with Friedrich Engels at that
        time. In the next stage, which was begun in 1857-58, he
        formulated the theory of the AMP in a general way; there
        he initiated the inquiry into the internal economic relations
        in a specific way, and at the same time, he set forth the
        legal and political forms in their economic relations. . . .
        Moreover, he had not yet found the characterizing term
        to be applied to this conception; this he introduced in
        1859, when he wrote for the first time of the Asiatic mode
        of production. This theory was further developed in the
        1860s, in the writings which culminated in the first volume
        of 'Capital'. There the theory is found in a strict sense
        but it is not systematized, being expressed here and there
        in the text.
    Consequently for Krader, in Marx's 'theory in the strict
    sense' neither oriental despotism nor stagnation play a part
    (ibid., pp. 152, 185, 275, 292). While Godelier's conclusions
    have very much the same import for contemporary discussions,
    the notions of stagnation and oriental despotism are thought
    to remain in Marx's post-1858 concept of the AMP but from
    the perspective of contemporary scholarship should be con-

sidered the 'dead parts' of the Marx and Engels legacy
(Godelier, 1970, pp. 135-7). However, in terms of the continuity between Marx and Engels and earlier writers, Godelier
would stress the continuity between these dead parts of
Marx's concept of the AMP and the seventeenth-century
travellers and the Enlightenment, whereas Krader would
restrict the sources of the concept of the AMP to Adam Smith,
Hegel, and Richard Jones (Krader, 1975, p. 7).

2   The rise of the absolutist states in Eastern and Western
Europe and the contrast between these states and the state
of the Ottoman Empire and China is the subject of Perry
Anderson's 'Lineages of the Absolutist State' (1974). Anderson sets out to challenge Marx's and Engels's descriptions
of absolutism 'as a state system representing either an
equilibrium between bourgeoisie and nobility, or even an
outright dominance of capital itself' (ibid., p. 17). Proceeding by means of a regional typology, Anderson, following
Althusser (1959), sets out to prove the eminently feudal,
rather bourgeois, character of absolutist states, distinguishing the Western absolutist form, which develops as 'a
compensation for the disappearance of serfdom' (ibid., p. 195)
from its Eastern counterpart, 'a device for the consolidation
of serfdom' (ibid.). The two geneses of these forms of
absolutism are said to be distinct, both of an entirely different
'species' from Chinese and Islamic civilizations in the early
modern period; the latter represent 'two patently divergent
morphologies of state and society' (ibid., p. 547). Somewhat
less concerned with cultural boundaries and typology, Wallerstein (1974) examines the role of the state at the beginnings
of the capitalist world system, considering the arguments
concerning the extent of state involvement in capitalist enterprise, the economic impact of its involvement and, finally,
its class content (Anderson's object of study). When coming
to this final problem, Wallerstein reframes what he calls the
'not too helpfully abbreviated' distinction between the aristocracy and the bourgeoisie, as one between capitalist landlords and capitalist merchants (ibid., p. 157). By the sixteenth century, not only had members of the nobility become
entrepreneurs, but there was a transition from the status of
the successful bourgeois to land-owner and noble. Despite
this occupational mobility, the strength of the land-owning
class did not disintegrate. Wallerstein is less interested in
defining this renewed landlord class as feudal landlords than
in their role in the expansion of the capitalist world system
(ibid., p. 87).

3   Machiavelli's contrast between European and oriental political
forms is found in 'The Prince' (1537) and 'Discourses on the
First Ten Books of Titus Livius' (1531). On Machiavelli's
distinctions, see Chabod (1964).

4   See Koebner (1951) for an extended examination of Aristotle's
'Politics' and its translations at the end of the Middle Ages.

5   See Anderson (1974, pp. 50-1) for the view that Bodin's
    theory of sovereignty accurately reflected a fundamental
    contradiction between the royal monarch's absolute power
    and its limits by natural or divine law.
6   The influence of Bodin's environmentalist explanation of
    political and ethnic diversity, as well as his opposition
    between 'royal monarchy' and 'seigneurial monarchy', are
    evident in Montesquieu's 'L'Esprit des lois' (Stelling-Michaud,
    1960-1, p. 331).
7   Stelling-Michaud (1960-1, pp. 330-1) recalls the terror in-
    spired by Turkish expansion in the Balkans and the Danube
    plain, a terror exploited by the church, reformers, and the
    civil authorities, contributing to the spread of the belief in
    a total opposition between Christian Europe and the Islamic
    East. However, Stelling-Michaud goes on to contrast this
    Turkophobia with a current of curiosity and admiration of
    certain qualities and virtues of the Turks - their tolerance
    of the Christian minority, their discipline and their sobriety -
    and cites Postel's 'De la Republique des Turcs' (1560) as
    a case in point.
8   See Hodgen (1964) on these writers' 'anthropological' inter-
    ests.
9   For an extensive bibliography of the accounts of seventeenth-
    century travellers, see Krader (1975). It was the views of
    François Bernier, physician and follower of Descartes's
    opponent, Pierre Gassendi, which came to prevail through-
    out most of the eighteenth century. Bernier served several
    years as a physician in Mogul India in the 1660s and published
    a series of reports, letters, and diaries of his travels.
10 Krader (1975, pp. 26-7) makes the case that J.B. Tavernier,
    a merchant who became quite wealthy by trading in the East,
    and J. Chardin were perhaps the only seventeenth-century
    writers to qualify the notion of the Mogul emperor as the sole
    proprietor. 'The Mogol emperor was, according to Tavernier,
    the absolute lord of all the lands; he was not their owner,
    but their master . . . he was landlord, not land-owner,
    accordingly, receiving income from them in his public, not
    private capacity' (ibid., p. 27). See J.B. Tavernier, 'Les
    six Voyages qu'il a fait en Turquie, en Perse, et aux Indes'
    (1679); and J. Chardin, 'Voyages en Perse et autres lieux
    de l'Orient' (1711).
11 Montesquieu's belief in the role of commerce as a prevention,
    or as a development entailing the suppression of despotic
    rule, has recently been noted by Hirschman (1977, pp. 70-81).
    Montesquieu 'hailed the bill of exchange and arbitrage as
    auxiliaries of the constitutional safeguards and as bulwarks
    against despotism and les grands coups d'autorité; and there
    can be little doubt that these passages on the favourable
    political consequences of economic expansion constitute an
    important, and hitherto neglected, contribution to his central
    political thesis, just as they represent a basic justification of
    the new commercial-industrial age' (ibid., p. 78).

12   See Krader (1975, pp. 29–32) on the sources of Voltaire's notion of the patriarchal nature of Chinese society.
13   This attempt to provide a positive connotation for the concept of despotism failed.
     The use of the term by the economists was so paradoxical that it gave rise to all manner of polemics and violent discussions, while men like Turgot, very close to the physiocrats in their philosophy, refused to accept it and even condemned it. The result was that it never became current in politics – also because of the political defeat suffered by the physiocrats – but remained the symbol of a lost cause and vain attempt (Venturi, 1963, p. 136).
14   Schumpeter credited Quesnay as the first to see the 'economic' as a coherent whole of interdependent parts. Dumont (1977, pp. 50–67) puts forward the idea that such a holistic view could only have been derived from 'the projection of a general conception of the universe as a totality on to the economic level' (ibid., p. 51). This general conception in turn drew upon the traditional view of China.
15   Anquetil-Duperron has been hailed as the thinker who 'opposed the real Orient to Montesquieu's mythical Orient' (Althusser, 1958, p. 596). Similarly, Venturi (1963) and Stelling-Michaud (1960-1) have cited the French orientalist as the bitter critic of colonialism. However, Anderson (1974, p. 466) has gone into Anquetil-Duperron's later works, particularly 'L'Inde en rapport avec Europe' (1782), where he appears more as 'a disappointed French patriot, chagrined by the success of British colonialism in ousting its Gallic rival from the Carnatic, and the subcontinent . . . [calling for] "the French flag to fly with majesty once again in the seas and lands of India" '.
16   See Venturi (1963, p. 140) and Krader (1975, p. 195). The latter criticizes Venturi for having mistaken Anquetil-Duperron's subject of attack for John Dalrymple. Alexander Dalrymple was one of the first to develop the theory of communal property in India and was criticized by Anquetil-Duperron who maintained that private property in land existed in India.
17   Adam Smith's solution to the question of land-ownership in the Orient was that the sovereign there collected rent on his land in his private capacity and tax in his public capacity. Smith's idea of the specific relation 'the tax-rent couple' was taken up by Marx (see Krader, 1975, p. 38n;
18   There is some dispute as to whether Adam Smith's mention of canals was intended to illustrate the role of the state in irrigation or in communication (see Krader, 1975, 38n; Wittfogel, 1957, p. 372).
19   For Hegel's debt to the Enlightenment, see D'Hondt (1974). For the continuities between Voltaire's and Hegel's view of oriental history see Krader (1975, p. 30).
20   Mill believed that the basic economic structure of oriental

despotism should be retained, i.e. government appropriation of the surplus. The reason for Mill's rather radical
approach to this question was his fear that the creation of
'strong' private-property rights in India would lead to the
creation of an unproductive and reactionary class of landed
aristocrats. . . . Mill argued that although government
appropriation of the surplus had in the past led to stagnation, it need not in fact retard the economy if the surplus
were calculated scientifically, according to the utilitarian
theory of rent (Sawer, 1974, p. 1512).

21 Jones advocated the development of a landed aristocracy in
India to overcome the political ills which had resulted from
the lack of these intermediary bodies. Jones sought the
explanation of the government monopoly of land in the fact
that oriental societies were 'conquest states' (see Sawer,
1974, pp. 1516-23).

22 See Godelier (1970, pp. 33-5) on the impact of F. Schlegel,
'Uber die Sprache und Weisheit des Indien' (1808); R.C.
Rask, 'Essay on the Origin of the Ancient Scandinavian or
Icelandic Tongue' (1818); and J. Grimm, 'Deutsche Gram-
matik' (1822). Godelier remarks that linguistics 'était la
première des sciences historiques à démontrer l'existence
de lois d'évolution en histoire et à fonder la possibilité de
reconstruire le passé en remontant à partir de survivances
présentes ou de formes derivées connues vers des formes
originaires inconnues' (ibid.).

23 The principal works in question are: 'The Fifth Report'
(1812); Metcalfe's appendix to the 'Report from the Select
Committee in the House of Commons' (1830); M. Elphinstone's
'History of India' (1839); M. Wilks's 'Historical Sketches of
the South of India' (1810); and G. Campbell's 'Modern
India' (1052).

24 For Maine's sources on India see Dumont (1966, pp. 80-9);
for Maine's sources on the South Slavs see Krader (1976,
p. 144).

25 The major mid-eighteenth-century works on the village
community in northern Europe and Russia were: Olufson,
'Bidraq til Oplysning om Danmarks indvortes Fortfatning i
de aildre Tider, isaer i det trettende Aahundrede' (1821);
A. von Haxthausen, 'Studien über die innern Zustande
das Volksleben und insbesondere die landlichen Einrichtungen
Russlands', 3 vols (1847-52); 'Die landliche Verfassung Russlands' (1862); G.L. von Maurer, 'Einleiteung zur Geschichte
der Mark- Hof- und Stadt-Verfassung' (1854); 'Geschichte
der Markenverfassung in Deutschland' (1856); 'Geschichte
der Dorfverfassung in Deutschland', 2 vols (1865-6).

26 'Ancient Law' (1861); 'Village Communities in the East and
West' (1871); 'Lectures on the Early History of Institutions'
(1875).

27 Among these were M. Kovalevsky's 'Obsinnoe Zemlevladenie'
(1879) and J.B. Phear's 'The Aryan Village in India' which

Marx consulted at the end of his life (see Krader, 1972, and 1975).

28   The theory of agrarian communism is widely held and widely disputed. At the basis of the theory is the premise that the tillers of the soil in primitive societies and in the early periods of civil society held their land in common; the social unit that held the land in common was bound together by ties of kinship or proximal residence, or both. The form of this social unit was a village, or kin-village, gens, sib, clan, phratry, tribe. The landholdings took the form of ownership by the social unit as a whole; individual and family relations to the soil were in the form of possession, which is distinct from property, whether common or individual. . . . Theory and practice of agrarian communism (i.e. Utopian communities) are a part of the social and political struggles of the past century as well as of their scientific and philosophical reflection. It has frequently happened that those involved in the struggle, in particular, those who called the theory of agrarian communism in question, have purported to argue on scientific grounds alone, but this has made their argument no less politically intentioned (Krader, 1976, pp. 123-4).

29   See Venturi (1966) on the social history of populism in Russia. Venturi illustrates the role that political activity among the peasantry in the 1870s played in the changing populist conception of the Russian peasantry:
Here and there they saw the peasants' mentality in a new light not at all as they had imagined. One day, for instance, Aptekman was describing to a crowded group what social life would be like when the people owned its own lands, woods and waters. He was interrupted by a peasant who shouted, 'That's grand. We'll divide the land and I'll take two workers and then I'll be in a fine position.' N. Morozov too saw for himself how far the patriarchal collectivism and spirit of equality in village life were being undermined by the rise of richer and stronger elements (ibid., p. 505).

30   See Walicki (1969, pp. 132-47).

31   On Marx's movement from a more abstract philosophical anthropology to a more concrete ethnological pursuit see Krader (1973).

32   For Marx's readings on Asia at this time, see Krader (1975, p. 80); and Ruben (1953-4).

33   Marx is here drawing directly upon Sir Stamford Raffles, 'History of Java' (1817).

34   Engels to Marx, 6 June 1853 (Marx and Engels, 1965, pp. 82-3).

35   Engels stressed the importance of artificial irrigation for agriculture in the East, but did not exclude the possibility that irrigation works were carried on by local or regional

organizations (see Marx and Engels, 1965, p. 82).
36 Marx criticizes the British, the most recent conquerors, for having neglected the department of public works, allowing previously tilled land to go to waste. This passage of Marx on the role of the state in irrigation in the East has been interpreted by Wittfogel as evidence of Marx's support of the 'hydraulic hypothesis'. Marx here, however, is not seeking the origins of the state in Asia, but describing functions taken on by the state rather than by voluntary associations as in Europe. Although the search for origins appears in Engels's letter to Marx (6 June 1853), Engels is seeking the origins of the absence of private property in land. In 'Anti-Dühring' (1878) Engels would in fact propose that the state in Asia had developed out of the function of providing irrigation.
37 There is some controversy as to which source Marx was using when he enumerated these occupations. According to Krader (1975, p. 83), Marx took this list from Mark Wilks, 'Historical Sketches of the South of India' (1820).
38 The presence of inequality within these communities had been remarked upon by Marx in his article 'The Duchess of Sutherland and Slavery' ('New York Daily Tribune', 9 February 1853) where Marx compared the Scottish clan with 'Asiatic family communities' (Marx, 1975, p. 144).
39 In his letter of 14 June 1853, Marx wrote to Engels:
Your article on Switzerland ('New York Daily Tribune', 17 May 1853) was of course a direct smack at the leading articles in the 'Tribune' (against centralization, etc.), and its Carey. I have continued this hidden warfare in the first article on India, in which the destruction of the native industry by England is described as revolutionary. This will be very shocking to them. As for the rest, the whole rule of Britain in India was swinish, and is to this day (Marx and Engels, 1965, pp. 84-5).
40 A two-volume edition published by the Foreign Languages Publishers, Moscow, under the editorship of the Marx-Engels-Lenin Institute. The first volume included the 'Introduction' previously published by Kautsky in 'Die Neue Zeit' (1903, XXI, p. 1) and seven notebooks covering the chapters on money and capital. The second volume contained an index of the notebooks written from August 1857 to March 1858, a draft of the 'Contribution to the Critique of Political Economy' (1859), referate on the various notebooks, a précis of what became the first section of 'Capital', vol. I, extracts from notes on Ricardo with index, and a sketch of Bastiat and Carey (previously published by Kautsky, 'Neue Zeit' (1904, XXII, p. 2) (Rubel, 1956, p. 136)). A photo-offset was issued by Dietz Verlag in 1953. According to Rosdolsky, only three or four copies of this edition reached the West (1968, p. 7, cited by Nicolaus (1973, p. 7)). The section, the 'Formen', was

published separately three times between 1939 and 1940
in the Soviet Union (see Baron, 1975, p. 140).

41 The continued existence of common property in India by
the mid-nineteenth century was considered to be a minority
phenomenon by Marx (Marx and Engels, 1965, p. 86). On
Marx's appreciation of the complexity of Indian land tenure, see
Avineri, 1969, pp. 313-16 (Marx's article 'Land Tenure in India',
'New York Daily Tribune', 7 June 1858).

42 A presupposition of wage labour, and one of the historic
preconditions for capital, is free labour and the exchange
of this free labour for money, in order to reproduce and
to realize money, to consume the use value of labour not
for individual consumption, but as use value for money
(Marx, 1957-8a, p. 479).

43 For Marx's changing usage of the concepts of property
and possession with reference to the AMP, see Krader
(1975, pp. 13-16, 101, 176).

44 It will be seen on closer inspection that all these processes
of dissolution mean the dissolution of relations of production in which: use-value predominates, production for
direct consumption; in which exchange value and its
production presupposes the predominance of another form;
and hence that, in all these relations, payments in kind
and services in kind predominate over payment and money-
services. But this only by the way. It will likewise be
found on closer observation that all the dissolved relations
were possible only with a definite degree of development
of the material (and hence also the intellectual) forces of
production (Marx, 1857-8a, p. 502).

45 Marx and Engels in 'The German Ideology' had proposed a
four-fold periodization: patriarchal tribal, corresponding
to primitive forms of economy (hunting, fishing, horti-
culture), ancient communal and State ownership (Ancient
Greece and Rome), feudal society and bourgeois society
(1969, I, pp. 21-4).

46 Marx makes a distinction between pastoralists and peoples
living by hunting and fishing:
For example, with pastoral peoples (mere hunting and
fishing peoples lie outside the point where real development begins). Certain forms of tillage occur among them,
sporadic ones. Landed property is determined by this.
It is held in common and retains this form to a greater
or lesser degree according to the greater or lesser degree of
attachment displayed by these peoples to their traditions ...
1973, p. 107).

47 'The Ethnological Notebooks', ed. L. Krader (1972), contain
Marx's notes and comments on L.H. Morgan, 'Ancient Society'
(1877), Sir John Phear, 'The Aryan Village in India and
Ceylon' (1880), Sir Henry Maine, 'Lectures on the Early
History of Institutions' (1875) and Sir John Lubbock, 'The
Origin of Civilisation' (1879). Marx's notes and comments

on M.M. Kovalevsky, 'Obščinnoe Zemlevladenie' (1879) (Communal possession of land) are found in L. Krader, 1975, pp. 343–412. Marx's letter to the Editorial Board of 'Otechesvenniye Zapiski' (November 1877) was first published by the populists in 1886 (Marx and Engels, 1965, pp. 311–13). Marx's letter to Vera Zazulich (8 March 1881) was first published by B. Nikolaevesky in 1924. The three drafts of the letter are reprinted in CERM (1970).

48   John Austin, 'The Province of Jurisprudence Determined' (1832).

PART II
THE FATE OF THE
AMP FROM PLEKHANOV TO
STALIN

EDITORS' INTRODUCTION

In the fifty years after Marx's death, evolution became a matter
of increasing political importance. An understanding of the
social evolution of particular nations and the characterization of
the current situation in evolutionary terms was considered the
theoretical precondition for formulating an appropriate national
and international strategy. Before the Russian Revolution of
1917, the evolutionary concerns of Marx's followers centred
on the necessity of a capitalist transformation of pre-capitalist
forms for the achievement of socialism. Whereas Marx had
waivered on the necessity of this sequence in his correspondence
with V. Zazulich and Mikhailovsky, Russian Marxists (Plekhanov,
Lenin, and Trotsky) surmounted the question of such a necessary sequence as far as Russia was concerned by demonstrating
the degree of capitalist development there.

Plekhanov's 'Our Differences' (1881) is his stock-taking
of the divergences between 'scientific socialism' as advanced
by Marx and Engels, and the Narodniks' Utopian fantasies.
Plekhanov had shared in these fantasies as a member of the
populist movement before he went into exile in 1880. In his
view, not only had the communes - which the Narodniks por-
trayed as the basis for a future socialist society - disintegrated
with the growth of exchange and the development of private
property, but their existence in Muscovite Russia had provided
the foundations of Russian absolutism. It is no wonder that
Plekhanov refrained from publishing both Marx's letter to
'Otechestvenniye Zapiski' sent to him by Engels in 1884 and
Marx's reply to Zazulich. In Plekhanov's polemics first with
the Narodniks and later with Lenin, the Asiatic state exploiting
village communes was held up as a negative spectre to dis-
courage idealizations of Russia's past and to argue against the
Russian Social Democrats' adoption of a policy of land nationalization in their agrarian programme.

In addition to his polemics with the Russian populists and
his organizational activities in the nascent Russian Social
Democratic Party, Plekhanov dedicated his years of exile to
the study of the works of Hegel, Marx, and Engels and con-

temporary ethnologists. For Plekhanov, Marx's historical
materialism was a form of geographic and technological determinism. According to the 'father of Russian Marxism':
    The characteristics of *social* man are determined at every
    time by the development of the productive forces, because
    on the degree of the development of the productive forces
    depends the entire structure of social union . . . this
    structure is determined in the long run by the character-
    istics of the geographical environment, which affords men
    a greater or lesser possibility of developing their produc-
    tive forces (Plekhanov, 1961, p. 739).
Plekhanov's review of Léon Metchnikoff's 'La Civilisation et
les grands fleuves historiques' combines his mechanistic interpretation of historical materialism with an attack on anarchism.
While he agrees with Metchnikoff's analysis of the geographical
and environmental causes of oriental despotism, he is quick to
criticize any notion of 'possibilism' on the part of the author,
tarnished by the idealistic brush of anarchism. For the reviewer,
socialism, like oriental despotism, is equally determined by the
logic and development of the productive forces.
    Lenin, like his teacher Plekhanov, dedicated his early writings
to a critique of Russian populism and to a Marxist analysis of
nineteenth-century Russia. He emphasized the increasing dominance of capitalist relations of production and circulation which
had effectively dissolved pre-capitalist relations within the
Russian commune. In describing the pre-capitalist phase of
Russian history, Lenin does not draw sharp distinctions between
patriarchal, Asiatic, and feudal relations. Within the context of
his clash with Plekhanov over the agrarian programme (1906) he
reluctantly accepted the latter's depiction of Muscovite Russia
as exemplifying the AMP, but he argued against the assumption
that the nationalization of land automatically entailed the restoration of the AMP. Unlike Plekhanov, Kautsky, or Luxemburg,
Lenin demonstrated no specific interest in pre-capitalist societies
and their evolution per se. As for Russian history, he was more
concerned to document the penetration of capitalism than to
ponder its pre-capitalist past. There is no indication that he
questioned the validity of the concept of the AMP (as opposed to
its relevance for Russian history) as evidenced by his references
to Marx's 1859 'Preface' (Lenin, 1960-70, vol. 24, p. 56) and
his notes on the Marx-Engels correspondence (Lenin, 1959).
    Among Marx's followers, P. Lafargue, H. Cunow, and R.
Luxemburg pursued ethnological studies, drawing widely on
the ethnological and ethnographic accounts of their day. In
works dedicated to ancient societies or contemporary primitives,
they employed concepts of primitive communism, agrarian communism, oriental society, oriental despotism, and the Asiatic
mode of production. Oriental despotism was considered a form
of 'natural economy', i.e. one where production was governed
by use-value, and as one of the earliest forms of class society.
In these ethnological works neither oriental society nor oriental

despotism was particularly controversial. Furthermore, there
was neither debate nor consensus on a unique or necessary
form or stage following the dissolution of primitive communism.
However, when these concepts were employed in the analysis
of societies simultaneously being examined with respect to the
necessity of passing through capitalism before achieving socialism, oriental despotism or the AMP took on immediate political
relevance.

For the leadership of the Second International, the question
of the necessary phase of capitalism lay behind their debates
over the correct socialist attitude to be adopted towards European colonialism and the location of the principal revolutionary
forces in the international struggle for socialism. Both Kautsky
and Luxemburg took a stand against the reformists Bernstein
and Van Kol, who espoused the idea that capitalism was a
necessary phase through which the colonies must pass. According to Kautsky, the task of socialists was to liberate the masses
of the whole world and to bring civilization to the colonial
peoples through friendship and assistance from the European
proletariat. The latter were considered the avant-garde of the
international socialist revolution.

But both the First World War and the Bolshevik Revolution
dampened the idea of a world-wide revolution led by the European proletariat. The 'nationalization' of revolution had its
parallel in the 'nationalization' of evolution, whereby the history
of nations was interpreted as a succession of modes of produc
tion. In the Soviet Union of the 1920s, both Trotsky's analysis
of the specificities of the development of Russia and Plekhanov's
emphasis on the external forces that had shaped Russian history
were attacked by the increasing 'official' history of Poktrovsky
and his school. The latter minimized the differences between
Russian and Western European development, depicting an
internal dynamic of Russian society as it moved from feudalism
through capitalism to socialism.

In the late 1920s and early 1930s, there was an intense debate on the nature of China's past among Soviet scholars and
strategists. The debate was intimately linked to the formulation
of Comintern policy for contemporary China, and specifically,
to the question of whether or not Soviet policy should be
directed at consolidating the alliance between the Kuomintang
and the Communists. This strategy was grounded in the idea
that the Chinese would have to undergo a bourgeois-democratic
revolution or overthrow the remnants of feudalism before pre-
cipitating a socialist revolution.

Both Varga and Riazanov analysed contemporary China in
terms of an articulation of the capitalist and Asiatic modes of
production. Varga disputed the anti-feudal nature of the
Chinese urban bourgeoisie, arguing that membership of the
urban bourgeoisie and the rural landlord classes largely over-
lapped. In 1928, the Sinologist Mad'iar published his study of
Chinese rural economy, in which he tried to introduce the

concept of the AMP as specifically relevant to Chinese agrarian history. In his foreword to Kokin and Papaian (1930), extracts from which are translated here, Mad'iar traces the development of Marx's thinking on the AMP.

A conference of historians, held in Tiflis in 1930, discussed the appropriateness of the concept of the AMP to the history of Turkey and Persia. Berin, who advocated the theory of a specific Asiatic mode of production, was accused of dogmatically following Marx's writings; a similar accusation of Mad'iar and his followers was made by Godes at the 1931 Leningrad discussions, excerpts from which are reproduced here. The major proponents of the AMP, Riazanov and Mad'iar, were to disappear during the purges of the mid-1930s. The concept thereafter came to be interpreted as an Asiatic variant of slavery or feudalism.

Stalin's 'Dialectical and Historical Materialism' (1938) listed primitive communism, slavery, feudalism, capitalism, and socialism as the five main types of relations of production known to history. Blame for the suppression of the concept of the AMP and the implantation of a unilineal evolutionary scheme is frequently assigned to Stalin. The tendency towards a mechanical vulgar materialist conception of Marx's historical materialism certainly pre-dates Stalin. The merging of world history and national histories as a sequence of universal stages is perhaps partly a product of the nationalization of revolution.

The assignment of the blame to Stalin only serves to obscure the implicit unilinealism and mechanicism of some of Marx's previous followers.

OUR DIFFERENCES*

G. Plekhanov

Listening to our Narodniks one could really think that the
Russian village community is an exceptionally enduring organization.
Neither the internecine struggles during the period of the
independent principalities, the Mongol yoke, or the bloody
period of Ivan the Terrible, nor the years of unrest during
the interregnum, nor the reforms of Peter and Catherine
which introduced into Russia the principles of West European
culture, nothing shook or changed the cherished institution
of peasant life,
says one of the most easily excitable Narodniks, Mr K—n, in a
book on 'the forms of land tenure among the Russian people';
'the serfdom could not obliterate it, its abolition could not be
brought about by the peasants leaving voluntarily for new lands
or by forcible expulsions,' etc., etc., in a word,
The ages went by, all strived to be happy,
In the world all repeatedly changed,
but the Russian village community remained unchanged and un-
changeable. Unfortunately, this glorification, despite all its
indisputable eloquence, proves nothing at all. The village communities display indubitable vitality as long as they do not
emerge from the conditions of natural economy.
The simplicity of the organization for production in these
self-sufficing communities that constantly reproduce themselves in the same form, and when accidentally destroyed,
spring up again on the spot and with the same name - this
simplicity supplies the key to the secret of the unchangeableness of Asiatic societies, an unchangeableness in such
striking contrast with the constant dissolution and refounding
of Asiatic states, and the never-ceasing changes of dynasty.
The structure of the economical elements of society remains
untouched by the storm-clouds of the political sky (Marx,
1867, vol. I, pp. 338-9).
But that same basic element of the barbarian societies which
stands firm against the storms of political revolutions turns out

to be powerless and defenceless against the logic of economic evolution. The development of money economy and commodity production little by little undermines communal land tenure.(1) Added to this there is the destructive influence of the state which is compelled by the very force of circumstances to support the principle of individualism. It is set on this road by the pressure of the higher estates, whose interests are hostile to the communal principle, as well as by its own ever-growing needs. The development of money economy, which in its turn is a consequence of the development of the productive forces, i.e. of the growth of the social wealth, brings into being new social functions, the maintenance of which would be unthinkable by means of the former system of taxes levied in kind. The need for money compels the government to support all the measures and principles of social economy which increase the flow of money into the country and quicken the pulse of social and economic life. But these abstract principles of social economy do not exist of themselves, they are only the general expression of the real interests of a certain class, namely that of trade and industry. Having emerged partly from the former members of the village community and partly from other estates, this class is essentially interested in mobilizing immovable property and its owners, since the latter are labour-power. The principle of communal land tenure is an obstacle to both of these aims. That is why it first arouses aversion, and then more or less resolute attacks on the part of the rising bourgeoisie. But neither do these blows destroy the village community at once. Its downfall is prepared by degrees. For a long time the outward relations of the members of the community apparently remain completely unchanged, whereas its inner character undergoes serious metamorphoses which result in its final disintegration. The process is sometimes a very lengthy one, but once it reaches a certain degree of intensity it cannot be stopped by any 'seizures of power' by any secret society. The only serious rebuff to a victorious individualism can be given by those social forces which are called to being by the very process of the disintegration of the village community. Its members, who were once equal as far as property, rights, and obligations went, are divided, thanks to the process referred to, into two sections. Some are attracted towards the urban bourgeoisie and try to merge with it in a single class of exploiters. All the land of the village community is little by little concentrated in the hands of this privileged class. Others are partly expelled from the community and, being deprived of land, take their labour-power to market, while others again form a new category of community-pariahs whose exploitation is facilitated, among other things, by the conveniences afforded by the community organization. Only where historical circumstances elaborate a new economic basis for the reorganization of society in the interests of this lower class, only when this class begins to adopt a conscious attitude to the basic causes of its enslavement and to the essential conditions of its emancipation, only there and only then can one 'expect' a new social revolution without falling into Manilovism. This new process also takes place gradually, but once it has started it will go on to its logical end in just the same way with the relentlessness of astronomic phenomena. In that case the social revolution does not rely on 'possible' success of conspirators but on the certain and insuperable course of social evolution.

Mutato nomine de te fabula narratur, we may say addressing the Russian village community. It is precisely the recentness of the development of money economy in Russia that explains the stability which our village community has shown until recently and which still continues to move weak thinkers. Until the abolition of the serfdom nearly all the communal - and to a great extent - economy of Russia was a natural economy, highly favourable to the maintenance of the village community. That is why the community could not be destroyed by the political events at the time of the principality and veche system and the Moscow centralization, of Peter's reforms and the 'drum-beating enlightenment' of the Petersburg autocrats. No matter how grievous the effect of these events was on the national welfare, there is no doubt that in the final account they themselves were not forerunners of radical upheavals in the public economy, but only the consequence of the mutual relations existing between individual village communities. The Moscow despotism was based on the very 'ancient foundations of the life of the people' that our Narodniks are so enthusiastic over. However, both reactionary Baron von Haxthausen and the revolutionary agitator Bakunin understood this clearly. Were Russia isolated from the economic and political influences of West European life, it would be difficult to foresee when history would undermine at last the economic foundation of the Russian political set-up. But the influence of international relations accelerated the natural, though slow, process of development of money economy of commodity production. The Reform of 19 February was a necessary concession to the new economic trend and in turn it gave it new strength. The village community did not, and indeed could not, adapt itself to the new conditions. Its organism was over-strained, and one must be blind not to notice the signs of its disintegration now. Those are the facts.

NOTES

* From G.V. Plekhanov, Our Differences (1881), 'Selected Philosophical Works', London, Lawrence & Wishart, 1961, vol. 1, pp. 271-5.
1 The influence of money economy on the decline of primitive communism is wonderfully described by Mr G. Ivanov (Uspensky) in the family community (from a Village Diary, 'Otechestvenniye Zapiski', September 1880, pp. 38-9):

At present there is such an immense accumulation of in-
soluble and difficult tasks in the life of peasant families
that if the big peasant families (I mean those near the
towns) still stand fast, it is only, so to speak, by observ-
ing the exterior ritual; but there is already little interior
truth. I fairly often come into contact with one of these
big peasant families. It is headed by an old woman of 70,
a strong woman, intelligent and experienced in her way.
But she derived all her experience under the serfdom and
in an exclusively agricultural household, all of whose
members contribute their labour, the whole income going
to the old woman and she distributing it at her discretion
and by general agreement. But then a high road was built
and a barrel of cabbage sold to the carters began to bring
in so much that it was more profitable than a whole year's
labour on the ploughland of, say, one man. This is already
a clear violation of the equality of labour and earnings.
Then the machine came, calves began to get dearer and
were needed in the capital. One of the sons became a
coach-driver and in half a year he earned as much as the
whole family in the country in a year. Another brother
became a *dvornik* in Petersburg and got fifteen rubles a
month - more than he sometimes got in a whole year. But
the youngest brother and the sisters barked trees the
whole spring and summer and did not earn a third of
what the coachman earned in two months. . . . And
thanks to this, although everything appears to be well
in the family, and each one contributes 'equally' by his
labour, it is not really so: the *dvornik* concealed four
red notes from his mother and the coach-driver still
more. And how could they do otherwise? The girl worked
her fingers raw with the tan the whole summer for five
rubles while the coachman got twenty-five in a single
night for driving gentlemen round Petersburg from mid-
night till dawn. Besides, the old woman's authority would
have still meant quite a lot if the family's earnings had
been only the result of agricultural labour. In this matter
she is in fact an authority, but the question is: what does
she know about a *dvornik*'s, a coachman's or other new
earnings and what a piece of advice can she give on the
matter? Her authority is, therefore, purely fictitious and
if it means anything it is only for the women who remain
at home; but even the women know quite well that their
husbands only appear to have a respectful and submissive
attitude to the old woman: the women have a very detailed
knowledge of their husbands' earnings and know whether
a lot is hidden from the old woman and by whom, and they
themselves keep those secrets as close as possible. The
authority of the head of the family is fictitious and so are
all the family and communal relations: each one hides
something from the old woman who is the representative

of those relations, and keeps it for himself. If the old woman dies, the large family will not remain as much as two days in its present state. Each one will wish for more sincere relationships and this wish will inevitably lead to something else - the desire for each to live according to his income, to enjoy as much as he gets.

CIVILISATION AND THE GREAT HISTORICAL RIVERS*

G. Plekhanov

What causes affect the internal development of human societies?
Why do some races remain almost at a complete standstill while
others unite into great state bodies, with the beginning of a
political presence, where learning, literature, applied and fine
arts and great productive forces, in a word, where everything
we call civilisation develops? And why, at different times, are
different peoples the bearers of civilisation? Why in a given
historical epoch are the principal forces of civilisation concentrated in a certain locality or localities only to move later into
a new territory, whose sovereignty is then contested by new
rivals? Why did civilisation's centre of gravity shift from the
banks of the Nile, Tigris, and Euphrates to Greece and Rome,
then to Central Europe and will eventually perhaps be transferred to North America? Is there any uniformity in this
apparently arbitrary progress of civilisation? Finally, and this
is perhaps the most interesting point: even in civilised countries
not everything is perfect. Historical development is bought at a
very high price. At least for a very long time, it admittedly
leads to the division of society into classes and ranks, to the
domination by a small number of privileged people, the oppression of the mass of people, the degradation of the human individual and very frequently to the worst and most obnoxious form
of despotism. So what drives humanity into the harsh school
of civilisation? Which force drives some races to abandon their
original social conditions and to bind together in new political
groups 'enserfed' to historical development?
    Of course, it would be very easy to answer such questions
by an allusion to divine decree, i.e. it thus pleased God in
His infinite wisdom. Such an answer leaves no more room for
doubt. But in science, this answer can have no meaning at all.
As long as we refer to a divine will we have no right to speak
of regularity in phenomena. It is the task of science to discover this regularity. Therefore, rigorous scholarship could
never coexist peacefully with religion. Gradually science has
driven religion from all its positions. In our time no one who

entertains any notion of scientific thought ever refers to divine
will as the cause of natural phenomena or social development.
However, it is not enough to forge references to a divine will
in order to acquire a strictly scientific way of thinking. In
social science, we still come across completely arbitrary explanations of facts and completely unscientific interpretations
of phenomena. In such explanations, there is not always a
felt obligation to apply logic. When there is a question about
the origins of this or that phenomenon, people, who are quite
serious and hardly stupid, are content with solutions which
answer absolutely nothing, since they are no more than the
repetition of the question in another form. For example, ask
a 'scholar' one of the above-mentioned questions: Why do some
peoples develop so amazingly slowly while others cover the road
to civilisation very quickly? The 'scholar' will answer you without hesitation that this fact can be explained by the characteristics of the different races. Do you understand the meaning
of this reply? Some peoples develop so slowly because it is a
characteristic of their race to develop slowly; others, on the
other hand, become civilised very quickly because the chief
characteristic of their races is that they can develop quickly.
If you want to compare the civilised peoples with one another,
you will also observe with what astonishing ease all the characteristics of their history can be explained by their racial features.
In the hands of skilled people 'race' becomes a key to solving
all the questions of social science. In the face of the wonder-
ful power of this magical word one might expect that the concept
connected with it would be distinguished by its total clarity
and perfect precision. Actually, the reality is quite different.
 In a very successful section of his book, 'Civilisation and
the Great Historical Rivers', (1) Léon Metchnikoff subjects the
modern doctrine on race to a strict analysis. He first emphasises
the remarkable fact that the very people who like to point to
racial characteristics to explain social phenomena are the
Darwinians, followers of the principle of evolution. In theory
they completely agree that the characteristics of a particular
type or a particular species are only the result of its adapta-
tion to its surroundings in its struggle for existence. L.J.
Metchnikoff then brilliantly shows how imprecise and erroneous
the whole concept of race is and quotes the evidence of the
eminent anthropologist Topinard. The latter maintains that due
to the already long and continuous mixing and crossing of the
various races it is no longer possible to speak of any pure race.
'When we speak of an Indo-Germanic, Latin, German, English
or Slavic race, the word race can only have meaning as a
political label, merely signifying a chance accumulating of
various anthropological elements.' L.J. Metchnikoff adds that all
great civilisations were the result of very motley mixtures of
various ethnic elements. One can at no time distinguish with
any degree of approximation the relative significance of individual elements within these mixtures. For example, it is hard

to say which of the three races, the black, the yellow, or the
white, did most for the civilisation of Ancient Egypt. The
history of the Chaldees in fact shows that the black race, the
so-called Cushites, was ahead of all the others in its civilisation. The black scholars of that time could therefore have
considered their race the most capable of progress and looked
down sympathetically on the white and yellow races who, in
those days, lived in the deepest ignorance. Now the position
has changed. Now the white race is ahead of all the others,
while the black race is noted for its backwardness. But if
we wanted to assume that the white race is more capable of
development than all the others, this would hardly explain the
difference in the historical fates of the peoples. We should not
know 'why ethnic groups as related to one another as the Kurds
and the Germans, the English and the Afghans, all belonging
to the Aryan branch of the white race, have nevertheless
played such different roles in history' (Metchnikoff, 1889,
p. 98). Clearly this variation in their roles was not determined
by the characteristics of their race, but by other causes which
those who are accustomed to attribute everything to race have
hardly considered.

But where must we look for these other causes? Metchnikoff
replies that they are to be sought in the geographical conditions
of human societies. It is widely known that the nature of geographical conditions varies greatly from area to area. Thus,
the fates of races which experience the powerful effects of
these differences are likewise distinct.

The idea that the geographical environment influences the
historical fate of humanity is far from new in science. In our
time it has been developed by Montesquieu, Hegel, Buckle,
Ritter, and many others. Nevertheless, the influence of geographical environment on society's historical fate has often
been very unclear and even quite wrongly understood. Hippo-
crates, for example, found that Asian peoples were altogether
more cowardly than Europeans. He tried to explain this by
the influence of the geographical environment. 'In Asia', he
said, 'the temperature of the various seasons does not vary to
the extent that it does in Europe. In Asia the temperature is
more constant. Thus the human spirit is not subjected to such
shocks or the body to such harsh changes. Such changes are
more suited to creating an indomitable, passionate and independent character than an unvaried life. Just such changes
and variety awaken the human spirit and prevent it from remaining at a standstill.' From the same, but more detailed,
perspective, Montesquieu, in his famous book 'The Spirit of
Laws', also speaks of the influence of climate on the human
character (1877, 234/Book XIV, Ch. 2):(2)

  A cold air constringes the extremities of the external fibres
  of the body; this encreases their elasticity and favours the
  return of the blood from the extremities to the heart. It
  contracts those very fibres; consequently it encreases also

their force. On the contrary, a warm air relaxes and
lengthens the extremes of the fibres; of course it diminishes
their force and elasticity. People are therefore more vigorous in cold climates. . . . This superiority of strength must
produce a great many effects; for instance, a great self-
confidence, that is more courage; a greater sense of superiority, that is, less desire of revenge; a greater opinion of
security, that is more frankness, less suspicion, policy
and cunning. . . . The inhabitants of warm countries are,
like old men, timorous; the people in the cold countries
are, like young men, brave.

Proceeding from the more or less probable physiological effects
of climate on man, Montesquieu undertakes a construction of
the character of the peoples who live in hot and cold lands by
purely deductive means. According to the deductively constructed characters, he advances conjectures about the 'spirit
of law' most suitable for the peoples of various countries.
Wherever there is really something in the laws of some country
which corresponds to the supposition he has made, the theory
appears to be strongly proven. Buckle also frequently reasoned
in a similar fashion. The basic error of these researchers lies
in the fact that they sought the signs of geographical and
environmental influence in the psychology or even the physiology
of different peoples, rather than in their social conditions. Under
the influence of the geographical environment, they said, this
or that character of a certain people develops and its history
is determined by this character. But obviously such a statement leads us back to the characteristics of race. The only
difference is that in this case the origin of the supposed
characteristics of the race are explained rather more thoroughly.
The researchers along these lines forgot that man lives in
society which influences his character and customs in an in-
finitely greater way than the direct influence of nature. Thus,
in order to assess correctly the effect of the geographical
environment on the historical fate of man, one must examine
closely how the natural environment affects the form and
characteristics of the social environment. The latter above all
else determines a person's character and inclinations. Metchnikoff has made a detailed examination of the influences of 'the
great historical rivers', the Nile, the Tigris, and the Euphrates,
the Yang-tse-Kiang and the Hwang-ho, on the development of
ancient oriental civilisations. These civilisations were not merely
ancient, but indeed the most ancient. Obviously, then, a
correctly undertaken study of these civilisations will enlighten
us on which geographical conditions contributed most to a
decisive appearance of humanity on the road to civilisation.
The most important characteristics of 'the great historical
rivers', according to L.J. Metchnikoff (1889, p. 189), can be
reduced to the following:

The lands watered by them can become either rich granaries
from which the people working for some days can obtain

their sustenance for the whole of the year – or they can
become graveyards filled with innumerable corpses, the
victims of floods, hunger and infectious diseases. To make
the most of the land under the normal conditions created
by these rivers, communal efforts of strictly disciplined
workers are required. These workers are recruited from
the population on its banks, a population usually consisting
of diverse groups, among whom there are great differences
in language, race, appearance and customs. The canals of
the Yang-tse-Kiang and the dams of the Hwang-ho are in
all probability a result of skilfully organised communal work
over many more generations than the pyramids and the
temples of Egypt. The slightest negligence in the digging
of a ditch or in the support of a dam, the least carelessness
or egotistical step by one person or a group of people in
the maintenance of the common water wealth becomes the
course of social evil and wide reaching misfortune in such
unusual conditions. Consequently such a source of irrigation
requires, under threat of death, a close and lasting solidarity between the masses of the population who are often
strangers or even enemies. It condemns everyone to such
works, whose common usefulness only becomes apparent as
time goes on, and whose plan very often remains totally
incomprehensible to the ordinary person. Here is the actual
cause of the timid and idolising worship of the river, this
god which nourishes and commands, kills and gives life,
entrusting its secrets only to a few chosen ones and demanding blind obedience from ordinary mortals.
Indeed, a hymn to the Nile, translated by H. Maspéro and
quoted by Metchnikoff (1889, pp. 215-16), leaves no room for
doubt on the reason for this deification:
    Hail to you, oh Nile, you who manifest yourself on the
    earth and appear peacefully to revive Egypt. The creator
    of corn, maker of barley, he who perpetuates the course of
    the seasons. . . . His working gives peace to millions of
    unfortunate people. When he sinks the gods in heaven are
    overturned and the people perish . . . when he swells,
    the earth is filled with pleasure, every stomach rejoices,
    every creature receives its food, all teeth work . . . even
    the sacrificial offerings only exist thanks to him . . .
But this mighty and mysterious god, the creator 'of all good
things', also created the despotism which was so characteristic
of ancient Egypt. Lenormand writes about the Egyptians in the
following manner:
    No single people went so far in the veneration of the power
    of the sovereign, elevating it and idolising it to so great an
    extent. This happened because nowhere did the conditions
    of material life and the production of the bare necessities
    of existence so clearly demonstrate to the people the neces-
    sity of this power.
Careful observation also shows that the strength of the Chaldean

kings rested upon the material conditions of life, on the production of the bare necessities. The kings themselves apparently
realised this. 'For the good of the people I investigated the
secrets of the rivers . . .' says one of them; 'I led the river
water into the deserts; I filled the dried out ditches with
it . . . I watered the desert plains; I gave them fruitfulness
and abundance; I made them into the abode of happiness'
(Metchnikoff, 1889, p. 258). Generally, despotic power had the
same economic significance in China, at least in the region
between the Yang-tse-Kiang and the lower Hwang-ho. 'There
we again find a land which richly rewards the people for their
efforts, but which forces them on pain of annihilation to complete solidarity and to strict and constant discipline extending
to even the smallest details of life' (ibid., pp. 353-4). Finally,
in India too, where the geographical characteristics of the
land hindered the growth of a unified state, despotism nevertheless grew into an inevitable, and for a time useful, form
of government.

The material that science now has at its disposal is sufficient
to elucidate the socio-political conditions reigning among the
Indian Aryans before the beginning of the despotic body
politic. Characteristic of these conditions were independent
communities with elected chiefs at their heads. As a result of
geographic conditions, the communities could remain independent
as Indian Aryans only so long as they continued to live in the
rich valleys of Kashmir, 'where the age-old hymns of the
Rig-Veda originated, and in the area of the Hindu-Kush. The
abundance of rain and the fertility of the soil favoured a
pastoral life or the free cultivation of the land by small inde
pendent groups' (ibid., p. 303). If the Aryans had not explained
their settlements in India they would have been able to maintain
their original free institutions for ever; however, they would
have played no part in history at all. But they pressed on
unceasingly into the interior of the country where other geographical conditions awaited them. The broad strip of land
which is called the 'middle land', 'Madja-Deza' in Mann's law
book, possesses all the characteristics favourable to historical
development. Agriculture, which there too rewards a man's
work most richly, is unthinkable without the planned organisation and work-discipline that created the ancient civilisations
and simultaneously caused the rise of the old despotic empires.
The same is true of the whole Ganges Valley. A great irrigation
system is absolutely essential there; the slightest negligence
in the maintenance of this system threatens the land with much
more misfortune than the most devastating enemy attack. The
central power administering this system acquires tremendous
might and becomes despotic.

But the ancient civilisations of the East are only the first
great steps of humanity on the path of historical development.
For further steps a different geographical environment, one
which favours other socio-political conditions, is required. The

development of technical skills and the growth of wealth create
the need for international commerce or exchange. Civilisation's
centre of gravity moves to the coasts of the Mediterranean.
New people come on to the historical scene: Phoenicians, Greeks,
Romans, etc. Those peoples who, like the ancient Egyptians,
for one reason or another, cannot adapt to the new demands
of civilisation, step into the background and begin to decline.
However, for the peoples who know how to adapt to these
conditions a new florescent period begins. For example, this
was the fate of Chaldea which first appeared as an exclusively
agricultural country, but later played an extremely important
role in the Persian Gulf trade. . . . But gradually international
commerce develops to such an extent that the Mediterranean
becomes too narrow for it. Civilised humanity now enters the
epoch of the civilisation of the oceans of the world, settling
those countries best suited to world sea trade in the first
place.

From a socio-political point of view, the ancient civilisations
of the rivers, as we have already said, are characterised by
the reign of great despotic powers. A new principle enters
the world in the period of Mediterranean civilisation, a period
in which our author also places the civilisation of medieval
Europe. The predominant form of government is the feudal
republic which is based on slavery and therefore is always
more or less oligarchic (ibid., pp. 44-5). Finally, the direction of the most recent epoch of civilisation, the period of
the world oceans, is hostile to both despotism and oligarchy.
It intends to make the human rights proclaimed by the great
revolution into reality, to destroy the rule of one class over
the other, and actually to carry out the principles of freedom,
equality, and fraternity (ibid., pp. 51-2).

Therefore, in the given natural conditions for the existence
of humanity, despotism, the division of society into classes,
the oppression of one class by another, and many of the other
dark sides of civilisation were necessary for its further
development. But the success of this development led gradually
to the negation of these dark sides. The time is not distant
when a new period of free and sensible social conditions will
begin. L.J. Metchnikoff comes to this conclusion and, of course,
we are in perfect agreement. However, it seems to us that the
writer occasionally does not express himself altogether accurately.
At the end of the first chapter of his book, he claims, among
other things, that 'sociological progress stands in inverse
relation to compulsion and force, and in direct relation to the
absence of any restriction of the will to freedom or to anarchy'
(op. cit., p. 28). In his opinion, this has already been proved
by Proudhon. We definitely disagree with such a claim. First,
the thesis advanced by the author is refuted by the content
of his own book. When one compares the social institutions which
grew up in the 'river civilisations' period with those of the
original human societies, one realises that 'compulsion' and

'force' played a much greater role in the former than in the
latter. But can one say that the 'river civilisations' were
behind the original societies in the sense of 'sociological
progress'? One can sympathise more with the free life of the
savages than with a civilised despotism, but why confer on the
word progress such an arbitrary and paradoxical meaning,
according to which primitive peoples who have hardly developed
would have travelled further along the path of progress than
the peoples who made a long historical journey. Above all, we
believe that abstract formulae are always damaging in social
science. Owing to their abstractness they are always extremely
one-sided and never embrace the rich content of real life and
history. To classify different societies according to the extent
of their 'progress' while at the same time only considering one
side of their situation, is as unjustifiable today as is the
classification of plant types, species and families according to
the number of stamen filaments, for example. In such a procedure, errors and contradictions soon appear in vast numbers. There was more freedom in original human societies than
under oriental despotisms; but the latter had progressed much
further because their populations exercised a greater control
over nature than the free savages. One cannot speak of human
progress without a consideration of the degree of control over
nature. The quality of gaining such control is the most important criterion in distinguishing human being from animal. If
man did not possess it, and if the extent of his control over
nature could not be increased when circumstances made it
possible, there would be no history and no human progress
(in whatever sense of this word). We wish furthermore to
make the following comment. As has already been said, we have
no doubt that humanity will be led by history to freedom and
equality. But will it be led to anarchy? Can one really state
that social development continues in the direction indicated
by Proudhon? Certainly not. Proudhon's 'Anarchy' is so satu-
rated with the spirit of petty bourgeois individualism that if
one day humanity were really permeated by Proudhon's ideas,
any planned organisation of co-operative production on any
basis would be impossible least of all on the basis of freedom
and equality. Of course, in the social forms which civilised
humanity is now clearly approaching, there can be no place
for either despotism or political domination of one class by
another, or for the apparently democratic government which
now serves to support the bourgeoisie's economic control over
the proletariat in some countries. But will society not retain
control of its own productive forces? Of course it will, since
for a society to give up this control would be synonymous
with suicide. Therefore, society will not be anarchical, for
anarchy means the total absence of all control and leaves an
unlimited space not for freedom, unthinkable without organisation, but for the caprice of individuals, making any organisation
impossible. Above all, we wish to note that Metchnikoff's state-

ments on anarchy were to the great detriment of his book -
because they prevented a systematic development of his completely correct and fertile ideas. Sometimes he seems to shrink
from his conclusions which stand in direct antithesis to anarchist views and he attempts to soften them by various
objections which contradict his theory. Here is an example.
After he has clearly proven that production in ancient Egypt
could only be organised with the aid of despotic power, he
considers it necessary to moderate the sense of his words. He
refers to E. Réclus, in whose opinion the geographical conditions of the Nile Valley left its inhabitants the choice between
despotism and association, dependent upon the equality of all
(ibid., p. 227). Admittedly, our author comments that the
words 'progress' and 'development' would make no sense if
the ancient Egyptians could have started with free association.
But if this is true, the ancient Egyptians only had one way
out - despotism, and therefore one cannot say, as Metchnikoff
does, that 'the geographical conditions of the Nile Valley even
from the beginning in no way forced despotism on the inhabitants', but 'only advised them to practise solidarity'. Again
we repeat that Metchnikoff damaged his own scholarly results
by such objections. It seems that this tendency comes from
his predilection for 'anarchy'. This predilection was also
injurious to our writer in another way.

It prevented him from applying the materialistic view of
history consistently in his book. The anarchistic theory is
extremely idealistic: Anarchists talk a lot about freedom, the
rights of the individual, the harmfulness of all control, etc.,
but they never take into consideration the objective conditions
of human development, conditions which are not only created
without human will, but which, on the contrary, influence and
determine the direction of human will and actions. Indeed, the
task of today's materialism in its application to social science
is to show how human development progressed under the
influence of the objective conditions of existence, independent
of human will. In other words, today's materialism stands in
harsh antithesis to the spirit of anarchist theory. A person
who only partly sympathises with anarchism, but does not take
this theory very seriously, will be carried away in spite of
himself by idealistic interpretations of history, even if he is
fundamentally a convinced materialist. For example, the main
theses of Metchnikoff's philosophy of history are totally materialistic. But in the details he frequently strays into idealism. We
find such details in almost every section of his book. In the
chapter 'The Indus and the Ganges', he says that a very
special task in the history of humanity fell to India and that
it 'had begun the history of justice with the disgraceful institution of castes' (ibid., p. 281). He attaches so much significance to castes because, in his opinion, they imposed certain
limitations on royal power which in the other centres of the
river civilisations was unlimited.

In ancient times, only India was able to bring about such
a social order which was regulated by its inner mechanism,
independent of the compulsion of any personal and arbitrary
power. In order to solve this task of world history, India
exaggerated to absurd, even monstrous lengths, the class
differences prevalent everywhere (ibid.).

All this takes on an even more idealistic hue since we are never
shown how the Indian castes actually arose or how the author
connects their origin with the peculiarities of geographical
environment. Respected researchers cast strong doubt even on
the existence of these castes as Metchnikoff has imagined them.
Henry Maine, whose competence no one will deny, writes (1881,
pp. 56–7):

I am aware that the popular impression here is that Indian
society is divided, so to speak, into a number of horizontal
strata, each representing a caste. This is an entire mistake.
It is extremely doubtful whether the Brahminical theory of
caste upon caste was ever true . . . it is even likely that
more importance has been attached to it in modern than
ever was in ancient times. The real India contains one
priestly caste, which in certain, though a very limited,
sense is the highest of all . . . caste is merely a name for
trade or occupation . . .

We will not argue over whether Maine is right or not, although we believe that he is right. In any case Metchnikoff
should have proved that Maine's view was incorrect before he
went on to speak of the special historical task of India. But
our author neglected the demands of criticism in the interest
of a biased idealistic view which in no way accords with the
main thesis of his historical theory.

Before indicating other such wanderings into idealism, we
should like to say that, in the work in question, the geographer
leaves the historian altogether too little scope. This may be
because the writer had no time to revise the work so as to
relate individual elements to one another correctly. But whatever the reason this failing is very noticeable in the book. The
geographical characteristics of the countries where the 'river
civilisations' ran their course are conveyed in such detail that
they leave nothing to be desired. But as soon as there is any
mention of the character of social conditions created by these
characteristics, the description is very brief. It becomes too
brief as soon as the writer touches on the internal development of the social order which arose on the banks of the
geographical rivers. This is why Metchnikoff's theory suffers
from a geographical formalism. It will be pointed out that modern
scholarship up to now offers very little material for a thorough
study of the character and development of the social conditions
in the ancient Orient. This is quite true, but at least all the
material available should be used; something which Metchnikoff
did not do. So if it no doubt follows from his book that the
geographical environment affects man chiefly through the

economic conditions arising under its influence, then these
conditions are given very little consideration by the writer.
If he had taken more notice of them he might have found the
solution to that phenomenon which he had to explain as simply
the 'exhaustion' or diminution of the 'life fluid', etc. of a
given historical people. The necessity for humanity to cross
over from one period of civilisation to another would have
shown up more clearly. The characteristic features of the
different periods would have shown up better, and the origin
of these features would be more comprehensible.

It seems to us that if L.J. Metchnikoff wished to speak
about the characteristics of the Mediterranean civilisations
he should at least have indicated how these characteristics
arose from the geographical environment using economics in
his demonstration. Now it remains unclear why these civilisations, as he calls them, were oligarchic, i.e. why they were
based on slavery. Furthermore, he needs to explain more
thoroughly why and to what extent he can compare the life
of medieval continental Europe with that of the Phoenician,
Carthaginian, Athenian, and other republics. The fact that
the masses in both situations were subjugated is by no means
sufficient to justify such a comparison. Furthermore, the
subjugation of the masses in the Mediterranean civilisations
has its own history. To explain this history, it is not enough
to refer to the role of the Mediterranean in general; the
internal economic history of the societies that arose on its
shores must be scientifically examined. Metchnikoff maintains
that medieval feudalism with its institution of serfdom was no
more than a rustic complement to the urban republics of
Carthage, Athens, and Rome (op. cit., p. 50). But rustic of
course means almost entirely agricultural and for agricultural
peoples, the sea, according to Metchnikoff's own theory, has
no significance at all. Such peoples even frequently depart
from it or 'turn their backs on it' instead of seeking it. So
in what way could the Mediterranean have a decisive influence
on the pattern of the social conditions in medieval agricultural
Europe? It is clear that the whole course of history cannot be
explained merely by the effect of geographical environment.
Of course, the environment played its part; but the social
conditions that arose under its influence also have their own
inner logic, which can often even conflict with the requirements of the geographical environment. In recent European
history, just as in any other case, one could count many
examples of such contradictions. The study of the inner logic
of social, especially economic, conditions, is no less necessary
than the study of the geographical basis of world history.
These researches complement each other and under their joint
effect, the innermost secrets of history will gradually be revealed. In the last forty years, Marx, especially, and several
of his followers have done much for the study of the inner
logic of social conditions. Unfortunately L.J. Metchnikoff

ignored almost all their results. In general his work reaches
the same conclusions as the Marxists.(3) However, he would
have achieved a balanced view and consistency had he used
the historical views of Marx and Engels. One has but to
examine Metchnikoff's above-mentioned views, according to
which historical development leads to a transformation of social
conditions tending towards freedom and equality. What is this
idea based on? - on the general consideration that man must
learn with time to organise his work without a despot's stick
and without the self-seeking supervision of the employer. This
consideration has strong probability. It is so strong that our
author really had no need to support it with analogies from
nature, since all such analogies are very forced and arbitrary.
But, however probable Metchnikoff's viewpoint, it remains only
probable. It would have only taken on the character of certainty
had he shown in a few words how the logic of the internal
conditions leads modern civilised countries to this end. Marx's
school did this by giving the necessary attention to the growth
and properties of the modern productive forces, as well as to
the conflict between these forces and modern conditions of
production. Marx showed that socialism necessarily follows from
capitalism. It is a pity that L.J. Metchnikoff did not think it
necessary to deal with Marx's doctrine. This doctrine would
confer on his theory of progress the character of something
strictly worked out and credible. The Marxist school has its
followers in Russia too. N. Sieber's book 'The Outlines of
Primitive Economic Culture' would have been a great help in
Metchnikoff's researches.

Despite these remarks, we do not forget for a moment that
'criticism is easy, while art is difficult'. Rare are the works
for which even those critics most well disposed to the author
would offer no criticism. Nor have we forgotten that L.J.
Metchnikoff's book cannot be regarded as a completed work,
since death prevented him from revising it properly.

NOTES

* From Die Zivilisation und die grossen historischen Flüsse,
  'Die Neue Zeit', Year 9, vol. 1, 1890-1, pp. 436-48. Translated by J. Gordon-Kerr.
1 L.J. Metchnikoff, 'La Civilisation et les grands fleuves
  historiques', Preface by Elisée Réclus, Paris, 1889.
2 Voltaire comments on this in his 'Dictionnaire philosophique':
  One must be on one's guard against such generalisations.
  Nobody has ever seen a Lapp or Samoyed go to war and
  the Arabs conquered in eighty years a greater territory
  than the Roman empire contained. The Spaniards with
  smaller numbers defeated the north German warriors at
  Mühlberg [1547 in the war of Schmalkalden]. This statement of the author's is as wrong as the others which refer
  to the climate.

3 Editorial note by K. Kautsky (eds): Twenty years before
   Mr Metchnikoff, Marx in 'Capital' had already noted some
   of the essential bases of the 'river civilisations'. At one
   point he says, 'The necessity for predicting the rise and
   fall of the Nile created Egyptian astronomy, and with it
   the dominion of the priests as directors of agriculture' and
   later, 'One of the material bases of the power of the state
   over the small disconnected producing organisms in India,
   was the regulation of the water supply' (Marx, 1867, p. 481).
   Permit me to note here that, stimulated by these passages,
   I found that what was true of the valleys of the Nile and
   the Ganges was also true for the Euphrates and the Tigris,
   the Yang-tse-Kiang and the Hwang-ho and that the material
   basis not only of the Egyptians and Indians, but also
   the Chinese and the Mesopotamian Empire was formed partly
   by the necessity of controlling the river and to which
   oriental despotism can to some extent also be partly attributed. Without knowing anything of Herr Metchnikoff, and
   before his book appeared, I developed these theories in
   an essay on Modern Nationality, 'Die Neue Zeit', 1887, V,
   pp. 392ff.

BIBLIOGRAPHY

Maine, H., 1881, 'Village Communities in the East and West',
    London, John Murray.
Marx, K., 1974, 'Capital', vol. I, Lawrence & Wishart.
Montesquieu, Baron de, 1877, 'The Spirit of Laws', London,
    S. Crowder, C. Ware, T. Payne.

A LETTER TO THE ST PETERSBURG WORKERS*

V.I. Lenin

What were Plekhanov's arguments in favour of municipalisation? In both his speeches he laid most stress on the question of guarantees against restoration. This curious argument runs as follows. Nationalised land was the economic basis of Muscovy before the reign of Peter I. Our present revolution, like every other revolution, contains no guarantees against restoration. Therefore, in order to prevent the possibility of restoration (i.e. the restoration of the old, pre-revolutionary regime), we must particularly shun nationalisation.

To the Mensheviks this argument seemed particularly convincing, and they enthusiastically applauded Plekhanov, especially for the 'strong language' he used about nationalisation ('Socialist-Revolutionary talk', etc.). And yet, if one ponders over the matter a little, one will easily see that the argument is sheer sophistry.

First of all, look at this 'nationalisation in Muscovy before the reign of Peter I'. We will not dwell on the fact that Plekhanov's views on history are an exaggerated version of the liberal-Narodnik view of Muscovy. It is absurd to talk about the land being nationalised in Russia in the period before Peter I; we have only to refer to Klyuchevsky, Yefimenko and other historians. But let us leave these excursions into history. Let us assume for a moment that the land was really nationalised in Muscovy before the reign of Peter I, in the seventeenth century. What follows from it? According to Plekhanov's logic, it follows that nationalisation would facilitate the restoration of Muscovy. But such logic is sophistry and not logic, it is juggling with words without analysing the economic basis of developments, or the economic content of concepts. In so far as (or if) the land was nationalised in Muscovy, the economic basis of this nationalisation was the Asiatic mode of production. But it is the capitalist mode of production that became established in Russia in the second half of the nineteenth century, and is absolutely predominant in the twentieth century. What, then, remains of Plekhanov's

argument? He confused nationalisation based on the Asiatic
mode of production with nationalisation based on the capitalist
mode of production. Because the words are identical he failed
to see the fundamental difference in economic, that is, production relations. Although he built up his argument on the
restoration of Muscovy (i.e. the alleged restoration of Asiatic
modes of production), he actually spoke about political restoration, such as the restoration of the Bourbons (which he
mentioned), that is, the restoration of the anti-republican form
of government on the basis of capitalist production relations.

Was Plekhanov told at the Congress that he had got himself muddled up? Yes, a comrade who at the Congress called
himself Demyan said in his speech that Plekhanov's 'restoration'
bogy was an out-and-out fizzle. The logical deduction from
his premises is the restoration of Muscovy, i.e. the restoration
of the Asiatic mode of production - which is a sheer absurdity
in the epoch of capitalism. What actually followed from his conclusions and examples is the restoration of the Empire by
Napoleon, or the restoration of the Bourbons after the great
French bourgeois revolution. But first, this sort of restoration
had nothing in common with pre-capitalist modes of production.
And secondly, this sort of restoration followed, not on the
nationalisation of the land, but on the sale of the landed
estates, that is, a measure that was arch-bourgeois, purely
bourgeois and certainly one that strengthened bourgeois, i.e.
capitalist production relations. Thus neither form of restoration
that Plekhanov dragged in - neither the restoration of the
Asiatic mode of production (the restoration of Muscovy), nor
restoration in France in the nineteenth century, had anything
at all to do with the question of nationalisation.

What was Comrade Plekhanov's reply to Comrade Demyan's
absolutely irrefutable arguments? He replied with uncommon
adroitness. He exclaimed: 'Lenin is a Socialist-Revolutionary.
And Comrade Demyan is feeding me a new brand of Demyan
hash.'

The Mensheviks were delighted. They laughed till their
sides ached at Plekhanov's sparkling wit. The hall rocked
with applause. The question whether there was any logic in
Plekhanov's argument about restoration was completely shelved
at this Menshevik Congress.

I am far from denying, of course, that Plekhanov's reply
was not only a superb piece of wit, but, if you will, also of
Marxist profundity. Nevertheless, I take the liberty of thinking
that Comrade Plekhanov got himself hopelessly muddled up over
the restoration of Muscovy and restoration in France in the
nineteenth century. I take the liberty of thinking that 'Demyan
hash' will become a 'historic term' that will be applied to Comrade
Plekhanov and not to Comrade Demyan (as the Mensheviks,
fascinated by the brilliance of Plekhanov's wit, think). At all
events, when Comrade Plekhanov, in speaking about the seizure
of power in the present Russian revolution, was tickling his

Mensheviks with a story about a Communard in some provincial town in France who munched sausage after the unsuccessful 'seizure of power', several delegates at the Unity Congress
remarked that Plekhanov's speeches were like a 'Moscow stew',
and that they sparkled with 'sausage wit'.

As I have already said, I was the first reporter on the
agrarian question. And in winding up the debate, I was not
the last to be given the floor but the first, preceding the
other four reporters. Consequently I spoke after Comrade
Demyan and before Comrade Plekhanov. Hence I was unable to
foresee Plekhanov's brilliant defence against Demyan's arguments. I briefly reiterated these arguments and concentrated
on the question of restoration as such, rather than on revealing the utter futility of the talk about restoration as an
argument in favour of municipalisation. What guarantees against
restoration have you in mind? - I asked Comrade Plekhanov.
Is it absolute guarantees in the sense of eliminating the economic foundation which engenders restoration? Or a relative
and temporary guarantee, i.e. creating political conditions
that would not rule out the possibility of restoration, but
would merely make it less probable, would hamper restoration?
If the former, then my answer is: the only complete guarantee
against restoration in Russia (after a victorious revolution in
Russia) is a socialist revolution in the West. There is and can
be no other guarantee. Thus, from this aspect, the question
is: how can the bourgeois-democratic revolution in Russia
facilitate, or accelerate, the socialist revolution in the West?
The only conceivable answer to this is: if the miserable Manifesto of 17 October gave a powerful impetus to the working-
class movement in Europe, then the complete victory of the
bourgeois revolution in Russia will almost inevitably (or at all
events, in all probability) arouse a number of such political
upheavals in Europe as will give a very powerful impetus to
the socialist revolution.

Now let us examine the 'second', i.e. relative guarantee
against restoration. What is the economic foundation of restoration on the basis of the capitalist mode of production, i.e. not
the comical 'restoration of Muscovy' but restoration of the type
that occurred in France at the beginning of the nineteenth
century? The condition of the small commodity producer in any
capitalist society. The small commodity producer wavers between
labour and capital. Together with the working class he fights
against the survivals of serfdom and the police-ridden autocracy.
But at the same time he longs to strengthen his position as a
property-owner in bourgeois society, and therefore, if the
conditions of development of this society are at all favourable
(for example, industrial prosperity, expansion of the home
market as a result of the agrarian revolution, etc.), the small
commodity producer inevitably turns against the proletarian who
is fighting for socialism. Consequently, I said, restoration on
the basis of small commodity production, of small peasant

property in capitalist society, is not only possible in Russia,
but even inevitable, for Russia is mainly a petty-bourgeois
country. I went on to say that from the point of view of restoration, the position of the Russian revolution may be expressed in the following thesis: the Russian revolution is strong
enough to achieve victory by its own efforts; but it is not
strong enough to retain the fruits of victory. It can achieve
victory because the proletariat jointly with the revolutionary
peasantry can constitute an invincible force. But it cannot
retain its victory, because in a country where small production
is vastly developed, the small commodity producers (including
the peasants) will inevitably turn against the proletarians when
they pass from freedom to socialism. To be able to retain its
victory, to be able to prevent restoration, the Russian revolution will need non-Russian reserves, will need outside assistance. Are there such reserves? Yes, there are: the socialist
proletariat in the West.

Whoever overlooks this in discussing the question of restoration reveals that his views on the Russian revolution are extremely narrow. He forgets that France at the end of the
eighteenth century, in the period of her bourgeois-democratic
revolution, was surrounded by far more backward, semi-feudal
countries, which served as the reserves of restoration; whereas
Russia at the beginning of the twentieth century, in the period
of her bourgeois-democratic revolution, is surrounded by far
more advanced countries, where there is a social force capable
of becoming the reserve of the revolution.

To sum up. In raising the question of guarantees against
restoration, Plekhanov touched upon a number of most interesting subjects, but he explained nothing at all on the point
at issue and led away (led his Menshevik audience away) from
the question of municipalisation. Indeed, if the small commodity
producers, as a class, are the bulwark of capitalist restoration
(this is what we shall for short call restoration on the basis,
not of the Asiatic, but of the capitalist mode of production),
where does municipalisation come in? Municipalisation is a form
of land-ownership; but is it not clear that the forms of land-
ownership do not alter the main and fundamental features of
a class? The petty bourgeois will certainly and inevitably serve
as the bulwark of restoration against the proletariat, no matter
whether the land is nationalised, municipalised or divided. If
any sharp distinctions between the forms of land-ownership can
be drawn in this respect, it can, perhaps, only be in favour
of division, since that creates closer ties between the small
proprietor and the land - closer and, therefore, more difficult
to break.(1)

NOTES

*   From Report to the Unity Congress of the RSDLP (1906),
    'Collected Works', Moscow, Foreign Languages Publishing
    House, 1962, vol. 10, pp. 331-5.

1   We say 'perhaps', because it is still an open question
    whether these closest ties between the small proprietor and
    his 'plot' are not the most reliable bulwark of Bonapartism.
    But this is not the place to go into the details of this con-
    crete question.

THE LEGITIMACY OF THE AMP*

L.I. Mad'iar**

Marx's teaching on social formations embraces the totality of all
productive relations of mankind in the course of all historical
development. The development of productive forces and the
division of labour, the origin of private property, the formation of classes, the means of exploitation and class struggle,
the formation of the state, and an analysis of all progressive
eras of economic formation in society, the economic basis and
the ideological superstructure in various social formations,
the origin, development and disappearance of various modes
of production and so on, all are included in Marx's teachings
on social formations.

The enormous theoretical and practical significance of all
these questions is indisputable. The complexity, diversity,
and variety of these questions forces us to concentrate in this
brief essay on only a few controversial questions.

We have set for ourselves a relatively modest and narrow
problem. We would like to present only a few of the indisputable teachings of Marx, restore their true meaning, purge
them of eclectic, revisionist, and mechanistic distortions and
then outline some of the most pressing problems for the solution
of these questions.

Marx's work on social formations is itself a product of historical development. This aspect of Marx's work developed together
with his entire system. And, if we wish truly to understand
Marx, then obviously we must outline, however cursorily, the
origin and development of Marx's teaching.

We must begin, undoubtedly, with Hegel. With Hegel the
dialectic 'is standing on its head. It must be turned right
side up again, if you would discover the rational kernel within the mystical shell' (Marx, 1867, p. 29).

In his study of social formations, Marx turned the Hegelian
dialectic right side up. Engels draws attention to this in his
article on Marx's book 'A Contribution to the Critique of Political Economy'. . . .

It is precisely the Hegelian understanding of history which

is the direct theoretical predecessor of the new materialistic
understanding of history developed by Marx. Marx turned
Hegel's dialectic right side up, not only in the sphere of
philosophy but also in the sphere of historical understanding.

Hegel gave his basic conception of history in his work
'Lectures on the Philosophy of World History'. In our view
Marx used this work of Hegel's not only for his criticism of
Hegel's 'Philosophy of Right', but also in developing his theory
of historical materialism. Marx surmounted, stood on its head,
the conception Hegel developed in this work. Here we cannot,
and perhaps need not, set forth even briefly, Hegel's conception of the autonomous development of the absolute spirit in
the historical process. For our purposes, it will suffice to
note that, according to Hegel, the absolute spirit passes
through a series of steps in its development. Hegel divides
world history into the following periods:
1   Oriental World: China, India, Persia, Western Asia, and
Egypt are included.
2   The Greek and Roman World: The history of Ancient Greece
and Rome is subsumed under this period.
3   The Germanic World: This is subdivided into three periods:
the first extending from the Byzantine Empire to Charlemagne,
the second, the Middle Ages, the third and final, the modern
era which is covered in a separate part of the book.

So if we remove the mystical shell, we find that Hegel
divides world history into four eras, namely:
1   The period of the origin and development of oriental
societies.
2   The ancient world.
3   The Middle Ages, i.e. the era of feudalism.
4   Bourgeois society.

We find in Hegel a number of theses concerning the Orient
which Marx and Engels turned right side up. . . . Among
other things, Hegel said of China . . . 'Since equality prevails in China, but without any freedom, despotism is necessarily the mode of government' (1830a, p. 124). In relation to
India he stated (ibid., p. 161) that '. . . while we found a
moral despotism in *China*, whatever may be called a relic of
political life in *India*, is a despotism *without a principle*, without any rule of morality and religion, (ibid., p. 154):

In respect to property the Brahmins have a great advantage,
for they pay no taxes. The prince receives half the income
from the lands of others; the remainder has to suffice for
the cost of cultivation and the support of the labourers. It
is an extremely important question, whether the cultivated
land in India is recognized as belonging to the cultivator,
or belong to a so-called manorial proprietor. The English
themselves have had great difficulty in establishing a clear
understanding about it. For when they conquered Bengal,
it was of great importance to them to determine the mode in
which taxes were to be raised on property, and they had

to ascertain whether these should be imposed on tenant
cultivators or the lord of the soil. They imposed the tribute
on the latter; but the result was that the proprietors acted
in the most arbitrary manner: drove away the tenant cultivators, and declaring that such and such an amount of land
was not under cultivation, gained an abatement of tribute.
They then took back the expelled cultivators as day-labourers,
at a low rate of wages, and had the land cultivated on their
own behalf. The whole income belonging to every village is,
as already stated, divided into two parts, of which one
belongs to the Rajah, the other to the cultivators; but
proportionate shares are also received by the Provost of
the place, the Judge, the Water-Surveyor, the Brahmin
who superintends religious worship, the Astrologer (who
is also a Brahmin, and announces the days of good and ill
omen), the Smith, the Carpenter, the Potter, the Washerman, the Barber, the Physician, the Dancing Girls, the
Musician, the Poet. This arrangement is fixed and immutable,
and subject to no one's will. *All political* revolutions, therefore, are matters of indifference to the common Hindoo,
for his lot is unchanged.

Here are a few of Hegel's views on Persia (ibid., pp. 183-
4):

In those times one City constituted the whole Empire -
Nineveh for example: so also Ecbatana in Media. . . . These
cities arose in consequence of a twofold necessity - on the
one hand that of giving up the nomad life and pursuing
agriculture, handicrafts and trade in a fixed abode: and
on the other hand of gaining protection against the roving
mountain peoples, and the predatory Arabs. Older traditions indicate that this entire valley district was transversed
by Nomads, and that this mode of life gave way before that
of the cities. Thus Abraham wandered forth with his family
from Mesopotamia westwards, into mountainous Palestine. . . .
The land around Babylon, was intersected by innumerable
canals; more for purposes of agriculture - to irrigate the
soil and to obviate inundations - than for navigation.

All the land and all the water belonged to the Great King
of the Persians. 'Land and Water' were the demands of
Darius Hystaspes and Xerxes from the Greeks. But the
King was only the abstract sovereign: the enjoyment of the
country remained to the nations themselves; whose obligations were comprised in the maintenance of the court and
the satraps, and the contribution of the choicest part of
their property (ibid., p. 190).

On Egypt, Hegel wrote (ibid., p. 204):

The Egyptians were, like the Hindoos, divided into castes,
and the children always continued the trade and business
of their parents. . . . Herodotus mentions the seven
following castes: the priests, the warriors, the neatherds,

the swineherds, the merchants (or trading population
generally), the interpreters - who seem only at a later
date to have constituted a separate class - and, lastly, the
seafaring class. . . . Herodotus says of the priests, that
they in particular received arable land, and had it cultivated for rent; for the land generally was in the possession
of the priests, warriors and kings. Joseph was a minister
of the king, according to Holy Scripture, and contrived
to make him master of all landed property.

If we abstract these disparate ideas from their context,
cleansing them of their idealistic ingredients, we can see that
Hegel, within a mystical overview and barrage of idealistic
arguments, reached the following conclusions:
(a)   The form of the state in the Orient was despotism.
(b)   The ultimate owner of land and water is the state.
(c)   Irrigation, the control of rivers, and public works in
general played an enormous role.
(d)   The village commune, and the caste system, which
originated and developed on the basis of the division of labour,
a stereotyped division of labour, left its imprint on the entire
social structure.
(e)   The entire structure of society was stagnant.
These ideas we later find in Marx and Engels in an entirely
different framework, in an entirely different interpretation.
Marx and Engels took the materialistic kernel, developed it,
and discarded the mystical shell and idealistic components.

In various works of Marx and Engels we find extremely
valuable remarks about the absence of private land-ownership
and the significance of irrigation in the Orient, oriental despotism, castes and guilds, the stereotyped division of labour,
the role and significance of the commune, the division of
labour in the commune, the connection between the necessity
for artificial irrigation, and the absence of private land-
ownership. Comparing these remarks with Hegel's statements,
we can clearly see that Hegel has become old-fashioned and
antiquated and understand the giant step forward taken by
Marx and Engels in precisely this area, in this particular
science.

It is interesting to note that one finds far fewer germs of
historical materialism in Hegel's writings on ancient (Greek and
Roman) society, feudal (German), and bourgeois society than
in those on the Orient. We cannot pursue this point here. For
our purposes, it suffices to note that Marx began from Hegel,
surpassed Hegel, and put the Hegelian teaching aright in
developing his conception of social formations. Marx began with
a critical review of Hegel's 'Philosophy of Right' and 'Philosophy
of History'.

The first stage in the development of Marx's conception of
social formations extended over the period 1844-7. In addition
to 'Contribution to the Critique of Hegel's Philosophy of Right',
this first stage is exemplified in the following works of Marx.

(This list does not pretend to be exhaustive.)
1   'The German Ideology'.
2   Letter to Annekov of 28 December 1846.
3   'The Poverty of Philosophy'.
4   'Wage-Labour and Capital'.

The sheer enumeration of these works illustrates that we can only indicate Marx's basic arguments on social formations in the present context. . . .

We cite, in our view, the decisive passage for this period (Marx, 1849, p. 28):

In production, men not only act on nature but also on one another. They produce only by co-operating in a certain way and mutually exchanging their activities. In order to produce, they enter into definite connections and relations with one another and only within these social connections and relations does their action on nature, does production, take place.

These social relations into which the producers enter with one another, the conditions under which they exchange their activities and participate in the whole act of production, will naturally vary according to the character of the means of production. With the invention of a new instrument of warfare, firearms, the whole internal organisation of the army necessarily changed; the relationships within which individuals can constitute an army and act as an army were transformed and the relations of different armies to one another also changed.

Thus the social relations within which individuals produce, the social relations of production, change, are transformed, with the change and development of the material means of production, the productive forces. The relations of production in their totality constitute what are called social relations, society, and, specifically, a society at a definite stage of historical development, a society with a peculiar, distinctive character. *Ancient* society, *feudal* society, *bourgeois* society are such totalities of production relations, each of which at the same time denotes a special stage of development in the history of mankind.

In comparing this passage with later statements by Marx on the same topic in the more classic and comprehensive thesis of the 'Introduction* to the Grundrisse' we can see that Marx has already developed his conception of the historical process by 1847. Furthermore, Marx had already distinguished three social formations or three societies in a concise and clear-cut fashion: the ancient, the feudal, and the bourgeois are each considered as constituting a particular stage in the development of the history of mankind. On this point there is and

---

*Ed. note: Mad'iar cites the publication of this work in 'A Contribution to the Critique of Political Economy'.

can be no doubt. . . .

From 1848 to 1859, Marx further developed his theory of
social formations and produced his final formulations. These
final formations are given in the 'Introduction to the Grundrisse'
and in the 'Preface to a Contribution to the Critique of Political
Economy'. We now know that the notebook in which Marx wrote
the 'Introduction' is dated 23 August 1857, and Marx dated the
'Preface' January 1859. We have not quoted these basic documents on the Marxist teaching of social formations. . . . They
have become classics, and are widely known. In them, Marx
not only formulates his method, but in brief outline summarizes
his own views on economic development as a natural historical
process. The connection between the 'Introduction' and the
'Preface' and Marx's earlier works on the same topic is so
obvious that the existence of this connection requires no proof.
The 'Introduction' and the 'Preface' expand, refine, complete,
give final formulation to and fill with living content those same
ideas which we find in embryo, in undeveloped form, in the
works of the young Marx. Marx himself points out this connection: 'The salient points of our conception were first outlined
in an academic, although polemical, form in my *Misère de la
philosophie* . . . this book which was aimed at Proudon
appeared in 1847' (1859, p. 22).

Once more he indicates this connection in his lectures on
wage labour. However, a comparison of the 'Introduction'
and the 'Preface' with Marx's earlier works reveals not only
an identity in the basic propositions, and a common methodology,
but also the enormous development which Marx himself had
achieved in the elaboration of historical materialism, in the
application of dialectical materialism and the formulation of the
theory of social formations. In the 'Introduction', Marx wrote
the following statement (1857-8a, pp. 105-6):

Since bourgeois society is itself only a contradictory form
of development, relations derived from earlier forms will
often be found within it only in an entirely stunted form,
or even travestied. For example, communal property. Although it is true, therefore, that the categories of bourgeois
economics possess a truth for all other forms of society,
this is to be taken only with a grain of salt. They can
contain them in a developed, or stunted, or caricatured
form etc., but always with an essential difference. The so-
called historical presentation of development is founded, as
a rule, on the fact that the latest form regards the previous
ones as steps leading up to itself, and, since it is only
rarely and only under quite specific conditions able to
criticize itself - leaving aside, of course, the historical
periods which appear to themselves as times of decadence -
it always conceives them one-sidedly. The Christian religion
was able to be of assistance in reaching an objective understanding of earlier mythologies only when its own self-
criticism had been accomplished to a certain degree, so to

speak. Likewise, bourgeois economics arrived at an understanding of feudal, ancient, oriental economics only after
the self-criticism of bourgeois society had begun.

So, in 1858, Marx already distinguished four social formations: oriental, ancient, feudal and bourgeois society. Here
for the first time, Marx declares the existence of oriental
society as a completely distinct form on the same basis as
ancient, feudal and bourgeois.

Two years later, in 1859, Marx formulated the same idea
in the following fashion: 'In broad outline, the Asiatic, ancient,
feudal and modern bourgeois modes of production may be
designated as epochs marking progress in the economic development of society' (1859, p. 21).

Here we once more see that Marx distinguished four modes
of production, that is four social formations, namely oriental,
ancient, feudal and the bourgeois modes of production, as
progressive formations of society.

So we can see that in 1857 and 1859, Marx, with complete
clarity, without any room for doubt, distinguished four modes
of productions, four social formations, four societies. Oriental
society, the Asiatic mode of production, was presented by Marx
as a progressive era in the economic formation of society, on
the same basis as ancient, feudal, and bourgeois societies.

And now the following questions arise perfectly naturally:
1   Hegel spoke, in mystical terms, of four societies. Marx
knew Hegel extremely well. How did it happen that from
approximately 1845 to 1848 Marx noted perfectly clearly only
three social formations, but we find that he made no mention
of oriental society or the Asiatic mode of production?
2   What precisely moved Marx to a 'recognition' of oriental
society or the Asiatic mode of production?
3   Did Marx change his view of oriental society and the
Asiatic mode of production? Did he not later refute his own
theses that oriental society represented a whole era in the
economic formation of society?
4   Finally, was Marx mistaken when he considered oriental
society, the Asiatic mode of production, as a separate social
formation?
We will briefly consider these questions.

1   Marx did not mention the existence of oriental society,
in spite of Hegel's clear indications, until 1853, because in
that period he was not yet able to put aright the Hegelian
conception. He had not found the key to the Orient; he had
not yet found the key to the productive relations of oriental
society. He had not found the fundamental difference which
separated oriental society from all other societies. This is a
difference not from the viewpoint of absolute idealism, as is
found in Hegel, but a difference from the standpoint of the
aggregate of productive relations. It is not difficult to show
that during this period, Marx was little concerned with the

Orient, knew relatively little about the Orient, and in any
event, had not yet found the key to the Orient. . . .

2   Marx was moved to a recognition, or a discovery of a
distinct oriental society through a study of the Orient.
We leave it to Marx's biographers to determine which books,
journals, and articles Marx read on the Orient in the period
1848–57. Marx himself records the results of these endeavours
in his letters and articles, written in the period 1853–7.
These letters of Marx and Engels give us an answer to the
question of what moved Marx to recognize, to discover oriental
society and the Asiatic mode of production. During this same
period, Marx's articles on India appeared, in which the same
basic propositions are formulated as in these letters.
It is true that Marx and Engels do not give a detailed
analysis of oriental society and the Asiatic mode of production
in these letters and articles. But neither do they give a detailed analysis of ancient society or feudal society. Marx has
left a detailed analysis of only bourgeois society and the
capitalist mode of production, and even that analysis he did
not succeed in finishing. Engels entirely correctly observed
that:

> Political economy, however, as the science of the conditions
> and forms under which the various human societies have
> produced and exchanged and on this basis, have distributed
> their products – political economy in this wider sense has
> still to be brought into being. Such economic science as we
> possess up to the present is limited almost exclusively to
> the genesis and development of the capitalist mode of pro-
> duction (1878, p. 181).

It seems to us that this remark of Engels has been little
considered, or not considered at all in the discussion of political economy. Marx and Engels considered that political economy
should study not only the capitalist mode of production, but
the conditions and forms of production, trade, and distribution
in various human societies. Such a political economy has not
yet been founded, but it is essential. For even to understand
bourgeois economy, it is necessary to study the forms which
preceded or coexist along with the bourgeois economy. Engels
was correct that: 'Such an investigation and comparison has up
to the present been undertaken, in general outline, only by
Marx, and we therefore owe almost exclusively to his researches
all that has so far been established concerning pre-bourgeois
theoretical economy' (1878, p. 182).
In addition, it can be said that not only Marx, but Engels
as well, gave us much for the understanding of the pre-
bourgeois economy, the pre-capitalist modes of production.
But it is only possible to find the basic principles of pre-
capitalist social formations in general outline in the works of
Marx and Engels.
If we now compare Hegel's writings of oriental society with

Marx's statements, it becomes clear why Marx did not mention
oriental society in the period 1845-53, and why he recognized
it as a separate social formation only after 1853. This happened
because only in 1853 did he find the key to the oriental
heavens, to the entire religious and political history of the
Orient, to an understanding of the productive relationships
of oriental society.

3   We maintain that Marx did not change his views on oriental
society and did not revise his teaching on the existence of
four social formations throughout the rest of his life. We maintain that in all of the authoritative works of Marx and Engels,
the concept of the existence of four class social formations
may be consistently found. We maintain that Marx, in the discussion of pre-capitalist social formations, almost always treats
the Orient as a separate problem.

First of all it is necessary to clear up what may be called
the philological aspects of the question. Is it true that Marx
speaks of oriental society and the Asiatic mode of production
only in the 'Introduction', but in his later works denies even
the formulation itself? No, this is not true. We find the term
in the later works of Marx. For example in 'Capital', analysing
the role of commodities in pre-capitalist social formations,
he writes (1867, p. 83):

> In the ancient Asiatic and other ancient modes of production,
> we find that the conversion of products into commodities,
> and therefore the conversion of men into producers of
> commodities, holds a subordinate place, which, however,
> increases in importance as the primitive communities approach
> nearer and nearer their dissolution.

Marx left the term in the last German edition of 'Capital'
which appeared during his lifetime. Furthermore, in the last
French edition of 'Capital' which appeared during Marx's lifetime, the translation of which was thoroughly edited by Marx,
we find the following formulation: 'Dans les modes de production de la vieille Asie, de l'antiquité en général le transformation
du produit en marchandise ne joue qu'un role subalterne' and
so on.

In 'Capital III', in the chapter 'Historical Facts about
Merchant's Capital', Marx, referring to China and India, states
that 'unlike the English, the Russian commerce . . . leaves
the economic groundwork of Asiatic production untouched'.
In the context, he mentions 'the obstacles presented by the
internal solidity and organisation of pre-capitalistic, national
modes of production' (1894, pp. 333-4).

However, we consider these conclusions purely formal, and
we would not have mentioned them if some individual comrades
in their ardour had not tried to deny that which is written
plainly in black and white.

We consider a much more important and authoritative conclusion the simple and unarguable fact that Marx, in the

analysis of almost all economic categories of capitalist society,
separately considers the question of the role and significance
of these categories in other social formations, and in particular
under the Asiatic mode of production.

We have already indicated that in his analysis of commodities
Marx emphasized its subsidiary role in ancient Asian, ancient,
etc. modes of production and blames bourgeois political economy
for treating 'forms of social production that preceded the
bourgeois form . . . in much the same way as the Fathers
of the Church treated pre-Christian religions'. In pre-
capitalist, social formations, commodity production is not found
'in the same predominating and characteristic manner as now-
a-days. Hence its Fetish character is comparatively easy to be
seen through' (1867, pp. 85–6).

When Marx analyses the question of co-operation, he treats
separately the question of co-operation in oriental societies.
His discussion of this question is essentially an expanded
formulation of the ideas expressed in the correspondence of
Marx and Engels, and in the articles on India (ibid.,
pp. 351-16):

> The colossal effects of simple co-operation are to be seen
> in the gigantic structures of the ancient Asiatics, Egyptians,
> Etruscans, etc. . . . This power of Asiatic and Egyptian
> kings, Etruscan theocrats, etc., has in modern society
> been transferred to the capitalist, whether he be an isolated,
> or as in joint-stock companies, a collective capitalist.
>
> Co-operation, such as we find it at the dawn of human
> development, among races who live by the chase, or, say,
> in the agriculture of Indian communities, is based, on the
> one hand, on ownership in common of the means of produc-
> tion, and on the other hand, on the fact, that in those
> cases, each individual has no more torn himself off from the
> navelstring of his tribe or community, than each bee has
> freed itself from connexion with the hive. Such co-operation
> is distinguished from capitalistic co-operation by both of
> the above characteristics. The sporadic application of co-
> operation on a large scale in ancient times, in the middle
> ages, and in modern colonies, reposes on relations of
> dominion and servitude, principally on slavery. The capital-
> istic form, on the contrary, pre-supposes from first to
> last, the free wage-labourer, who sells his labour power
> to capital.

When Marx analyses the problem of the genesis and accumulation of capital, he considers the genesis of capital in various
social-economic formations separately. . . .

He indicates that
> *under Asian forms, usury can continue a long time, without
> producing anything more than economic decay and political
> corruption.* Only where and when the other prerequisites
> of capitalist production are present does usury become one
> of the means assisting in establishment of the new mode of

production of ruining the feudal lord and small-scale producer, on the one hand, and centralising the conditions
of labour into capital, on the other `. . . (1894, p. 597,
emphasis L. Mad'iar).

When Marx analyses the question of land-ownership, he
notes its dependence upon and relation to social formations.
The arguments he puts forth in 'Capital' strongly recall the
ideas found in an embryonic form in 'The German Ideology'.

Marx also indicates that land-ownership developed under
particular conditions in ancient society.

The legal view itself only means that the land-owner can
do with the land what every owner of commodities can do
with his commodities. And this view, this legal view of
free private ownership of land, arises in the ancient world
only with the dissolution of the organic order of society,
and in the modern world only with the development of
capitalist production.

Marx points out that the concept of private ownership of land
'has been imported by Europeans to Asia only here and there'
(Marx, 1894, p. 616). (Of course, since Marx wrote these
lines, the situation has completely changed, and now land is
private property in just those parts of Asia where free ownership of land did not exist in his lifetime.)

When Marx discusses the theory of rent, he warns of the
three main mistakes which can dim the analysis. The first of
these is the confusion of various forms of rent which exist
at various stages of development in the social process of
production (ibid., p. 634):

Whatever the specific form of rent may be, all types have
this in common: the appropriation of rent is that economic
form in which landed property is realised, and ground-rent,
in turn, presupposes the existence of landed property, the
ownership of certain portions of our planet by certain
individuals. The owner may be an individual representing
the community, as in Asia, Egypt, etc.; or this landed
property may be merely incidental to the ownership of the
immediate producers themselves by some individual as under
slavery or serfdom; or it may be a purely private ownership of Nature by non-producers, a mere title to land; or,
finally, it may be a relationship to the land which, as in
the case of colonists and small peasants owning land, seems
to be directly included - in the isolated and not socially
developed labour - in the appropriation and production of
the products of particular plots of land by the direct producers.

This *common element* in the various forms of rent,
namely that of being the economic realisation of landed
property, of legal fiction by grace of which certain individuals have an exclusive right to certain parts of our
planet - makes it possible for the differences to escape
detection.

We see here that Marx enumerates separately Asia, Egypt, and so on, that is oriental society, slave-owning society, serfdom, and capitalism.

Analysing the question of the genesis of capitalist ground rent, Marx once more considers the question of ancient society and the Orient separately. He demonstrates that in natural economy, the product and surplus product of large estates consists not only of the products of agricultural labour, but emphasizes (ibid., p. 787) that the surplus product

encompasses equally well the products of industrial labour. Domestic handicrafts and manufacturing labour, as secondary occupations of agriculture, which forms the basis, are the prerequisite of that mode of production upon which natural economy rests - in European antiquity and the Middle Ages as well as in the present-day Indian community, in which the traditional organisation has not yet been destroyed.

Here we once more see a list of three pre-capitalist formations. When Marx indicates that under pre-capitalist relations, the surplus labour for the land-owner can be extracted only by extra-economic compulsion, he once more treats the Orient separately (ibid., p. 791):

Should the direct producers not be confronted by a private land-owner, but rather, as in Asia, under direct subordination to a state which stands over them as their landlord and simultaneously as sovereign, then rent and taxes coincide, or rather, there exists no tax which differs from this form of ground rent. Under such circumstances, there need exist no stronger political or economic pressure than that common to all subjection to that state. The state is then the supreme lord. Sovereignty here consists in the ownership of land concentrated on a national scale. But, on the other hand, no private ownership of land exists, although there is both private and common possession and use of land.

When Marx raises the question of the genesis of capitalist land rent, he indicates three problems (1905, p. 42):

. . . should now be set forth: 1   The transition from feudal land-ownership to a different form, commercial land rent, regulated by capitalist production, or, on the other hand, the conversion of this feudal landed property into free peasant property. 2   How rent comes into existence in countries such as the United States, where originally land has not been appropriated and where, at any rate in a formal sense, the bourgeois mode of production prevails from the beginning. 3   The Asiatic forms of land-ownership still in existence.

Here again the question of Asiatic forms of land-ownership is put forth as a separate problem. It is possible, as a matter of fact it is necessary, to continue to collect the opinions of Marx on the various economic categories as they apply to the various modes of production. With an absolutely remarkable

consistency, even for Marx, he applies his method of strict
definition and historical specificity to all problems of political
economy concerning various modes of production.

When Marx proposes the law of population for capitalist
society, he immediately indicates that the law is inapplicable
to other social formations.

The labouring population therefore produces, along with
the accumulation of capital produced by it, *the means by
which it itself is made relatively superfluous, is turned into
a relative surplus-population; and does this to an always
increasing extent. This is a law of population peculiar to
the capitalist mode of production*; and in fact every special
historic mode of production has its own special laws of
population, historically valid within its limits alone. An
abstract law of population exists for plants and animals only,
and only in so far as man has not interfered with them
(1867, pp. 591-2, emphasis L. Mad'iar).

When Marx points out the cyclical nature of industrial production under capitalism, he emphasizes that under previous
methods of production such cycles did not exist: 'This peculiar
course of modern industry, which occurs in no earlier period
of human history, was also impossible in the childhood of
capitalist production' (1867, p. 593).

It would not be difficult to prove that Marx sees the question of the interrelationship between the city and the countryside as one specific to the various social formations. However,
we have shown this elsewhere, and it is not necessary to
detail the evidence for this proposition here. Nor would it be
difficult to prove that Marx also viewed the problems of foreign
trade from the perspective of their role in the various social-
economic formations. However, it seems to us that this far
from complete list of the teachings of Marx is sufficient to
prove how correct Engels was when he asserted that in the
works of Marx, if only in general terms, we can find the
basic principles for a study not only of the capitalist mode
of production but generally of all modes of production. This
proves that Marx conducts his analysis of the various economic
categories as they apply to the four-class social formations
with the strictest consistency. It proves that Marx never refuted his thesis on the existence of oriental society, but, on
the contrary, developed it and gave it a firmer basis in his
later works.

4   Now we must answer the fourth question. Was Marx mistaken when he put forth his theses on the existence of the
four class-social formations, and in particular, on the existence
of oriental society and the Asiatic mode of production. After
all, certain comrades consider that Marx was mistaken, that
the Asiatic mode of production, as a separate social formation,
did not exist. They are ready to condescendingly forgive poor
Marx this 'mistake' in view of the fact that he knew little about

the Orient; that during the time Marx was working, the Orient
was in general little studied, and that Marx had read few books
on the Orient, etc. Such a formulation of the problem strikes
us as highly comical, but all the same we will consider the
question.

We know that Marx and Engels in their first works presented
the following characteristic features of oriental society:

1   An absence of private land-ownership.
2   A necessity for artificial irrigation and, connected with this,
a necessity for the organization of public works on a grand
scale.
3   The rural commune.
4   Despotism as form of government.

These characteristics do not by themselves uncover the
means by which the surplus product was extracted from the
immediate producers, but they provide a key to the discovery
of the most profound secret of the Asiatic mode of production.
These clues give the key to the discovery of the class relations
of oriental society. In order to find that key, we must start
from the commune, for from it, oriental society arose and
developed. We find a description of the commune in Hegel,
and in Marx's letter of 14 June 1853 (Marx and Engels, 1965,
p. 85; Marx, 1867, pp. 337-8). Marx characterizes the social
and economic organization of the commune in the following
fashion (1867, pp. 83-4):

> Those ancient social organisms of production are, as com-
> pared with bourgeois society, extremely simple and
> transparent. But they are founded either on the immature
> development of man individually, who has not yet severed
> the umbilical cord that unites him with his fellowmen in a
> primitive tribal community, or upon direct relations of
> subjection. They can arise and exist only when the develop-
> ment of the productive power of labour has not arisen
> beyond a low stage, and when, therefore, the social rela-
> tions within the sphere of material life, between man and
> man, and between man and Nature, are correspondingly
> narrow.

However, such communes existed among the Romans, the
Celts, the Germans; they existed, to use the expression of
Engels, from India to Ireland, and we should add that they
existed in America, Australia, and in all parts of the world.
The peculiarity of the development of these communes in the
Orient arose from the necessity for artificial irrigation. Marx
indicates this in 'Capital', when he emphasizes that 'One of
the material bases of the power of the state over the small
disconnected producing organisms in India, was the regulation of the water supply' (1867, p. 481).

Engels emphasizes the great significance of the necessity
of regulating the water supply from the viewpoint of the
origin of classes and the state in the following fashion (1878,
p. 179):

The state, which the primitive groups of communities of the same tribe had first arrived at only in order to safe-guard their common interests (e.g. irrigation in the East) and for protection against external enemies, from this stage onwards acquires just as much the function of maintaining by force the conditions of existence and domination of the ruling class against the subject class.

How did classes and the state develop from the commune? Engels answers this question in the following manner (1878, pp. 214-15, emphasis L. Mad'iar):

In each such community there was from the beginning cer-tain common interests, the safeguarding of which had to be handed over to individuals, true, under the control of the community as a whole: adjudication of disputes; repression of abuse of authority by individuals; control of water supplies, especially in hot countries; and finally, when conditions were still absolutely primitive, religious function. Such offices are found in aboriginal communities of every period - in the oldest German marks and even today in India. They are naturally endowed with a certain measure of authority and the beginnings of state power. . . . These organs which, if only because they represent the common interests of the whole group, hold a special position in relation to each individual community - in certain circum-stances even one of opposition - soon make themselves still more independent, partly through heredity of functions, which comes about almost as a matter of course in a world where everything occurs spontaneously, and partly because they become increasingly indispensable owing to the growing number of conflicts with other groups. It is not necessary for us to examine here how *this independence of social functions in relation to society increased with time until it developed into domination over society; how he who was originally the servant, where conditions were favourable*, changed gradually into the lord; how this lord, depending on the conditions, emerged as an oriental despot or satrap, the dynast of a Greek tribe, chieftain of a Celtic clan, and so on. . . . Here we are only concerned with establish-ing the fact that the exercise of a social function was everywhere the basis of political supremacy; and further that political supremacy has existed for any length of time only when it discharged its social functions. However great the number of despotisms which rose and fell in Persia and India, each was fully aware that above all it was the entre-preneur responsible for the collective maintenance of irrigation throughout the river valleys, without which no agriculture was possible there. It was reserved for the enlightened English to lose sight of this in India; they let the irrigation canals and sluices fall into decay, and are now at last discovering, through the regularly recurring famines, that they have neglected the one activity which

might have made their rule at least as legitimate as that
of their predecessors.

Thus we see how the necessity for artificial irrigation has
set apart a group of people from the commune who, in fulfilling their function as servants of the commune, were invested
with a certain authority, and how these servants of the commune
were transformed into the sovereigns of the commune, as they
set themselves up as a ruling class in relation to the commune.
Political power has been transformed from handmaiden into
mistress. By the same means, the former servants of the
commune became the masters of the commune - they became
its exploiters. In this sense, Marx spoke of the rule of the
priests in Egypt. In this same sense, in his letter to Bernstein
of 9 August 1882, Engels affirmed that 'the satrap of pasha
is the main Eastern form of exploiter'.

Furthermore, we must take into account still another important circumstance: pre-capitalist social formations are characterized by a tendency to

make trades hereditary; either to petrify them into castes,
or whenever definite historical conditions beget in the
individual a tendency to vary in a manner incompatible
with the nature of castes, to ossify them into guilds. Castes
and guilds arise from the action of the same natural law,
that regulates the differentiation of plants and animals into
species and varieties, except that, when a certain degree
of development has been reached, the heredity of castes
and the exclusiveness of guilds are ordained as a law of
society (Marx, 1867, p. 321).

In ancient and feudal societies, the stereotypification of the
division of labour never reached the point of forming castes.
However, in the majority of the ancient oriental countries,
the development of productive forces and the division of
labour led to the formation of the caste system. So it was in
Egypt.

Plato's Republic, in so far as the division of labour is
treated in it, as the formative principle of the State is
merely the Athenian idealisation of the Egyptian system of
castes, Egypt having served as the model of an industrial
country to many of his contemporaries also, amongst others
to Isocrates, and it continued to have this importance to
the Greeks of the Roman Empire (Marx, 1867, p. 346).

So it was in India, Assyria, and to a significant degree in
Persia. In China, there was an unsuccessful attempt to introduce the caste system under the influence of Buddhism, but
the division of society into literati, peasants, craftsmen, and
merchants played a rather important role even in ancient
China. As the division of labour was decreed as a law of
society, the prerequisites for the exploitation of one class by
another were laid down.

So the basic class division of oriental society is between
the fundamental masses of the peasants, joined together in

communes, and the former servants of the commune, who have separated themselves from the commune and formed themselves into a ruling class. (The priests in Egypt, the literati in ancient China, and so on.) The form of government is despotism. Private land-ownership is absent. The supreme owner of the land and water, the basic conditions for production, is the state. The basic economic form of exploitation is the tax, which coincides with rent. The ruling class exploits the commune, extracting the surplus product in the form of a tax or rent. The economic form of extraction of the surplus product by means of a tax which coincides with rent undoubtedly likens this form of exploitation to the feudal system. The absence of feudal property and a class of feudal lords constitutes the primary difference between oriental and feudal society.

In view of this it is necessary to emphasize that in oriental society the conduct of class war was confined to the struggle between the peasants of the commune and the state of rent and tax collectors. The struggle against merchant-usurer capital played an enormous role in oriental society, and there was frequently a struggle between the various castes. Given that the caste system and the whole social order was decreed not only as social but also as religious law, political struggle took place in these societies under a religious mask. In India, the struggle between Buddhism and Brahmanism was the struggle for the easing of the caste system. In Indonesia, Buddhism conquered Brahmanism simply because the masses of the people supported the struggle of Buddhism for the easing of caste divisions. Muhammadanism vanquished Buddhism because Muhammadanism generally does not recognize caste division. In China the despot emperor is the Son of Heaven, and so forth.

This, in most general outline, is the teaching of Marx on the class relations under the four-class social and economic formations. Of course, not a single one of these formations ever existed in pure form; in actual historical reality, each of these formations existed in endless variations and gradations. Needless to say, in each of these formations there were remains of previous formations and forerunners of future formations. . . .

It is impossible, however, to conclude this survey without mentioning that some bourgeois and petit bourgeois economists have tried to use Marx's teaching on oriental society to create the most absurd historical conceptions of supra-class states in the Orient, and the absence of class struggle in oriental society, etc. Should these pathetic, reactionary, and counter-revolutionary theories be refuted?

It seems to us that it is not worth while wasting time on such a task. And if some comrades who consider themselves Marxists have fallen under the influence of such ridiculous theories, it only proves that they consider themselves Marxists

in vain, and without any foundation. There is no need to
prove that Marx's theory of oriental society does not mean that
the Asiatic mode of production existed in all periods and in all
countries of the Orient. After all, Marx himself held that the
feudal mode of production existed in Japan in the nineteenth
century. . . .

Undoubtedly, at the present time, the agrarian order of the
great majority of oriental countries is witnessing conflict
between the peasantry and the landlord. It hardly bears repeating that one cannot speak of the absence of private land-
ownership in the Orient today; it hardly bears repeating, that
today the presence of landed property is the key to the
oriental heavens, to the understanding of the moving forces
of revolution in these countries. Nor is it necessary to re-
state that today, in all oriental countries, the anti-imperialist
and agrarian revolution is the order of history. Japan is the
only exception, an imperialist country where the proletarian
revolution, with its great number of bourgeois tasks, is the
order of the day. It hardly bears repeating that at the present
time, in all countries of the Orient, including Japan, the
product is extracted from the peasantry primarily in the form
of rent in kind, or secondarily, in the form of money rent,
as a transformed variety of rent in kind. Of course, merchant
usury, tax, and imperialistic exploitation of the peasant masses
play a tremendous role. But at the same time, it would be incorrect and would not correspond to the state of things to
search the social order of these countries, and the agrarian
order in particular only for remains and survivals of the feudal
order of exploitation and to mobilize the workers and the
peasants against them. Over relatively large territories of
China, land-owners' estates play a relatively subsidiary role
(Hopei, Shanhsi, some parts of Shantung, Kwanghsi, Shenhsi,
even Honan). In these areas, the surplus product is extracted
from the peasantry in the form of rent/tax. Even in those
areas of China where landlordship prevails, the exploitation of
the peasantry by means of the extraction of a rent/tax plays
an important role. It is no accident that the blows of the
peasant movement in China are directed against the payment
of rent, taxes, and usurers' interest. Nor is it accidental that
the slogan of the peasant movement is 'Down with land-owners
and gentry.' And it is absolutely impossible to explain the
gentry as a social category without a correct understanding of
Marx's teaching on oriental society. In significant parts of
India, where at one time the English conquerors created, in
the words of Marx, 'a caricature of peasant property', the
exploitation of the peasantry by means of a rent/tax played
and continues to play an enormous role; although under the
effects of merchant usury and imperialist exploitation, and
with the spread and intensification of mercantile monetary
relations, landlord ownership has arisen and developed quickly.
The English conquerors, even in those parts of the country

where they created 'a caricature of the system of large scale
English landholding' (Marx), and where they have turned the
Indian commune with its common landholding into a caricature
of itself, have declared themselves to be the supreme land-
holders and have become the receivers of the rent/tax or have
divided the rent/tax with the zamindars, talukdars, malguzars,
dzhagidars, etc. (as the Indian rent receivers are still called).
In India, the remains and survivals of the caste system still
play a significant role. In a whole series of colonies, where
the village or clan commune as an economic form, and along
with it communal land-ownership, have been preserved, the
first step of capital in reconstructing the social order is to
declare the country the booty of the supreme land-owner and
to introduce rent/tax. Without correct understanding of these
methods of exploitation, it is impossible to mobilize the masses
against all these types of exploitation. This is exactly what
the programme of the Communist International had in mind
when it stated that along with the remains of feudalism, the
remains and survivals of the Asiatic mode of production could
be found in the Orient.

NOTES

*   From Foreword to M. Kokin and G. Papaian, '"Tszin-Tian",
     agrarnyi stroi drevnogo Kitaia' ('"Tszin-Tian", the Agrarian
     Order of Ancient China'), Leningrad, 1930, pp. ii-xi, xvi,
     xviii-xl, lxvii-lxxi. Translated by Robert Croskey.
** Liudvig Ignat'evich Mad'iar was born in Hungary in 1891.
     He worked as a journalist throughout the First World War.
     He headed the press bureau under the Károly government
     and became an active supporter of the Hungarian Soviet
     Republic. After the fall of the Republic, he was imprisoned
     until 1922, when he was freed in an exchange of prisoners
     between Hungary and the Soviet Union. In the Soviet
     Union he worked for the Comintern and was in the Soviet
     diplomatic service in China in 1926-7.

THE ASIATIC BUREAUCRACY AS A CLASS*

M.D. Kokin**

The surplus labor or product was extracted by the bureaucrats
from the direct producer. I cannot call this group of people
other than bureaucrats, although it would be possible to find
another name for them. But this is not important. The term
is not important. Marx and Engels did not give a name to
this class. When mention is made of the ruling class in Asiatic
society, Marx speaks of the state. This is why I chose the
word bureaucrat as, in my opinion, the most appropriate name.
    The question may arise. Was every bureaucrat a member of
the ruling class? If the answer is yes, then is there not a
contradiction with the basic proposition that legal position is
only a result of the position in the process of production? In
any event, it seems to me that our sources provide ample
evidence for the conclusion that every bureaucrat was a member
of the ruling class in Asiatic society. But it should not be
forgotten that every bureaucrat was not only a screw in the
apparatus of oppression, but a part of the economic apparatus,
which controlled the productive process - the relations of
exploiter to exploited were in this case between bureaucrat and
commune member. This side of things must not be forgotten
when Asiatic society is examined. It is absolutely clear that
he who became a bureaucrat could only arise from those who
had a direct relationship to the caste of bureaucrats. Outsiders
did not enter that caste. It is worth mentioning a purely superficial but characteristic feature: in order to become a bureaucrat,
it was legally required to be literate first of all. The alphabet
in the countries of the Asiatic mode of production is a hiero-
glyphic alphabet. To achieve a command of this language, in
order to learn these hieroglyphics, long years, even decades,
are necessary. Who could devote years to the study of hieroglyphics except those who lived at the expense of others? Thus,
even in this respect, it is vividly clear that economic position
was decisive as to whether or not a person was master or
worked for the master.
    This is more or less what I wanted to say about the process

of class formation and the creation of the state in oriental society. It is perfectly clear that the classics of Marxism speak clearly of the specific characteristics of oriental society and in particular, of the oriental state; it is perfectly clear that these propositions do not contradict the teachings of Marx or Lenin on the state.

NOTES

* From 'Diskussia ob Aziatskom sposobe proizvodstva po dokladu M. Godesa' (A Discussion on the Asiatic Mode of Production According to the Report of M. Godes), Moscow-Leningrad, 1931, pp. 50-1. Translated by Robert Croskey.
** Mikhail D. Kokin (1906-39) was a Soviet specialist on China who worked in Leningrad. After 1931, like many Soviet sinologists, he turned his attention to modern Chinese history and began work on the Revolution of 1911.

THE AMP AND THE CLASS STRUGGLE*

E.S. Iolk**

Comrades, we face a stormy and difficult time in the intense
struggle for socialism in our country and in this period of
sharp class struggle throughout the world, especially in the
colonial East. In such circumstances, if we deem it possible
to discuss the question of the 'Asiatic mode of production',
then it is primarily because we are concerned with the political
conclusion to be drawn from the theory of the Asiatic mode
of production. I believe that the interest shown in our discussion by the public at present is obviously dictated by precisely these considerations, that is to say, considerations of
a practical nature.

We are not here because, as Comrade Kokin said, we 'need
at last to study the history of the Orient and include it in
world history'. This is not the reason we are discussing the
Asiatic mode of production, but in order to turn the practical
history now being created by the heroic struggle of the exploited masses of China, Indo-China, and India, to the interests
of the labouring masses by means of the correct method. This
must be the determining factor in our approach to the evaluation of one or another position in the present debate.

And so, if we approach things from this perspective, and
I think it is absolutely necessary to do so, then it is perfectly
clear that the interpretation of individual statements by Marx
and Engels on the Orient, presented to us in the guise of a
superficially perfected theory about a particular Asiatic mode
of production, is absolutely unacceptable from a political point
of view. . . .

In constructing a particular oriental social and economic formation, the proponents of the theory of 'Asiatic' mode of
production base themselves on the Marxist definition of the
essence of social formations, but they ground their 'formation'
on purely superficial remarks. The mechanism is very simple
here: individual statements by Marx and Engels are chosen
relating to the Orient, or to various aspects of life in the

Orient a new name, the Asiatic mode of production, is added, and a new formation is ready. The means by which the proponents of this theory define social formations, in particular the Asiatic mode of production, has been demonstrated before our very eyes in Comrade Kokin's presentation. . . .

I believe that today, in conditions of extremely sharp class struggle in the Orient, when the national liberation movement in the decisive colonial countries has reached its highest stage and the class contradictions are revealed as never before, it is not accidental that in our ranks the question of the class structure of oriental society lacks definition, and that there is a movement away from Marxist-Leninist positions. I do not wish to chide all of the named comrades for repeating bourgeois theories, but for me it is clear that their mistaken propositions reflect well-defined foreign influences. It is absolutely obvious that in such conditions any erroneous Marxist works can be widely used by our enemies. When the USSR - that widely recognized citadel of Marxist theory - 'exports' incorrect directives, even with the highly authoritative stamp of the Communist Academy, there is no doubt that this can have the most harmful of consequences for us. When, for example, the Chinese comrades are presented with a theory that the gentry are not an example, a survival, of the feudal order, that in China there were certain peculiar 'Asiatic' relations, and so on, this of course can disorient and confuse the ideology of the proletarian avant-garde.

This is why we have so vehemently criticized the theory of a separate Asiatic mode of production; this is why we consider the struggle against the theory as a deviation from Marxist-Leninist methodology, infused with mistaken political principles, particularly crucial.

NOTES

*   From 'Diskussia ob Aziatskom sposobe proizvodstva po dokladu M. Godesa' (A Discussion on the Asiatic Mode of Production According to the Report of M. Godes), Moscow-Leningrad, 1931, pp. 59, 63, 74. Translated by Robert Croskey.

**   Evgenii S. Iolk was born in Riga in 1900. He participated in the Civil War and later studied Chinese. He served in China in 1925-27 as a translator. In the late 1920s and early 1930s he held a number of academic positions. He served in the Soviet army from 1932 until 1942, the year of his death.

THE REAFFIRMATION OF UNILINEALISM*

M. Godes

The debates over the nature of the Chinese revolution which reached a peak in 1926-7 provide the basis for our discussion of the Asiatic mode of production and social formations in general.

The debates of that period are a vivid illustration of Comrade Stalin's notion that our theory lags behind practice, that theory is not catching up with practice. The Communist International has given a clear answer to the question of the nature of the social and economic order of China today, but in the present situation, we Marxists cannot boast that our theoretical work has anticipated revolutionary practice. Quite the contrary.

The directive of the Party and the Comintern on the nature of the Chinese revolution was disputed from two points of view. However, both led to the denial of feudalism in China. Radek, who has done so much for the discovery of the existence of feudalism in the ancient Chinese history, as a Trotskyist denied the presence of serious feudal survivals in present-day China, the social and economic order of which, in his opinion, is characterized by merchant capitalism. In 1930, Radek himself evaluated the error of his position in 1926-7 ('Problemy Kitaia' (Problems of China), 1930, no. 3, p. 11):

> During our discussion of 1926-7, my mistake in the evaluation of the driving forces of revolution consisted in not clearly understanding that the union of the land-owners' cabal with that of merchant capital was only possible on the basis of unliquidated feudal remnants. Therefore the burden on the peasant masses had, in my eyes, a more contemporary capitalist character.

Leaving aside the insufficient clarity of Comrade Radek's formulation, it is nevertheless apparent that he correctly sees his basic error in insufficient evaluation of the survivals of feudalism in contemporary China.

But attacks on the Party's estimation of the Chinese revolution as antifeudal, and on the Party's understanding of the nature of the economic order in contemporary China, were not only grounded in replacing feudalism by the so-called merchant capitalism. At the Fifteenth Party Congress, in the discussion of the report of our delegation to the Comintern, Lominadze gave a speech devoted to the Chinese revolution:

'I consider', he said, 'that the types of social relationships which are found in the Chinese countryside can only very conditionally be considered to be feudal, and then only with the qualification that they resemble very little the European Middle Ages. Survivals of the unique Chinese feudalism, which it would be better to call the Asiatic mode of production, as Marx did, are the reason for the continued and extremely sharp class struggles in the Chinese countryside' ('Stenografichiskii otchet KV s'ezda VKP(b)', p. 662). (Trans. note: Wissarion (Besso) Lominadze (1898–1934) specialized in Chinese affairs in the Comintern. He fell from favour in late 1930 over the collectivization question and committed suicide in 1934.)

At first glance, this statement might seem to be an innocent terminological correction. However, Lominadze, in this evaluation of the nature of the Canton Commune, gave a Trotskyist formulation, defining the tasks of the revolution as the tasks of a socialist revolution. The denial of the existence of feudalism in China, or the underestimation of its importance, seem to have always led to political mistakes, mistakes of precisely a Trotskyist nature. Incidentally, the evaluation of the social relationships in the Chinese countryside (Stenographic Report of the Fifteenth Congress of the VKP(b)), as characteristic of the Asiatic mode of production, was given in the first plan of the agrarian programme of the Chinese Communist Party, a plan which was later disavowed at the Sixth Party Congress.

A discussion of the Chinese revolution presents us immediately with a series of theoretical problems. It demands a more precise definition of feudalism and merchant capital; it requires an answer to the question of whether China and the Orient as a whole passed through a particular social formation which is not found in the history of European nations, or whether we find in the history of China, as in the rest of the Orient, a unique development of the same forms which are found in the history of Europe.

Beginning with the question of the Asiatic mode of production, the discussion grew into a dispute over social formations. The discussion originated with politics and its political import was evident throughout. Here, today, dealing with the question of its result, it is certainly apparent that we are not engaged in a purely academic debate. Considering the political nature of our discussion, we understand that success and the very possibility of specific research in the future depends on the correct solution to our methodological disputes.

Let's start with the school of Comrade Mad'iar. Although

not everyone who has written on the Orient and accepted the
concept of the Asiatic mode of production is a disciple of
Mad'iar, they all exhibit a natural attraction to that most
fruitful, and it must be said openly, most firmly based defender
of this theory. A characteristic of this entire school is an
acritical use of separate remarks of Marx on the Asiatic mode
of production and the Orient, frequently out of context, and
at any rate out of the context of Marx's general teachings on
social formations. It seems that the discussion of the Asiatic
mode of production and the pronouncements of partisans of
this theory have once more shown that quotations from Marx
do not constitute Marxist teachings, and that not everyone who
quotes Marx correctly applies his method, just as not everyone who repeats the Lord's prayer makes it to heaven.

What mode of production comes to replace the Asiatic - this
Mad'iar explained to us in his first book on China,* in which
he raised the question of the transition from the Asiatic to
the feudal mode. . . .

It would seem that when a Marxist describes the transition
from one class formation to another, he must first of all show
how the historical necessity and inevitability of transition to
the following stage of social development arose out of class
struggle within the old society. We find nothing of the kind in
the writings of our 'Asiatics', for them the question is decided
much more simply. They employ that relation always used by
the mechanicists to explain what they do not understand in
the historical process. (Trans. note: previously a similarity
to Bogdanov had been pointed out and deplored.) This relation
is merchant capital. For our mechanists, just as for their predecessors, merchant capital is the demiurge of history.

Therefore when speaking of the downfall of the Asiatic mode
of production, you defenders of the Asiatic mode of production
should state exactly how it fell. If, within the Asiatic mode,
forces capable of moving it to another stage of development
did not arise, then the intrusion of imperialism was truly the
force which moved the history of the people of the East from
dead centre. If the Asiatic mode of production, by means of
imminent laws within it, developed into the feudal mode of
production, then you must show that forces aided this transformation. Having raised this question, which requires a precise answer, I must honestly warn the audience that neither
my co-speaker (Kokin) nor his *alter ego*, Papaian, will give
us one. The weakest point in their theory is an incorrect
understanding of the class structure of oriental society and
consequently class struggle there. If they do not understand
this, they of course can understand nothing in the historical
process of the Orient.

---

*Ed. note: 'Ekonomika selskogo khozia'istva v kitaie' (A study
of the rural economy in China), Moscow, 1928.

Another question, for which I demand an answer from my opponents, and I hope to obtain one now, is how they view Mad'iar's statement that European capitalism found the Asiatic mode of production in the Orient. If the Asiatic mode of production ended in China in the Chou dynasty, then Mad'iar is incorrect in bringing it down to the nineteenth century. If Mad'iar is right when he considers that the European incursion found the Asiatic mode of production, then how does he explain the economic and social structure of contemporary China? None of us view contemporary China as a country with completely developed capitalist relations. If Mad'iar is right, and if the survivals of feudal relations predominate in China, then by what miracle, in an era of European intrusion, did the Asiatic mode of production evolve into the feudal mode of production? And if to the present day the survivals of the Asiatic mode of production dominate, then how do you connect this with the pronouncements of the Communist International on the question of the character of the Chinese revolution? These questions, comrades, you must answer.

Before the discussion, Comrade Papaian chided me that, in so far as the programme of the Comintern mentions the Asiatic mode of production, I am obliged to recognize it, just as I am obliged to pay Party dues. Later we shall speak of the programme of the Comintern. Here I wish only to say that I do not belong to the group of people who are ready to include among the Trotskyists all those who recognize the Asiatic mode of production. For me every theory is important, not according to what the author thinks of it, but because of where that theory leads, the conclusions that inevitably follow from its application to the contemporary Orient. The theory of the Asiatic mode of production in the form in which it is defended by the school of Mad'iar not only distorts the methodology of Marxism concerning the basic problems of history, but it is politically harmful as a theory since it obscures the question of feudal survivals in the contemporary Orient.

It is now time to abandon all discussion on the question of whether Marx in a series of his works recognized the Asiatic mode of production as a specific social formation. It is impossible to close one's eyes to the unquestionable statements by Marx, even if one does not accept the theory of the Asiatic mode of production.

In the works mentioned here (trans. note: Godes has cited 'Capital' and 'The Preface to a Critique of Political Economy'), Marx unambiguously speaks of the Asiatic mode of production, but does this mean that we are now obliged to apply mechanically these remarks to the contemporary Orient, or even apply them to other eras in the history of the Orient? People who are able to grasp only the letter of Marxism frequently forget its essence. The statements on the Asiatic mode of production cannot be expunged from the work of Marx. Our task is not to repeat those statements blindly. Given the fact that our present-

day conceptions of the history of oriental countries does not
support the existence of such a specific social formation, our
task is to explain how and why Marx, at this particular point
in the development of his theory and in the development of
historical science, expressed opinions on the social order of
the Orient, which have not always proven true. Regardless of
this, the essence of Marxism does not suffer. Marx once
expressed the opinion that India was the home of European
languages; I hope that no one will defend the Indo-European
theory solely on the basis of this remark.

'It is necessary to distinguish between the letter and the
essence of Marxism, between separate statements, and the
method of Marxism' (Stalin). Comrade Stalin himself during
the dispute with the Trotskyist opposition gave a brilliant
example of how to distinguish the letter from the essence of
Marxism.

Now I wish to answer the question of how we should understand the social relations of those countries which in our
discussion have been characterized as exemplifying the Asiatic
mode of production. Every attempt to give an affirmative
answer to this question must lead to only one answer, to a
recognition that the Asiatic mode of production is nothing other
than feudalism. The Orient, in a very unique fashion, went
through the same stages of social development as Europe. If
to this day, many comrades cannot recognize the existence of
feudalism in the Orient, this is due to the fact that their eyes
have been closed either by the Trotskyist fog, or simply by
a failure to understand the essence of feudalism.

Feudalism has been thus defined: land-ownership is the
basis for the accumulation of the surplus product; the direct
producer conducts his own independent economic operation;
in the relations between the owner of the means of production
and the direct producer, extra-economic compulsion prevails;
the hierarchy of land-ownership corresponds to the hierarchy
of political power. If this definition of feudalism is correct,
then obviously the question of feudalism in the Orient will
be decided on whether or not we find these characteristic
relations in the social order of these countries.

The study of the history of these countries consists in
demonstrating not the existence of feudalism in general, but
its specific manifestations. Of course, the Orient, with its
1,000-year history and its extraordinary variety, requires
maximum consideration for the unique nature of the different
countries. All those factors, which in the view of the 'Asiatics'
dictate a specific social formation, do not fulfil this requirement when contrasted with the definition of feudalism which
we have given, which is based on the classics of Marxism.
The Asiatic mode of production, according to the 'classical'
definition of Mad'iar, Kokin, and Papaian, includes the
nationalization of land, agricultural communes, artificial irrigation and, associated with the latter, a centralized despotic state.

Such an order in its pure form, with all these character-istics, can scarcely have existed; even if it did exist, this does not mean that one can speak of an absolutely separate formation, since all these characteristics, either individually, or taken together, do not fulfil the main requirement for the construction of a separate mode of production. In fact, the role of each of these features, characteristic of the history of the Asiatic peoples, has been strongly overemphasized.

Let me conclude with a few words about the political impor-tance of our discussion. Earlier I noted that the theory of the Asiatic mode of production could be used not only for the purposes of studying agriculture in the Chou dynasty, but has been used, and can still be used, to dispute the pronounce-ments of the Communist International on questions of the nature of colonial revolution. In this respect the concept of the Asiatic mode of production is the wet nurse for the theoretical position of Trotskyism. Not all defenders of this theory draw the Trotskyist tactical conclusions from their theory, but they are confused as to the question of the class structure of the colonial countries, particularly China. And if Comrade Mad'iar at present takes the proper position in questions of the Chinese revolution, this is because he obviously refuses to put his methodology in practice and rejects the theory of the Asiatic mode of production as it applies to contemporary China. Our comrades must now think about this side of the question. It would be better if they did not limit themselves to a shame-ful and silent refusal to apply their methodology to the present, but drew all the appropriate conclusions about a theory which no longer serves the revolution. For a theory which is useless in application to the present day must be discarded.

The theory of the Asiatic mode of production, which empha-sizes the exclusive specificities of oriental history, can easily play into the hands of nationalist elements in the Orient. They could hide under the veil of this exclusive nature and insist that the teachings of Marx and Lenin are inapplicable to the Orient. At the same time, this theory of exclusivity completely satisfies imperialism, since it is associated with the view that oriental society was stagnant and therefore that European capitalism played a messianic role.

The theory of the Asiatic mode of production, as a social formation specifically characteristic of only the Orient, not only cannot serve as a key to the oriental heavens. Today it has become a serious obstacle to further growth. It destines all who fall under its influence to utter futility. This theory, politically harmful and methodologically incorrect, must be discarded.

NOTE

* From 'Diskussia ob Aziatskom sposobe proizvodstva po
  dokladu M. Godesa' (Results of the Discussion on the
  Asiatic Mode of Production), Moscow-Leningrad, 1931,
  pp. 5, 6, 8, 19, 23, 27, 30-1, 34. Translated by Robert
  Croskey.

SELECTED FURTHER READINGS: PART II

S.H. Baron, 1958, Plekhanov's Russia: the impact of the West upon an oriental society, 'Journal of the History of Ideas', 19, pp. 388-404.

S.H. Baron, 1974, Plekhanov, Trotsky and the development of Soviet historiography, 'Soviet Studies', 26, pp. 380-95.

D. Boersner, 1957, 'The Bolsheviks and the National and Colonial Question', Geneva, Droz.

H. Carrère-d'Encausse and S. Schram, 1969, 'Marxism and Asia 1853-1964', London, Allen Lane.

G. Haupt and M. Reberioux, 1967, 'La Deuxième Internationale et l'Orient', Paris, Cujas.

R. Luxemburg, 1971, 'Introduction à l'économie politique' (1925), Paris, 10/18.

V.N. Nikiforov, 1970, 'Sovetskie isotoriki o problemakh Kitaia' (Soviet historians on the problem of China), Moscow.

G. Sofri, 1974, 'Il Modo di produzione asiatico, (1969), Turin, Einaudi.

E. Varga, 1964, 'Problems of the Political Economy of Capitalism', Moscow.

A. Vucinich, 1970, 'Science in Russian Culture 1861-1917', Stanford University Press.

A. Walicki, 1969, 'The Controversy over Capitalism', Clarendon Press.

K. Wittfogel, 1962, The Marxist view of China, 'China Quarterly', 11, pp. 1-20, 12, pp. 154-69.

PART III
THE WITTFOGEL WATERSHED

EDITORS' INTRODUCTION

Karl A. Wittfogel is perhaps the best known exponent of the theory of oriental despotism or 'hydraulic society' which he traces to Marx and to the writings of Enlightenment and nineteenth-century social and political thinkers, particularly Montesquieu and J.S. Mill. Wittfogel's theory has had considerable impact on anthropology and archeology and upon cold war ideology. His book 'Oriental Despotism' (1957) was a direct onslaught on the unilinealism of Stalin's five-stage theory. Following upon the initial steps towards de-Stalinization taken at the Twentieth Party Congress, Wittfogel's polemic provided a major impetus to the reopening of the debates on the AMP among Eastern and Western European Marxists who consider Wittfogel a renegade from Marxism.

In this selection we have tried to emphasize both the ideological and theoretical significance of Wittfogel's work and its repercussions among both Marxist and non-Marxist social scientists. At face value, the Marxists' charge that Wittfogel is a renegade would seem to warrant no further elaboration: a member of the German Communist Party in the late 1920s, Wittfogel broke away from the movement after the signing of the German-Soviet pact. He went on to play an instrumental role in McCarthy era witch-hunts of suspected Communists and fellow-travellers, while deploying his theory of oriental despotism in cold war polemics on the nature of Communist Russia and China. However, leaving aside politics and polemics for a moment, one should ask whether in fact there are signficant differences between Wittfogel's early and late work or between his so-called Marxist and renegade periods.

Throughout his career, Wittfogel, like Plekhanov, interpreted Marx's materialism as a form of geographical or ecological determinism, in which the forces of production determining a society's social and political organization are themselves determined by climate, soil, hydrography, etc. It is this aspect of Wittfogel's early, Marxist theorizing which is criticized by G. Lewin. Lewin, however, gives the impression that Wittfogel's interpretation of historical materialism as a kind of mechanistic

determinism, albeit of an ecological bent, represented a sharp
deviation from Marxist orthodoxy. In fact, Wittfogel's approach
(1929, 1931, 1932) is largely a refinement of Kautsky's and
Plekhanov's appreciation of historical materialism. Rather than
a renegade from the mechanical or vulgar materialism of Second
International Marxist orthodoxy, Wittfogel was surely its cham-
pion, and stands out as such when contrasted with Lukács or
Korsch.

Wittfogel's 1935 article on ancient China which we reproduce
here is perhaps an exception to his willingness to classify and
explain societies by reference to environmental differences.
Early Chinese history is periodized on the basis of a number
of elements, not the least of which are the changes in the form
of the exploitation of surplus labour. He depicts this period
of Chinese history as one encompassing a transition from clan
organization to feudalism and then to oriental absolutism. The
subsequent non-development of capitalism is explained as the
result of struggles between the state, merchant capital, and
landlords. Unlike his formulation of the hydraulic or irrigation
hypothesis in the 'Theory of Oriental Society', reproduced
here, or in 'Oriental Despotism', large-scale irrigation neither
precedes nor necessitates the development of the state; rather,
such irrigation is undertaken by the state in order to enlarge
the conditions of its own reproduction, a view similar to
M. Godelier's account of the role of irrigation among the Incas
(1973a, pp. 186-95).

The reviews of 'Oriental Despotism' included here are
intended as a partial documentation of its reception among
social scientists of various ideological and theoretical bents.
It is not surprising that distinct aspects of Wittfogel's theory of
the development, structure, and organization of oriental despotism are taken up by the different reviewers. Toynbee and
Eberhard question the intention and utility of a dualistic
model of social evolution opposing hydraulic to non-hydraulic,
or single-centred to multicentred societies. Levada, on the
other hand, concentrates on Wittfogel's polemics with the
Soviet state and his 'unorthodox' notion of class. Levada is
content to reiterate the 'orthodox' view from Engels's 'Origin
of the Family, Private Property and the State', that classes
and the exploitation of one class by another only arise with
private property. Wittfogel's call for a new sociology of class,
however, resounds in recent work by French Marxist anthropologists, P.P. Rey and E. Terray, who have come to challenge
the long-held view of primitive societies as classless societies.

Despite G.P. Murdock's willingness to expound upon the
political lessons of 'Oriental Despotism' in his review, the
anthropological and archaeological communities' reception of
Wittfogel's theory, particularly his hydraulic or irrigation
hypothesis, was noticeably apolitical, restricting the relevance
of Wittfogel's hypothesis to the origins of early civilizations.
Furthermore, it was paradoxically a-Marxist, at least until

1968 when Marvin Harris claimed Marx as a lineal ancestor for his cultural materialist strategy. Wittfogel, upon his arrival in America, had joined the self-proclaimed school of multilinear evolutionists led by Julian Steward, who stood in opposition to the 'general evolutionists', L. White and V. Gordon Childe. Multilinear evolution was also counterposed the 'unilinear' evolutionary theory of the nineteenth century. The idea that one of these nineteenth-century evolutionists, Karl Marx, was not a unilinear evolutionist - an idea that Wittfogel was putting forward in his polemical and marxological articles - was not taken up by the anthropological community until Marvin Harris's intervention.

Although Steward and Wittfogel both use the term multi-linear to refer to their evolutionary theory, they in fact address distinct issues. For J. Steward (1949) the differences between multilinear evolution and universal evolution concerned very basic attitudes towards the logic of science: multilinear evolution was empirical and inductive, universal evolution was philosophical and deductive. Wittfogel's theory of the development and non-evolution of oriental despotism involves a claim about the outcome of evolution. It has been interpreted as a bilinear theory based upon an initial dichotomy between non-hydraulic (multicentred) societies which evolve and hydraulic (single-centred) societies, the development of which depends upon external factors. As a set of conclusions about evolution, Wittfogel's theory is not strictly bilinear in so far as he has included diffusion as an alternative explanation to account for the development of oriental despotism in non-hydraulic areas. Perhaps the most contradictory aspect of Wittfogel's contribution to evolutionary theory is his shifting criteria for establish-ing whether or not a society can be considered to have evolved. Evolution is a matter of a combination of values and technology, unequally applied: the evolution of a multicentred type of society is gauged by the development of technology and strati-fication, while for a single-centred type to have 'evolved', it must have become a multicentred type.

The articles by Julian Steward and Barbara Price critically review much of the archeological and anthropological research spurred on by Wittfogel's work. Edmund Leach's analysis of a case of non-correspondence of large-scale irrigation and a centralized despotic state, brings up the idea of qualitatively different forms of state control in India and China.

No doubt both Wittfogel's polemical usage of the concept of oriental despotism and his environmental-cum-technological determinism follows in the tradition of Plekhanov. Wittfogel's interpretation of Marx on the AMP has given support to Marvin Harris's view of Marx as a cultural materialist. The introduction of a supposedly Marxist concept into a milieu where the theoretical problems of social evolution had no immediate political significance has had somewhat of a depoliticiz-ing effect. Questions about whether large-scale irrigation, population growth, warfare, trade, etc. directly influenced the development of the state in ancient Mesopotamia or Mexico are not asked from a perspective immediately concerned about whether contemporary Iran or Mexico must be capitalist before it becomes socialist, or holding up the spectre of oriental despotism to communist or anarchist opponents. Within the problematic of neo-evolutionary anthropology, Marx's ideas on the AMP have been equated with a theory of the origin of the state or of a particular kind of state.

THE STAGES OF DEVELOPMENT IN CHINESE ECONOMIC AND SOCIAL HISTORY*

Karl Wittfogel

1 HISTORICAL STAGES IN THE SETTLING OF CHINA

Whatever may have taken place in the Yangtze Valley and south of it during the second millennium and the first half of the first millennium BC, this region was not the cradle of Chinese civilization. It originated in the north, in the basin of the Huangho, and, as both written tradition and archeological discoveries testify, not in the northeast - in the alluvial plains of North China - but in the north central region and perhaps the northwest, in the territory of the modern provinces of Shansi, Shensi, Honan, and perhaps Kansu.

The action of the loess streams, which heighten and change their channels from year to year, made permanent habitation and agricultural activity in the eastwestern plain possible only if the streams were held within their banks by dykes of great size, a fact generally recognized by geographers(1). It removes any factual basis from a viewpoint, according to which the first development of Chinese agrarian society took place just where nature presented the greatest obstacles, insurmountable indeed by neolithic implements. Having discounted this view - which has few adherents(2) - we must regard the expansion into the northern plain as the second, and qualitatively extremely important, step in the economic, social and political evolution of China.

The third step, the organization of the central Chinese territory, occurred only at the end of the feudal epoch and took place in several stages. Elsewhere we have at length analyzed those natural factors which in the northeastern plain necessitated a specific kind of public works (embankments), as well as the 'oriental' conditions of soil and climate, which in the rice growing regions of Central and South China led to the construction of irrigation works on a large scale(3). The history of the settling of China is therefore essentially bound up with the development of the instruments and forms

of organization of social labor and of the state, determining it
and being in turn determined by it. Apart from the present
phase of development, which exhibits completely new phenomena,
a scientific economic history of China must constantly bear in
mind the process of settling, upon which the economic history
is based.

2 THE FIRST DIFFERENTIATION OF CHINESE AGRARIAN SOCIETY

An analysis of early Chinese history finds abundant sources
of evidence upon which to base a description of the original
forms of economic activity: excavations and written historical
traditions, old songs, traditional religious and profane customs.(4) Millet as the chief food,(5) hemp as probably the
first textile plant,(6) hunting and clearing of land done by
the men,(7) extensive hoe-culture(8) and possibly pig-raising(9)
done by the women,(10) collective forms of labor, perhaps on
the basis of the clan(11) (sib) and matriarchal forms of the
clan(12) – these are some of the leading features of primitive
Chinese society as revealed by the new deposits and the freshly
interpreted historical sources.

The transition to more intensive forms of cultivation, perhaps
already including the rudiments of irrigation,(13) the domestication of horses, oxen and sheep,(14) and the passing of the
dominant forms of production into the hands of the male, which
resulted in the development of a patriarchal system of kin-
ship(15) – all of these factors provided the socio-economic
prerequisites for the rise of the state. This may have come
about either in endogenous fashion, by the singling out of a
native ruling class to perform general administrative, juristic,
religious and special military duties; or it may have happened
exogenously through conquest, perhaps by nomads. But even
if conquest by nomadic, patriarchal tribes took place, it did
not determine the character of the society. It led rather to
an intermingling of the two different modes of production,
and the autochthonous agrarian economy predominated in
shaping both the organization and the prevailing ideology of
the resulting social order.

Certainly the old word for 'duke' (which Legge translates
as pastor)(16) indicates the importance of cattle-breeding to
the first Chinese ruling class. The word 'pa' used so frequently
during the Ch'un Ch'iu period for the 'leader of the feudal
states' really means 'father's brother'. But the latter word, as
a matter of fact, is, according to Conrady, of later origin.(17)
The ideograph for 'clan' emphasizes descent from a female
ancestor.(18) Early legendary history reports a female ancestry
for the Chou, the Shang and the Hsia dynasties,(19) as well
as an originally matrilineal succession.(20) And if the sheep and
the bull are prominent in the solemn state sacrifices during the

Chou period,(21) the great agricultural ceremony which the
'Son of Heaven' performs(22) likewise permits no doubt as to
the central position which agriculture occupied in the actual
economic structure as well as in the consciousness of the ruling
class. Whether or not agriculture was predominant from the
very start, it at any rate became so very early in the course
of history.

3 THE TRANSITION TO THE SECOND STAGE

In this first stage of socially differentiated Chinese agrarian
society, irrigation played a minor role at most; waterworks
certainly no essential part. Nothing, therefore, impeded the
elaboration of a feudal organization. Society was organized in
relatively small, decentralized states. The few ruling clans
dwelt among their military followers in fortified settlements
(towns), which they held as ficfs. The peasants of the surrounding villages, who were bound to the soil, were obligated to pay
taxes directly to the ruling clans. There is ample proof that
this was the form of social organization during the Chou period.(23)
It existed, in its ripest development and even in a state of growing crisis, from the eighth century at the very least. There can
therefore be no reason to doubt that the ruling class of the pre-
Chou period, too, was feudally organized. The 'noblemen' of
whom the bone-inscriptions of the Shang period speak(24) are
quite plainly feudal retainers.(25) If in the Hsia society was
already taking the form of a state, we must not lightly repudiate
the evidence of the Shu Ching, which indicates that the organization of this state was, at least embryonically, of a feudal
character(26) - especially since such an early feudal phase is
fully in harmony with our general knowledge of economic and
social history.

Although the first two dynasties seem to have been clearly
feudal, and the feudal structure of the Chou state is very pronounced, the old agrarian order early acquired a new feature -
public waterworks - which at first affected but slightly the
social and political organization. But in the course of a long
development, full of contradictions, its importance grew until
it led to grave convulsions which impressed a fundamentally new
stamp upon the entire social order.

We do not know the particulars of this transition, and it is
questionable how much we shall ever learn of it in detail. But
we do know - and this is fully sufficient - the point of departure
and the result, as well as the essential causes of the process.
The beginnings of a socially differentiated agrarian history of
China took place outside of the swampy plains of the vast northeastern river delta, although there may have been isolated cultural
oases in the more elevated regions or round about in the hill
districts of the East. During the later Chou period, however,
when Chinese feudalism was in open crisis, a progressive advance

is evident of the Chinese cultural area into the great eastern
alluvial plain. (27) The subjugation of the northern part of this
plain, formerly uninhabitable because of the ever-changing loess
streams, would have been impossible without the construction of
river dams of great magnitude. A technology first developed in
small river valleys(28) was now applied on a giant scale.

The history of the Chou period is filled with reports of so-
called public works(29) as well as of the activity of the directing
Minister of Works. (30) The origin of this post goes back into the
beginning of the Chou period, perhaps even much further. When
the Chou state was founded, the Ministry of Public Works was
one of the two ministries reserved for members of the ruling
family. (31) The inscriptions of the An Yang bones indicate that
the technique of embankment existed even during the Shang
period. (32) An ode of the 'Shih Ching', dealing with life in the
state of Chou during Shang times, speaks of public works and
also of one Ssi Kung (Minister of Public Works) as their direc-
tor. (33)

The value of the legendary tale of the 'labors of Yü' consists
in the fact that it gives information on the beginnings of public
works in very ancient times - though in a quite legendary manner.
We certainly do not need the document in order to prove the
development of this institution in the course of feudal times.
That such a development occurred is beyond all doubt. Moreover, it is quite impossible that the huge waterworks attributed(34)
to Yü by the legend could at that time have actually been executed within a single generation. Even if the 'labors of Yü' were
very much smaller than tradition maintains, they must have been
spread over a long interval - just as in the case of the preparatory
labors for the Great Wall and the Great Canal. Yü and the saga
spun about him remain, nevertheless, extremely noteworthy;
for they do prove that the beginning and development of the
system of public works in ancient China was felt to be the beginning of a new epoch of agricultural settlement and production
and thus the beginning of a new function of the ancient Chinese
state. (35)

4 THE DISSOLUTION OF THE OLD AGRARIAN COMMUNITY

Closely connected with the crisis of feudalism in the Chou period,
underlying and profoundly influencing it, is the crisis in the
ancient Chinese agrarian community. In the study of this phenomenon too much attention has been devoted to the question of the
geometric form of the village community. In discussing the
possibility and form of a schematic 'well-field system' (nine equal
fields, with a well in the central field, about which were grouped
the huts of the villagers), it was often forgotten that this question was by no means the only one or even the most important.
A purely philological approach, which disregards all evidence
save that from one or two sources (the 'Shih Ching' and Mencius),

which it then attempts to discredit by emphasizing their 'lateness',(36) is equally unsatisfactory and offers no assistance to the advancement of a material analysis.

During the fourth century BC, Shang Yang, prime minister of the state of Ch'in, replaced by a new system the agrarian organization which had prevailed up to that time. In the new system the individual peasant families (no longer large families, but small ones), became permanent possessors of the land which they cultivated.(37) The existence of an earlier state-regulated 'bound' system, up to the end of the Chou period, is therefore beyond doubt.(38)

The old agrarian order, which lasted certainly until the time of Shang Yang, must itself have been the result of a long and complicated development. Strange remains of tradition point to a prefamily condition of economic life.(39) Such a condition, as comparative anthropology has established, often appears in the beginnings of agrarian development in the form of an agrarian communism of the clan, generally combined with mother-right.(40) Because of its collective labour it was essentially different in its socio-economic structure from the 'well-land' system which followed. In the latter the village land was periodically redivided among several large patriarchal families. Each of these families now constituted an economic unit whose members tilled the soil in common. Such a system constitutes an important form of transition from the original agrarian communism to full-grown private property and to the small family.(41) It is noteworthy that Shang Yang, when he dissolved the old bound community – thereby, certainly, only aiding a development which had begun long before his time – decreed also the breaking up of the great peasant family.(42) By the time of Chou the original, presumably clanlike character of the village family had already changed considerably. With the dissolution of the clan into large families the original organization gave way, at least in part, to regional organization. Even before Shang Yang, there was a tendency in various states toward uniting villagers, apparently no longer organized on a clan basis, into groups of five as administrative units.(43) This is an interesting sign of the transitional character of the village community of those days.

Because of the diversity of development in different areas, the form of the agrarian community at the end of the Ch'un Ch'iu period (eighth to fifth century BC) was not at all uniform. The study of material facts shows that there was no single type of village community but rather a whole series of types which can by no means be reduced by philologists to a fantasy invented by Mencius. Mencius, who dealt with the well-land system, lived about 300 BC. The 'Chou Li', which according to Liang Ch'i-Ch'ao and Chen Huan-Chang contains old materials at least in part,(44) describes two completely divergent forms of the old well-system.(45) The account given by Ku Liang, commentator on the 'Chu'un Ch'iu', likewise differs materially from that of Mencius.(46) A statement in the 'Tso Chuan' relating to the

well-system makes clear that this system prevailed only in flat
country.(47) Thus, as was to be expected, the form of the
village community varied greatly, both temporally and spatially.
One feature, however, seems to be common to its various aspects.
The payment rendered to the feudal lords was generally made in
the form of labor, which was performed on a special 'public' or
'state' field.(48)

Why was there this change, toward the end of the Chou period,
from collective labor in the public field to exaction of a land
tax(49) from individual families, or, speaking in economic terms,
from labor rent to rent in kind? We find the explanation in the
'Lü Shi Ch'un Ch'iu': it was because the public fields were cultivated less carefully than the private ones.(50) A form of taxation
which was useful with extensive methods of cultivation proved
no longer suitable in view of the increasing intensity of labor.
New forms of taxation had to be found. They were found.

The material bases for the dissolution of the village commune
were prepared by the increase in the productivity of labor
through the introduction of metal implements,(51) and especially
by irrigation, which was coming more and more into use also in
north China at the end of the Chou period.(52) We cannot trace
this trend in detail. We merely wish to establish that the development of public forms of labor (dyke and canal building), which
originally grew up out of its private forms, now reacted in a
decisive manner upon the private forms of agricultural production. Although irrigation was first employed, in the central and
western sections of China, in an incidental and local manner,
the development of public works in the northeast (at first mostly
dykes) led to an irrigational agriculture supported by public
canal construction, which eventually rose to be the ruling form
of agricultural production.(53)

The end of the Chou period witnessed an ever expanding construction of canals, even in those northern regions which mainly
produced millet and wheat. The economic and political importance
of this canal construction is exhibited most clearly in the history
of the state of Ch'in. This state, which changed its agrarian
system earlier and in a more thoroughgoing manner than most of
the neighbouring states, ultimately raised itself to be the master
of all the 'China' of those days. Ch'in was of course not the only
state which moved in this direction. It was a time filled with
attempts at transition to a new agrarian and social order. Ch'in
achieved a more decisive transformation than its political rivals
and thereby gained the advantage leading to victory in the final
struggle for power over the empire. Ssu-Ma Chien's report on
the end of the Chou period reflects the profound impression
which the agrarian policies of the state of Ch'in made upon contemporaries and posterity.(54) The changing of the agrarian
order, canal-building on a great scale, both linked with ruthless
destruction of the old feudal structure, enabled the young
bureaucratic, centralized state to defeat the other states, which
were still feudal, and to enter into the heritage of the Chou
dynasty.

5 CHANGES IN THE ECONOMIC AND SOCIAL STRUCTURE

Hand in hand with the development of agricultural production, aiding it and being aided in turn, went a powerful development of handicraft and commercial activity. During the Chou period the idea had arisen that just as the peasant was bound to the soil from birth and as the feudal lord inherited his fief and posts of state, so handwork and commerce were likewise hereditary trades. But in actual practice this part of the old order, like the other two, had already been relaxed. (55) As production grows, commerce increases, developing its means of exchange (money)(56) and its channels of exchange (markets and wholesale trade). (57) In the second half of the Chou period we hear of merchants who have considerable wealth and stand in relations of a contractual nature to the feudal prince. (58)

The growth of public works - dykes, canals, fortifications, luxurious buildings - requires a new type of state officials. There now arises the class of wandering philosophers, political counselors and non-hereditary officials, stemming mostly from the poorer members of the old feudal class, and in lesser numbers also from the other classes. (59) They offer their services to the various high officials and dukes, on a frankly commercial basis. (60) Alongside of the ideal of the feudal knight, as he is honoured in the remains of old songs and in other traditions, there now appears the ideal of the literarily educated administrative official. Confucius, though not the first to do so, formulated this ideal in a way heavily fraught with historical consequences.

The placing of posts of state on a non-hereditary basis signified a break with the feudal view of the state. In this way the central power demonstrated plainly that it no longer needed the feudal retainers. The new tasks of the state made such a break objectively necessary, while the growing economic, military and political power of the central government made it technically possible.

When former social bonds are destroyed, new institutions and forces are needed in their stead, to be used as material and moral supports to the changing state. In addition to the new type of officials, and the great businessmen (commercial, industrial and bank capitalists, who could not be altogether relied on), (61) this support was sought from the structure of the family - primarily the peasant family, which with the development of agricultural production was evolving out of its earlier forms.

During the close of the Chou period, the head of the then maturing small family was the father. He embodied in the peasant family the accumulated experience (62) of increasingly skillful methods of agricultural production. The prestige derived from his economic leadership made him eminently fitted to exercise an almost unlimited moral and political authority over members of his family. (63)

Confucius's great rival, Mo Ti, agrees with him in this, that neither feudal heredity nor knightly qualities should determine fitness for state post.(64) In his demand that the members of the new ruling class should be professionally trained, his radicalism surpassed even that of Confucius.(65) But it was just these demands which led in the long run to his seeming, in the eyes of the new officials, a less suitable representative than Confucius, who did not hesitate to grant his 'gentleman' dignity and a carefully limited brilliance.(66)

What tipped the scales in the historical victory of Confucius over Mo Ti was the former's conception of the small family organization as a necessary basis for the new society. Mo Ti, with his demand for universal love, according to which children should love their neighbours' parents as their own,(67) obviously reflects the mood of a developing commodity economy with its abstract equality of all produced commodities in the market. If China's feudalism had been resolved into a capitalist economy, its ideology would have been transformed in accordance with the ideas of Mo Ti or of some similar thinker. To be sure, even the Confucian ideas did not prove to be entirely fitted to the later developments. But at all events they corresponded to this development incomparably better than any competing ideological patterns. The epoch of political transitional struggles, from the Ch'in period to the Sui (202 BC to AD 618), witnessed the struggle of Confucianism with other rival philosophies, namely, Taoism and Buddhism. The final establishment of 'oriental' society during the T'ang period brought the final victory of Confucianism and its definitive formulation in the Sung philosophy.(68)

6 THE FORMATION OF 'ORIENTAL' ABSOLUTISM

(a) From labor rent to rent in kind and money rent
Originally the feudal lords of ancient China took the surplus product of agricultural activity largely in the form of labor rent (work in the public fields); toward the end of feudalism, they began to substitute payment in kind. The new state, which expressed a new social system,(69) made rent in kind general, as being better suited to a more highly developed mode of production.

With the growing commodity economy and especially after the completion of the Great Canal, the state exacted part of the rent payment in the form of money. The principle, nevertheless, remained the same. The peasant should retain just enough of his crop to enable him to reproduce both his family's labor power and his own.(70) As the history of agrarian crises in China shows, this principle, when not modified by other factors (tenantry, commercial capital, and usury capital), was rarely departed from to ease the pressure but frequently to tighten it.(71)

(b) Further stages in the history of Chinese economy and
settlement
The growth of the Chinese state within the Yangtze Valley
compelled it to undertake public works on a new scale. The
great rice districts situated around the Yangtze required a very
extensive state activity in the construction of water-works. The
giant canals built in the great plain not only served agricultural
production, but assisted in the economic and military, i.e. the
political mastery by the bureaucratic state of all the important
centers of the enlarged area of production.(72) When in the
sixth century AD the Great Canal was unified by the linking of
its separate sections, in part very old, it represented a kind
of artificial Nile. This huge canal made it possible for the capital
which for military reasons had to be situated on the edge of the
economic key area, to maintain control of this area and, thereby, of the economically essential parts of the empire.(73)
    China's artificial Nile necessarily gave an enormous impetus
to both private and state economy, in the direction of extension,
intensification, and commercialization. The growth of money
economy during the T'ang period(74) and its fiscal, military and
political consequences are no accident. It represents the necessary result of the development and modification of the material
foundations of 'Asiatic' China.

(c) The struggle over land rent in the 'Asiatic' system of
production
The original form of the Asiatic system of production in China,
in consequence of the historical causes depicted above, had a
feudal point of departure which complicates the picture. The
bureaucratic centralized state developed in China on the foundation of an agrarian community which disintegrated in proportion
to the growth of the new society.
    The higher form of the Asiatic system of production displays,
alongside of simple commodity production, the beginnings of a
developed commodity production.(75) A commercial and money-
lending capital springs up within the framework of the bureaucratic state(76) based on a countless number of 'free' peasant
holdings. The state had to favor commodity economy and trade
as a means of making the land rent mobile.
    The ruling official class, which as representative of the common-
wealth struggled against the feudal land rent on behalf of the
state form of land tax, saw itself confronted by new opposing
forces. Because the latter were linked by necessity to the new
society, they were not so easy to get rid of as the outmoded
old ones. A vicious economic-political circle resulted: accumulation
of private wealth of the new type and accumulation of private
possession of land in the hands of officials, 'gentry' and great
merchants, reduction of land tax, enfeebling of the state,
agrarian crisis, internal crisis, external crisis - invasions(77) -
state crisis. Although this vicious circle could be periodically
smoothed over by the fall and rise of 'dynasties', it could never
be really overcome.

The details of this contradictory process can only be hinted at here. The new 'Confucian' ruling class, either active as officials or living privately as 'gentry', may originally have arisen out of the old feudal aristocracy. Its economic position was now determined essentially by its specific relation to the land and the products thereof. Because of this relation, it was vitally interested in the new form of land rent received as tax and disbursed by the state treasury.

The land rent which flows into the state treasury must be large. This calls for agronomic activity by the state, especially for widespread irrigation works. It demands economic and political struggle against private land-owners who seize the rent for themselves. This struggle takes the form of regulation of inheritances to effect the splitting up of landed estates. (78) It results in repeated attempts to destroy or limit large land-ownership. These measures prove effective so long as they are aimed against obsolete feudal strongholds. (79) They succeed only temporarily or not at all, or react against their initiators, (80) when they are directed against social-economic powers which adhere necessarily to the new economic system.

In order that land rent may be controlled, it must be mobile. Roads and canals are useful in this connection, and a certain measure of monetary development is also serviceable. We have already pointed out what the establishment of the Great Canal during the seventh century AD meant in this connection. (81)

Finally, land rent must also be capable of being centralized. Only then is a centralized official apparatus economically reproductive. A bureaucratic, oriental centralized state differs essentially from the administrative state of European absolutism growing up on the foundations of early capitalism. The dimensions of the territory of China had a specializing effect. Since the starting point and the means are different, the results must also be different. If the central power succeeds by and large in keeping its official apparatus in hand, even though with diminishing power in border districts it has succeeded in achieving the maximum degree of centralization which can be realized under the given circumstances.

The control of the economic key positions constitutes the fundamental prerequisite for the establishment of such control. Numerous institutions, such as the ideographical form of writing prevailing throughout the whole empire, (82) the introduction of the examination system (which, it is significant to note, received its final form shortly after the completion of the Great Canal), (83) and the regular and frequent shifting about of officials (84) - all of these are important expressions of the efforts of the central government to establish its power.

Investigators often overlook the methods of preventing the hereditary crystallization of high state positions and of large fortunes by the appointment of eunuchs to high political posts. It is not the 'discovery' or introduction of the technique of castration that is important historically. It is rather the evolution

of eunuchs from palace and harem servants to holders of high
political and sometimes even military posts.(85) Ma Tuan-Lin
is undoubtedly right in declaring that behind the change in the
status of eunuchs is concealed the struggle against feudal
inheritance.(86)

Along with the centripetal forces of the new economic and
political system there develop its centrifugal forces. There may
have been endeavours to hold in check the influential class of
great merchants - powerful since the end of the Chou period -
by means of the decimation of its wealth(87) and through social
chicanery.(88) Nevertheless, it could not be prevented from
drawing great profits from the necessary processes of circulation.(89) Furthermore, it could not be prevented from purchasing
with these profits the land of the impoverished peasants(90)
(who had been 'emancipated' to increase the land rent!). There
thus arose alongside of the remaining feudal land-owners and
officials holding land privately(91) a new class of town-dwelling
land-owners. But just as commercial capital created no new system
of production, so the new tenant relationships prevalent during
the 'Asiatic' epoch in China, even in so far as they were fixed
contractually, did not result in a new capitalist form of land rent.
They produced only marginal variants of the older type of land
rent, mainly preserving the form of payment in kind.

The Confucian official-state fought again and again for mastery
over private land-owning, i.e. for a monopoly of the land rent.
The antagonistic tendencies of the mandarinate itself (land purchase by officials), the indestructibility of the economic function
of commercial capital, and despite clashes of interest, the close
social connections between the two classes, prevented a real
victory in this struggle. Apart from a few reforms evoked by
special emergencies,(92) the fight against private land-ownership
was carried on without even the attempt of using the extreme
measures.

The efforts of the bureaucratic official-state to strike commercial
capital at its root, i.e. profits, were not much more effective.
The measures for the control of commerce by means of offices
for the equalization of prices, etc. were intended to weaken the
power of the commercial bourgeoisie and to strengthen that of
the state and its representatives.(93) The same end was served
by the millennium-old efforts of the mandarinate to control the
production or at least the sale of two of the most important
popular articles in agrarian China: salt and iron. Whereas other
and more sweeping monopolistic tendencies were frustrated (Wang
An Shi's plan to transfer village moneylending from private hands
to those of the state is interesting),(94) the control of salt and
iron continued in various forms from the close of the Chou
period.(95) It undoubtedly contributed greatly to the strengthening of the centripetal forces in the bureaucratic centralized
state.

CONCLUSIONS

Our survey, sketchy and incomplete as it unavoidably is, at all
events shows this: that the economic and social system of China
was by no means always the same. On the contrary, it went
through several clearly defined stages of development, before it
assumed that stationary shape in which Europeans learned to
know it.(96)
Insight into the rise and the dynamics of the economic system
of China is, it seems to us, not merely a condition sine qua non
for any deeper understanding of the cultural development of
China. It is, at the same time, an indispenable prerequisite to
an understanding of the forms which are today arising in China
in the course of the dissolution of the old order and the rise
of a new social and economic way of life.(97)
Attempted explanations based on metaphysical or racial considerations are evidently incapable of making intelligible why
Japan, at the end of the nineteenth century, could so promptly
evolve into industrial capitalism, while China has not even yet
been able to do so. But comparison of the socio-economic systems
of the two countries quickly shows that Japan, in contrast to
China, was not an 'Asiatic' country in our sense. It had indeed
an 'Asiatic' tinge (irrigational economy on a small scale), but
was nevertheless fundamentally more akin to the European
nations: her advanced feudal economy in the nineteenth century
had already taken the preliminary steps toward the evolution of
industrial capitalism.(98) The apparent riddle vanishes and its
solution, or, more exactly, the beginning of its solution comes
in sight.
In subsequent writings we shall attempt to work out this
solution in detail.

NOTES

* From The Foundations and Stages of Chinese Economic
   History, 'Zeitschrift für Sozialforschung', 4, 1935, pp. 35–58.
1 For the views of the earlier geographers see F.V. Richtofen,
   'China,' vol. I, Berlin, 1877, pp. 356ff. On the latest researches cf., among others, L.H. Dudley Buxton, 'China,
   the Land and the People,' Oxford, 1919, pp. 221ff., as well
   as G.B. Cressey, 'China's Geographic Foundations,' New
   York and London, 1934, pp. 159ff.
2 H. Maspero, Les origines de la civilisation chinoise, 'Annales
   de Géographie,' 1926, pp. 154ff. Also his 'La Chine antique,'
   Paris, 1927, pp. 21ff.
3 K. Wittfogel, 'Wirtschaft und Gesellschaft Chinas,' I, pp.
   189–206.
4 The evolutionary process analyzed by us remains comprehensible even if we exclude 'survivals' as auxiliaries to knowledge. But the fundamental scepticism with which Malinowski

regards such a use of survivals (article on Culture, in
'Encyclopaedia of the Social Sciences,' vol. IV, p. 624) is
untenable. A cultural element can indeed 'outlive its functions',
that is, its original function, and can continue to exist,
exercising changed functions – perhaps of lesser importance –
in a changed social milieu. Within a deeply traditional culture,
customs which have lost their original significance nevertheless may continue to operate with a new, derived importance,
that of maintaining the prestige of the ruling class. Malinowski
has well pointed out the inadequacy of the survival theory of
naive evolutionists. But instead of criticizing it constructively
from the standpoint of a superior concept of development, he
starts from an actually anti-evolutionary standpoint, and
destroys exactly those elements of accurate knowledge which
the naive survival theory does contain.

5   During the Shang period, 'the' harvest is the millet crop,
as the inscriptions teach us (Hopkins, Pictographic Recon-
naissances, 'JRAS,' 1926, p. 474). The little Hsia calendar
mentions millet and wheat. (Cf. Li Chi, trans. by R. Wilhelm
under the title of Li Gi, Jena, pp. 235, 238 and 240.) Confucius emphasizes the value of the 'Little Hsia-calendar';
within his time the calendar seems still to have existed (Lun
Yü, XV, X, 2; Legge, 'Chinese Classics,' 1, p. 297). Likewise 'Ssu-Ma Ch'ien,' ch. II, Chavannes, op. cit., I,
p. 171. The role of millet in the great sacrifices is so unambiguously represented in all old texts, from the 'Shih
Ching' to the 'Book of Etiquette,' – the 'I Li' – that there
can be no doubt cast upon the importance of this plant even
as far back as the beginning of tradition. This remains unchanged by the discovery of a grain of rice in a Yang Shao
shard (cf. on this the report of G. Edman and E. Söderberg
in 'Bull. Geolog. Survey China,' vol. 8, no. 4, 1929,
pp. 363ff.). Since according to the oldest calendar reports,
the climate has not been essentially different, from the earliest
times to the present, and because the soil relations would
have changed still less, rice, even if it were known in early
times could only have been sown in enclaves. As old funeral
customs show - later cowrie-shell money and rice were placed
in the mouth of the dead – it was an object of value ('I Li,'
ed. by J. Steele, London, 1917, p. 50). Rice was never a
food for the masses in the north, as far back as we can
pursue history. (Cf. 'Wirtschaft und Gesellschaft Chinas,'
I, pp. 72ff.)

6   J.G. Andersson, An Early Chinese Culture, 'Bull. Geolog.
Survey China,' p. 26. There is remarkable coincidence between this archeologically supported view and old sacrificial
and mourning customs. ('I Li,' II, pp. 9ff. and 183ff.)

7   Andersson, op. cit., p. 29. Also his 'Archäologische Studien
in China,' reprint from vol. LIV of 'Mitteilungen der Anthro-
pologischen Gesellschaft Wien,' 1924, pp. 66 and 70. The
square axes described here, which would be too weak for

clearing forests of the European type, are eminently suited
for clearing of bushes on the thinly wooded loess territories
of northwest China. On ground-clearing, see 'Shih Ching',
III, I, 7, Legge, 'Chinese Classics,' IV, 2, pp. 339ff.

8   Andersson, An Early Chinese Culture, p. 26
9   Ibid., p. 32. There is no doubt that the pig was the oldest
domesticated animal. Who reared it? We do not possess any
direct proofs as yet. The fact that pork is served at the
marriage ceremony by the young woman - at the ancestral
sacrifice by a man - (I Li, I., 29, and II, 131) is perhaps
ground for the belief that the woman had the duty of raising
the first domesticated animal. The offering of game and of
large, grazing domesticated animals during the sacrifices
lay exclusively in the hands of the men.

10 Andersson's guess that the men in this phase of development
were still 'hunters' (An Early Chinese Culture, p. 29), that
male and female activity thus were sharply differentiated,
finds support from a remarkable source. Davidson Black, on
the basis of his investigations of neolithic skeletal deposits,
declares that the great difference ascertained by him between
the male and female bones 'would seem to indicate that a
considerable specialisation of the respective work of the two
sexes obtained among the Yang Shao people'. Black, The
human skeletal remains, etc. In so far as the woman partici-
pates at all actively in the old sacrificial rites, she brings
vegetable gifts: millet, onions, wine, also rice ('Li Gi,'
Wilhelm ed., pp. 264, 361. 'I Li', I, pp. 32, II, pp. 134,
140, 167, 174 ff., 183, 187, 190. Cf. also 'The I Ching',
trans. by J. Legge, as 'The Sacred Books of the East',
vol. XVI, Oxford, 1882, Hexagram 54, p. 182).

11 Cf. for this W.H.R. Rivers, 'Social Organization,' London,
1924, p. 114; 'Wirtschaft und Gesellschaft Chinas', I,
pp. 348ff. Traces of clan communism have survived to the
present day in the very lately colonized south China. (Cf.
M. Volin and E. Wolk, 'The Peasant Movement in Kwangtung,'
I, Canton, 1927, pp. 52ff.)

12 We give below data on remains of the mother-right tradition.
On the importance of primitive hoe-culture by the woman for
the evolving of mother-right relations, cf. E. Müller-Lyer,
'Die Familie,' Munich, 1912, pp. 85ff.; H. Cunow, 'Zur
Urgeschichte der Ehe und Familie, Ergänzungsheft', no. 14
of 'Die Neue Zeit', Stuttgart, 1912, pp. 41ff. Also, stemming
from utterly different presuppositions: F. Grabner, 'Das
Weltbild der Primitiven,' Munich, 1924, p. 35; O. Menghin,
'Weltgeschichte der Steinzeit,' Vienna, 1931, p. 498; R.
Thurnwald, 'Die menschliche Gesellschaft,' Bd II, Berlin u.
Leipzig, 1932, p. 194.

13 If in the central and western parts of China climatic conditions resembled those of today only approximately, irrigation
must have been highly desirable for the production of
normal harvests of millet and wheat. This is true even if we

neglect the extremely problematic existence of a late neolithic cultivation of oryza sativa. (Cf. 'Wirtschaft und Gesellschaft Chinas,' I, pp. 209ff.)

14 The An Yang inscriptions report sacrifices of oxen (Hopkins, The Human Figure, 'JRAS' 1930, p. 105). One Shang ode mentions wagons - therefore, horses. ('Shih Ching,' IV, III, 3; Legge, 'Ch. Cl.,' IV, 2, p. 637.) The little Hsia calendar mentions horses and sheep. ('Li Gi,' pp. 237 and 239.) If we may rely on the relevant passages in 'Shu Ching,' the tradition of the 'shepherds' can be traced much further back. ('Shu Ching, Shun Tien,' 7 and 16, cf. also 9, Legge, 'Ch. Cl.,' III, 1, pp. 34, 42 and 35.)

15 If, as modern anthropology establishes, it is the woman who usually makes the transition to the early forms of agricultural production, the division of labor resulting therefrom is changed 'in most cases', when new tasks, like irrigation and terrace-construction or the yoking of domestic animals in the agricultural process, make the use of male labor power imperative. (Cf. F. Boas, article, Anthropology, in 'Encyclopaedia of Social Sciences,' II, p. 83.)

16 'Chinese Classics,' III, 1, p. 34 and passim.

17 A. Conrady, China, in 'Pflugk-Harttungs Weltgeschichte,' vol. Orient, Berlin, 1910, p. 490

18 See L. Wieger, 'Charactères chinois,' 4th ed. Hien Hien, 1924, Leçon 79 F, p. 206.

19 'Shih Ching,' III, II, 1 ('Ch. Cl.,' IV, 2, pp. 464ff.). 'Shih Ching,' IV, III, 3 (op. cit., IV, 2, p. 636). 'Annals of the Bamboo Books,' Part III (Legge, 'Ch. Cl.,' III, 1, Prolegomena, p. 117).

20 'Shu Ching,' I, III, 12 ('Ch. Cl.,' III, 1, p. 27). The statements made about the ancestresses of the first three dynasties belong to those 'reveries, extravagances and manifest falsities' against which the scholars of Tsin raised such indignant complaints when the 'Bamboo Books' were discovered. (Cf. Legge, 'Prolegomena to Shih Ching,' 'Ch. 'Ch. Cl.,' III, 1, p.106.) Franke justly says concerning this ancient motherright material that 'much in the Chinese world of ideas speaks against it' ('Geschichte,' 1, p. 74). Correct. But this is evidence for, not against, the genuineness of the mother-right legends. The rise of these legends - perhaps not of all, but in all events the nucleus of them - goes back to a time when the status of woman was fundamentally different from what it later became. That there was such a time is indicated by both the documents cited by us (and other similar documents) and many old songs which were preserved in the 'Shih Ching' collection.

21 'Li Chi,' IX, I, 1 and II, 2 ('The Sacred Books of the East,' vol. XXVII, pp. 417-28). Cf. also ed. Wilhelm, 'Li Gi,' pp. 255, 259.

22 'Li Chi,' XXII, 5, op. cit., vol. XXVIII, p. 239.

23 See 'Shih Ching,' III, III, 5; III, III, 7; IV, II, 4('Ch. Cl.,'

IV, pp. 537ff., 621ff.). Also the 'Tso Chuan,' passim.
24   Hopkins, op. cit., 1917, p. 806.
25   'Shih Ching,' IV, III, 5 (op. cit., IV, 2, pp. 643ff.).
26   'Shu Ching,' III, IV, I, 8 ('Ch. Cl.,', 1, p. 85).
27   We have hitherto avoided drawing upon the 'Yü Kung' in
     our analysis, as its authenticity is particularly disputed in
     recent times. The view that the work must be 'a post-fourth
     century production' (Ting, Prof. Granet's etc., p. 269)
     because it mentions Ssetchuan, presupposes a unity in the
     content of the 'Yü Kung' which is not certain. At the time
     depicted by the 'Yü Kung,' the largest part of the north-
     east plain must have been half or wholly waste ('Shu Ching,'
     III, I, III, 12ff. 'Ch. Cl.,' III, 1, pp. 99ff. Cf., too,
     'Wirtschaft und Gesellschaft Chinas,' I, pp. 279ff.). Towards
     the end of the Chou period this area as well as that of the
     Yangtze states was no longer in the condition which the
     'Yü Kung' describes, not only in what relates to the 'labors'
     but above all in connection with the yield of the soil. See
     note 3 on the development during the Chou period.
28   Cf. the highly instructive traditions about early waterworks
     in Shansi at the Fen: T'ae T'ae, as chief of the officers of the
     water 'cleared the channels of the Fun and T'aou and em-
     banked the great marsh, so as to make the great plain
     habitable ('Tso Chuan,' Duke Ch'aou, Year 1, 'Ch. Cl.,'
     V, 2, p. 580).
29   The 'Tso Chuan' continually tells of such public works, but,
     in accordance with the whole character of the writing, is
     really interested in them only so far as they are military and
     showy works. The statement in the Year 31 of Duke Seang
     ('Ch. Cl.,' V, 2, p. 566) proves that the dyke technique
     was well known at the time of Tso Chuan. That, by means
     of this technique, already then giant canals and dykes were
     constructed, is shown in the 29th chapter of the 'Shih Chi'
     of Ssu-Ma Ch'ien (Chavannes III, pp. 522ff.).
30   Cf. 'Tso Chuan,' 'Ch. Cl.,' V, pp. 110, 186, 251, 258, 275,
     282, 297, 310, 319, 388, 397, 409, 428, 439, 447, 469, 515,
     548, 564, 599, 641, 667.
31   'Tso Chuan,' Duke Ting, Year 4, 'Ch. Cl.,' V, 2, p. 751.
32   Hopkins, op. cit., 1917, p. 791
33   'Shih Ching,' III, 1, 3, ('Ch. Cl.' IV, 2, p. 439).
34   'Shu Ching,' III, I ('Ch. Cl.,' III, 1, pp. 92ff.).
35   'Wirtschaft und Gesellschaft Chinas,' I, p. 282.
36   See Hu Shih, 'Wen Ts'un' (trend of thought reported by
     P. Demiéville in 'Bulletin de l'Ecole Française d'Extrême
     Orient,' XXIII, 1923, especially, p. 495). Besides, Hu Shih
     admits the historicity of the existence of a public field
     (p. 496).
37   Cf. Ssu-Ma Ch'ien, 'Biography of Lord Shang,' trans. by
     J.J.L. Duyvendak ('The Book of Lord Shang,' London,
     1928, pp. 18ff.). Cf. Besides the additional sources given by
     Duyvendak, op. cit., p. 45.

38 Cf. 'Tso Chuan,' 'Ch. Cl.,' V, 2, pp. 447, 517, 558, 609,
671, 700, 773 and passim, also 'Shih Ching,' II, VI, 6
('Ch. Cl.,' IV, 2, pp. 374ff.). III, I, 3 (op. cit., pp. 438ff.).
Even if the passages of the 'Chou Li' dealing with the agrarian
relations of the Chou period were written, as Karlgren believes,
no earlier than the fourth century BC - which does not exclude the possibility that they may reflect older institutions -
the information conveyed in this work remains instructive
enough. It is important to establish the fact that the statements of the 'Chou Li' do not contradict the parallel dates
in the 'Tso Chuan' and 'Shih Ching' but - apart from certain
exceptions which cannot be discussed here - merely supplement them and make them concrete. Our view, moreover,
does not need support from the 'Chou Li.' It remains valid,
even if we put this source completely aside.
39 'Li Chi,' VII, I (Legge, op. cit., vol. XXVII, pp. 364ff.
R. Wilhelm ('Li Gi,' pp. 30ff.) who in distinction from Legge,
does not mention any public and common spirit, which is not
found in the text). Couvreur's translation which introduces
a 'chef de l'empire' is utterly fantastic ('Li Chi,' ed. Couvreur,
Jo Kien Fou, 1913, I, p. 497). Wilhelm's translation, 'At
the time when the great way ruled, the world was common
property' conveys exactly the sense of the nine Chinese
characters. It contains an idea which corresponds perhaps
very little to the views later dominant in China, but corresponds excellently with certain findings of modern anthropology.
Cf., moreover, the 'Lü Shi Ch'un Ch'iu,' ed. Wilhelm, p. 346.
40 'Nevertheless, there are facts pointing definitely to the close
connection between communal ownership and mother right,
on the one hand, and individual ownership and father-right
on the other hand' (Rivers, 'Social Organization,' p. 114).
41 M. Kowalewsky's conclusions concerning this ('Tableau des
origines et de l'évolution de la famille et de la propriété,'
Stockholm, 1890) receive confirmation in important respects
from recent investigations. Cf., besides Rivers (op. cit.),
R. Thurnwald, 'Die menschliche Gesellschaft' Bd. II, S. 26.
See also F. Engels, 'Der Ursprung der Familie, des Privat-
eigentums und des Staats,' 20th ed., Stuttgart, 1921. pp.
42ff. and p. 143.
42 The size of the average family is set at three adult males
capable of working, in the 'Ta Tai Li Chi' ('Li Gi,' ed.
Wilhelm, p. 70). Shang Yang sought to destroy this type
of large family. He issued an order 'forbidding fathers and
sons, elder and younger brothers, from living together in
the same house'. 'People who had two males or more [in the
family - addition by Duyvendak] without dividing the household, had to pay double taxes' (Duyvendak, op. cit.,
pp. 18 and 15).
43 Tso Chuan, Duke Seang, Year 31 ('Ch. Cl.,' V, 2, p. 558),
Duke Ch'aou, Year 24 (op. cit., p. 700), Duke Ting, Year
9 (op. cit., p. 773).

44   Chen Huan-Chang, op. cit., I, p. 35. Liang Ch'i-Ch'ao
'History of Chinese Political Thought,' p. 27.
45   First form: 'Chou Li' X, 8ff. (ed. Biot, I, p. 223). Second
form: 'Chou Li,' IX, 27 ff. and XV, 6ff. (ed. Biot, I,
pp. 206ff. and I, 340ff.).
46   Ku Liang's representation is reproduced in Legge, 'Prolegomena to Ch'un Ch'iu' ('Ch. Cl.,' V, p. 68), 'Mencius,'
III, A, 3 and I, B, 5 (ed. Wilhelm with the title 'Mong Dsi,'
Jena, 1916, pp. 51 and 16).
47   Wei Yen, the marshal of Ch'i, 'set about . . . defining the
meres; marking out the higher lands and the downs; distinguishing the poor and salt tracts; enumerating the
boundaries of the flooded districts; raising small banks on
the plains between dykes; assigning the wet low grounds
for pasturage; dividing the wide rich plains into tsings'
(Tso Chuan, Duke Seang, Year 25. Legge, 'Ch. Cl.,' V,
2, p. 517. Our italics).
48   Little Hsia Calendar (op. cit., p. 235); 'Shih Ching,' II,
VI, 9 ('Ch. Cl.,' IV, 2, p. 381). Besides the statement
cited in note 5, the well-system is mentioned several times
in the same work under Duke Seang, Year 31 (op. cit., 558).
It is to be assumed that where the well-system existed it was
linked with the institution of a special public field. The labor
of the population (townsmen as well as countrymen) on the
public fields is explicitly spoken of in the 'Tso Chuan,'
Duke Ch'aou, Year 18 (op. cit., p. 671). The two commentators
of the 'Ch'un Ch'iu,' 'Kung Yang' and 'Ku Liang,' both
speak of the 'anciently' performed labor in the public field
(Ch. Cl'. V, 1, Prolegomena, pp. 68ff.). According to
'Ku Liang,' wells and dwellings of peasants were situated
on the public field, an assertion which finds its support
perhaps in the 'Shih Ching' songs II, VI, 6 (op. cit., p.
375). Mencius speaks of the public fixing of boundaries as
a reality, but, on the other hand, about the nine-field
system as an earlier institution, whose re-installation he
recommends - but not for the fields in the neighborhood
of the city! The public fields in his time evidently no longer
had the form he recommended. He censures arbitrary drawing of boundaries as a device of evil officials to augment their
pay! When one remembers that during the time of Mencius
the abolition of the old 'boundaries' in the state of Ch'in
was deemed an extraordinary innovation, which the other
states by no means followed, it immediately becomes evident
that Mencius's suggestions for reforms were far from being pure
fantasies; they rather set out from existing relations, which
to be sure were visibly in complete disintegration. A very
complicated description is that of the Chou Li, which for
reasons of space we shall not treat of here. Maspero, who
points out the independent nature of the Chou Li, and who
likewise sees the variations among Kung Yang, Ku Liang and
Mencius - which he seeks to explain, not as real variations

but merely as philological variations of an unknown common
source - Maspero, at all events, stresses at the same time
emphatically that the well-system was undoubtedly an historical reality. ('La Chine antique,' p. 109, note.)
49 'Ch'un Ch'iu,' Duke Deuen, Year 15. There may be added
to this the two commentaries already cited, as well as a remark
of Tso Chuan evidently pointing to labor in the public field
('Ch. Cl.,' V, 1, pp. 327 and 320). The new tax, after more
than a hundred years, did not yet prove sufficient for the
state of Lu, which imposed it (cf. 'Lun Yü,' XII, 9, 'Ch. Cl.,'
p. 255). The state of Ts'i, in the sixth century, was not
contented with 20 per cent of the yield. It laid taxes which
left the population (peasant?) only a third of the product of
its labor. ('Tso Chuan Duke Ch'aou,' Year 3, op. cit.,
p. 589.) We have already told how Shang Yang's re-ordering
of the agrarian system was connected with a fundamental
transposition to a tax imposable on families.
50 'Lü Shi Ch'un Ch'iu,' trans. by R. Wilhelm with the title
'Frühling und Herbst des Lü Bu We,' Jena, 1928, p. 262.
51 The use of iron must have been widespread by the sixth
century (Tso Chuan, 'Duke Ch'aou,' Year 30, op. cit.,
p. 732). Mencius mentions iron implements in tilling of fields
(III, I, 4, Legge, 'Ch. Cl.', II, p. 248). Eberhard claims,
following recent Chinese investigations, that bronze implements were used in agriculture during the Chou period,
although late and infrequently (W. Eberhard, 'Zur Landwirtschaft der Han-Zeit, Mitteilungen des Seminars für
Orientalische Sprachen,' Berlin, XXXV, 1932. Ostasiatische
Studien, p. 88).
52 The 'Shih Ching' is already acquainted with irrigation. Cf.
II, VIII, 8, 'Ch. Cl.,' IV, p. 417; also IV, I, 5, op. cit.,
pp. 600ff. Ssu-Ma Ch'ien describes the extension of the
system of canals at the end of the Chou period with continual
reference to the importance of the constructions for irrigational purposes (Chavannes, III, pp. 522ff.).
53 The result of this development is described by the 'Lü Shih
Ch'un Ch'iu' (ed. Wilhelm, p. 159). Also, in the framework
of the rotation of the calendar: pp. 28 and 36. Cf. too 'Li
Chi' (Legge, XXVIII, pp. 264 and 286) as well as the 'Chou
Li' - admitted to be authoritative for what appertains to the
fourth century BC (Biot, I, p. 367).
54 Ssu-Ma Ch'ien, ch. XXIX (Chavannes, III, pp. 523ff.).
55 The fact that the dukes of the various feudal states were
enthroned by hereditary succession - cf. 'Shih Ching,'
III, III, 7, 'Ch. Cl.', IV, 2, pp. 516ff. - needs no proof.
It is true that at first the Shang and Chou rulers like the
Carolingians and the Merovingians may have become rulers
over the imperial territory by means of great military vic-
tories. There is much evidence for this. Still, corresponding
to the economic-military structure of this 'empire', there
very soon set in a relapse into an agglomerate of single

feudal states with relative independence. Respecting the
inheritance of high state posts, in 'old families', we possess
data which reach back to the middle of the Shang period; if
we trust the reports of the 'Shu Ching,' very much earlier
still. (Cf. 'Shu Ching,' P'an Keng. 'Ch. Cl.,' III, 1, p. 229.
For the Chou period cf. 'Tso Chuan,' pp. 429, 551, 580,
683.)

The hereditary character of the non-noble classes must
likewise have continued to exist until the time of the 'Spring
and Autumn Annals' (eighth to fifth centuries). In the year
582 we find the following praise of the state of Tsin: 'his
common people attend diligently to their husbandry; his merchants, mechanics and inferior employees know nothing of
changing their hereditary employments'. ('Tso Chuan,'
p. 410.) A statement from the end of the sixth century
implies that while non-official callings ought to be hereditary,
they are in fact no longer entirely so. ' Sons must not change
the business of their fathers - husbandry, some mechanical
art or trade' ('Tso Chuan,' op. cit., p. 718). The 'Lü Shi
Ch'un Ch'iu' speaks similarly (Wilhelm ed., p. 454) of the
ties of the peasant to his land and of the merchant to his
calling - now clearly as of an ideal.

56   The inscriptions of the Shang period take cognizance of
shell-money (Hopkins, op. cit., 1917, p. 382). We find in
the eighth century and following that the variety of means of
exchange has grown greatly. (See 'Tso Chuan,' pp. 12, 191,
224, 263ff., 282, 310, 427, 525.)

57   Primitive market traffic, perhaps still on a natural basis, is
described in the 'Yi Ching' (Hexagram 21, Legge, p. 383).
The 'Tso Chuan' knows highly developed market relations
('Ch. Cl.,' 467, 581, 589, 671, 681, 683, 712, 816, 843,
856). It is a characteristic fact of the growing bureaucratiza-
tion that we hear of the existence of markets often only
indirectly, namely, on the occasion of the naming of the
officials who supervise them. Large purchases of grain for
state purposes are reported. ('Tso Chuan,' pp. 111ff. Cf.
also pp. 21 and 167.) The question of taxing the merchants
on the borders or in the markets plays a considerable part
in the discussion of the later Chou period. (Cf. Mencius,
I, II, 5; Legge, 'Ch. Cl.,' II, p. 162; also II, 1, 5, pp.
199ff.; 'Ta Tai Li Chi,' 'Li Gi,' Wilhelm ed., p. 187; and
'Lü Shi Ch'un Ch'iu,' pp. 93 ff.)

58   'Tso Chuan,' 'Ch. Cl.,' V, 2, p. 664. In this case, as in the
first of the sales of grain mentioned in the above note, the
merchant is evidently of noble origin. The account of 'Tso
Chuan' proves that merchants normally ranked below the
officialdom and could be taken up into the latter class only
through very great services. 'Tso Chuan,' op. cit., p. 799.
That a travelling merchant who is in a position to offer im-
promptu a dozen oxen and a number of hides is wealthy
enough to mix in politics is to be seen from the report in

the 'Tso Chuan,' p. 224. Cf. also the somewhat divergent
account in the 'Lü Shi Ch'un Ch'iu,' pp. 246ff.

59 Both Confucius and Mencius belonged to poor or obscure
branches of noble families. (Cf. 'Lun Yü,' IX, VI, 3; Legge,
'Ch. Cl.,' p. 218; also Legge's preface to his edition of
Mencius, 'Ch. Cl.,' II, p. 15.) Confucius only underlined
a tendency already present in his time, in making the choice
of the pupils he educated for a political public career independent of their wealth. (See 'Lun Yü,' VI, IX, and especially VII, VIII, 'Ch. Cl.,' pp. 188 and 197.) The classic
figure of a very rich merchant who at the end of the Chou
Period attains the highest state office is Lü Pu We. (Cf.
Kao Yu's biography, trans. by Wilhelm, 'Frühling und
Herbst des Lü Bu We,' pp. 1ff.)

60 Confucius seeks to fix the amount of a suitable emolument
for the new type of official demanded by him. The emolument
is no longer in land, but in adequate payment for capable
officials. ('Lun Yü,' VI, III. 'Ch. Cl.,' pp. 185ff.) Also
Mo Ti demands that a competent officer should receive adequate payment. (Forke ed., under the title of 'Me Ti,'
Berlin, 1922, p. 197.) A chapter of the 'Li Chi,' whose
Confucian purity is thought doubtful, but which surely
expresses the main tendency of the transition period, contains the following advice: 'A superior man will not for words
of small importance receive great emoluments, nor for words
of great importance small emolument' ('Li Chi,' Legge, op.
cit., XXVIII, p. 345).

61 Leading roles as counselors and high officials under the Han
emperor, Wu Ti, towards the end of the second century BC,
were played by K'ung Chin, a wealthy iron manufacturer,
Tung-Kuo Hien-Yang, a great salt producer, and the
merchant's son Sang Hung-Yang. (Cf. 'Ssu-Ma Ch'ien,'
ch. 30, Chavannes, III, pp. 567ff.) Cf. also O. Franke,
'Staatssozialistische Versuche im alten und mittelalterlichen
China,' Berlin, 1931, pp. 5ff., and Chun-Ming Chang,
The Genesis and Meaning of Huan K'uan's Discussion on
salt and iron, 'Chinese Social and Political Science Review,'
XVIII, April 1931, pp. 21 and 25. On the growing importance
of the banker see ibid., p. 12.

62 It is experience that counts, not bodily strength. 'Man becomes a farmer by accumulated experience in farming'.
Hsüntze (trans. by Hu Shih, 'The Development . . .',
p. 156.) Cf. 'The Works of Hsüntze,' H.H. Dubs ed., London,
1928, p. 115.

63 'Wirtschaft und Gesellschaft Chinas,' I, pp. 141ff.

64 Confucius proclaimed as a principle for the selection of his
pupils: 'In teaching there should be no distinction of classes.'
('Lun Yü,' XV, XXXVIII, 'Ch. Cl.,' I, p. 305. Cf. too
'Lun Yü,' VII, VII and XXVIII, op. cit., pp. 197 and 204.)
He excluded from his teaching the old knightly subjects of
learning, chariot-riding and archery, as well as the military

arts. ('Lun Yü,' IX, II, 'Ch. Cl.,' I, pp. 200, 202, 294 and
216.) Whereas the old feudal songs extol the brave chariot-
driver, the good marksman and the reckless hunter (cf.
'Shih Ching,' I, VII, 4, 'Ch. Cl.,' IV, 1, pp. 129ff. and
passim), Confucius stresses, having perhaps in mind the cited
passages in the 'Shi Ching,' that he has no use for such men
of strength. ('Lun Yü,' VII, X, op. cit., p. 198.) Mo Ti's
attack on the choice of officials by the old feudal principle
of heredity pervade all his writings ('Mo Ti,' Burke ed., pp.
193, 202, 208, 211, 226 and passim).

65 Confucius: the high official needs no professional training
('Lun Yü,' XIII, IV, 'Ch. Cl.,' I, pp. 264ff.). Completely
opposed by Mo Ti, who lauds the technical administrative
and agricultural abilities of the idealized officials of the
past ('Mo Ti,' p. 194 and passim).

66 Confucius on eating, clothing and music. ('Lun Yü,' X,
VI; X, VIII; XIV, XIII, 'Ch. Cl.,' pp. 230ff., 232ff.,
279.) These observations in Confucius are accompanied by
a number of limiting statements. But actual development
considerably lessened the importance of these limitations.
Ideologically the weakening of these restraints was rendered
easier by the consciously aestheticizing character of the
Confucian cultural ideal (cf. 'Lun Yü,' VII, VIII, 'Ch. Cl.,'
I, p. 211 and passim). Opposed, Mo Ti and his school, who
advocated simplicity in eating, clothing and dwelling (op.
cit., pp. 293ff., 296), and were scornful towards music
('Mo Ti,' p. 188 and Mohists, pp. 399ff.). Mo Ti's crude belief
in spirits is as compatible with his early bourgeois social
ideas as Cromwell's clumsy belief in the Bible was with his
historical function.

67 The thesis that piety and obedience in the patriarchally
constituted family must be the basis of society constantly
recurs in Confucius's teachings, as is well known. Even
when the parents are in the wrong, one must obey them.
If the father steals a sheep, the son's duty is not to inform
on him but to shield him! ('Lun Yü,' IV, XVIII and XIII,
XVIII, op. cit., pp. 170 and 270). Sharply opposed is Mo
Ti. ('Mo Ti,' pp. 242ff., 254 and 262ff.)

68 We intend at a later date to elaborate this idea on the basis
of a concrete analysis of the underlying economic and social
development. At this point we rest content with referring
to the presentation of Hu Shih, who makes this phenomenon
obvious. (Hu Shih, 'Religion and Philosophy in Chinese
History,' Symposium on Chinese Culture, pp. 32ff.)

69 Cf. K. Marx, 'Das Kapital, III,' 2, 4th ed. Hamburg, 1919,
pp. 324 and 173ff.

70 The principle is already proclaimed by Mencius and Shang
Yang. (Mencius, III, I, 3, 'Ch. Cl.,' II, p. 210. 'The Book
of Lord Shang,' pp. 176 and 306ff.)

71 The tax is demanded in bad years too, and when the people,
after the whole year's toil, are still not able to nourish their

parents 'they proceed to borrowing to increase their means,
till the old people and children are found lying in the ditches
and water-channels' (Mencius, op. cit., pp. 211ff.). This
condition, which set in toward the close of the Chou period
(i.e. with the dissolution of the old agrarian commune),
always reappeared in the history of the bureaucratic centralized state as the expression of the approach and outbreak of
socio-economic crisis. A budget of a poor peasant family of
the Han period shows that the taxes were so high, even in
normal years, that if the peasant met all his obligations the
year inevitably closed with a deficit. (See the attempted
calculation reported by the Han Shu, translated by Duyvendak,
'The Book of Lord Shang,' p. 43. Cf. too Eberhard, 'Zur
Landwirtschaft der Han Zeit,' pp. 76ff. and 81.) The famous
memorial of Lu Chih of the late eighth century AD reports
on the status of the land-tax system in the T'ang period
after the modernization of the tax. (Cf. the translation of
S. Balazs, 'Mitteilungen des Seminars für Orientalische
Sprachen,' Berlin, 1933, Ostasiatische Studien, pp. 29ff.)
Ma Tuan-Lin's account gives detailed information on tax
conditions during the Sung period. (See Biot's 'Mémoire
sur la condition de la propriété territoriale en Chine,'
'Journal Asiatique,' 1838, pp. 306ff.) The procedure:
merciless exaction of taxes leading to the flight or committing
to tenantry of the pauperized peasants, continuing throughout
the Ming dynasty into the Manchu dynasty. (Cf. Mable Ping-
Hua Lee, 'The Economic History of China,' New York, 1921,
pp. 350ff., 407 and 411.)

72 See 'Wirtschaft und Gesellschaft Chinas,' 1, pp. 287-90.
Cf. also Ch'ao-Ting Chi, The economic basis of unity and
division in Chinese history, 'Pacific Affairs,' 1831, December,
pp. 387 passim. Chi's conception of the dynamics of what he
calls the 'economic key areas' in Chinese history is, in our
judgment, one of the most valuable contributions to scientific
understanding of the economic development of China.

73 This is one of the most important ideas in the work of Ch'ao-
Ting Chi ('The Economic Basis,' p. 302).

74 Cf. S. Balazs, 'Beitrage zur Wirtschaftsgeschichte der T'ang-
Zeit' ('Mitteilungen des Seminars für Orientalische Sprachen,'
Berlin, 1931 and 1932). Balazs, by the use of sources, has
for the most part accurately described the phenomena.
Employing Max Weber's methodology, however, he seeks the
explanation not in the development of the material basis
but in the conditions of circulation and of politics, commerce
and wars, 1932, pp. 22 and 32.

75 Human labor power also becomes a commodity, in the form of
wage labor (cf. on the beginnings of industrial capitalism
in ' "Asiatic" China', 'Wirtschaft und Gesellschaft Chinas,'
I, pp. 525ff.).

76 Ibid., pp. 701ff. and 726ff.

77 We have analyzed the problems of invasion, as related to the

internal crisis of the Chinese state, in Probleme der chinesischen Wirtschaftsgeschichte, 'Archiv für Sozialwissenschaft
und Sozialpolitik,' vol. 57, 1926, pp. 325 and passim.
78 The new regulations of inheritance, originating during the
close of the Chou period, were pushed forward energetically
after the establishment of the absolute state. Han Wu Ti in
128 BC enacted a law which had as its object the division
of feudal estates previously inherited. The final statute of
the regulation of inheritance is shown by the codex of the
last dynasty as P.G. Boulais has translated it under the
title of 'Manuel du Code Chinois,' Shanghai, 1924. (Cf.
ibid., pp. 198ff.; O. Franke, 'Die Hechtsverhaltnisse am
Grundeigentum in China,' Leipzig, 1903, pp. 46ff.) It is
known that the majority of the ranks of nobility, in so far
as they were bestowed in post-feudal times, usually dis-
appeared after a few generations (cf. 'Manuel du Code
Chinois,' p. 103).
79 The stifling of feudal power in 154 BC 'politically meant the
end of feudalism and the concentration of power in the hands
of the imperial government'. Economically it signified the
impoverishment of the feudal lords (Chun-Ming Chang,
op. cit., p. 12). Bestowing a benefice on some dignitary was
robbed of any material importance. Ma Tuan-Lin summarizes
by saying: 'We know from the accounts of Li Pi that the so-
called bestowal of title and land was only an empty appearance,
as no inheritance took place. The "fief-holders" were not
allowed to draw any direct income from the fief. It thus becomes only an emolument or gift, there is no question of
bestowal of the land. . . . But if there be no hereditary
bestowal of the land, the feudal system is finished' ('Wen
Hsien T'ung Kao,' p. 276, 11b. Trans. by Balazs,
'Beitrage . . . ,' p. 66). Traces of feudal land possession
later appear often, finally in the form of the 'banner-land'
of the Manchus, but it never regained its historic role.
The vast 'palace farms' which members and officials of the
court seized for themselves in the Ming period obviously
determined the relation between landlords and direct producers in the form of tenantry, i.e. in a post-feudal manner.
80 This is the fate of the attempted reforms of Wang Mang
(AD 9–23), as well as a number of similar later efforts up
to the time of Wang An Shih (twelfth century AD). The
slogan, 'Back to the old well-field system', which always
recurs, means economically and politically: 'Give the state
the exclusive power over peasant landrents!' (On the history
of these attacks and reforms cf. Hu Shih, Wang Mang, the
Socialist Emperor of Nineteen Centuries Ago, ' North China
Branch of the Journal of the BAS,' 1928, pp. 218ff.; Franke,
'Staatssozialistische Versuche,' etc.) Cf. too the report of
Wang An-Shih in 1058 upon the reform of the officialdom,
trans. by O. Franke, Berlin, 1932, significant at the same
time as an example of a pushing forward in the direction of

a really rational professionally trained officialdom. How
ineffectual all efforts to limit private land-possessions remain is shown by the sources of Chinese economic history
trans. by Mrs Lee. Cf. among others, pp. 171 and 187 (early
and late Han); 233 (T'ang); 317ff. (Sung); 352, 369, 384
(Ming); 420 (Tsin).

81 Cf. on this the conclusions of the high official Tung Hsün-
Ho in the last (Manchu) dynasty, cited by Chi, 'The economic
basis,' p. 389.

82 Chinese writing separated social classes but united regions.
The particular character of this writing as the instrument
of the ruling official class in manifesting its special social
position and reproducing it continually was of great importance for Chinese social history.

83 The beginning of the system goes back to the start of the
new order, i.e. in the Han period. Ma Tuan-Lin has exhaus-
tively treated its development and maturity. (See Biot's
essay based on Ma Tuan-Lin's work, 'Essai sur l'histoire
de l'instruction publique en Chine,' Paris, 1847. Cf. especially, pp. 254ff.)

84 Wang An Shih combated this frequent shifting about of
officials, which, as he rightly contended, prevented real
familiarity with local problems. (Cf. his report of 1058,
Franke, op. cit., p. 40.) Wang's reform efforts and similar
ones were failures. The main interest of the Mandarinate
was far more in having a stable state-apparatus than in
having a specialized and capable one. (Cf. the sharp regulations against officials who dare to stay in their district
or office longer than they were ordered to ('Manuel du
Code Chinois,' p. 128.))

85 This transition begins already in the Chou period. The
eunuchs mentioned in the 'Shih Ching' (III, III, 10, 'Ch. Cl.,'
IV, 2, p. 561) still belong entirely in the category of 'harem
and household affairs'. We find in the Tso Chuan alongside
of this type ('Ch. Cl.,' V, 1, p. 173; and perhaps V, 2,
p. 610) a definitely political type of eunuch who may be versed
in the art of chariot-driving (V, 2, pp. 535 and 843) and who
even leads the army on to the battle-field (V, 1, pp. 137
and 145; cf. further V, 1, p. 191, and V, 2, p. 525). The
high officials' hatred toward these 'castrated' politicians
was already bitter. (Cf. V, 2, p. 475; see too 'Lü Shi Ch'un
Ch'iu,' Wilhelm ed., p. 137.) Ch'in Shih Huang Ti, the
founder of the first post-feudal dynasty, had a favorite
eunuch who accompanied him in his chariot and was apparently
politically educated ('Ssu-Ma Ch'ien,' ch. VI, Chavannes,
II, p. 192). The political influence of eunuchs increased
exceedingly under the Han. Ho-Ti (89–105) 'installed them
by preference within the highest administrative posts' (Biot,
'Essai sur l'histoire de l'instruction publique en Chine,'
p. 188). We cannot here deal with the further development.
It may only be remarked that the greatest admiral China

possessed, Chang Ho, who led Chinese fleets as far as
Arabia, Africa and across peninsular India, was a eunuch.
86 Cited in E.T.C. Werner ('Chinese Descriptive Sociology,'
London, 1910, p. 55).
87 That merchants were to be robbed of their profits and kept
fearful was already the motto of Shang Yang (cf. 'The Book
of Lord Shang,' p. 177 and passim). The power of commercial
capital grew exceedingly under the Han. Its agents were
treated with servile respect by the members of the ruling
class. Ssu-Ma Ch'ien has written illuminatingly on how the
central power sought to defend itself against commercial
capital and to bring the latter's wealth into the possession
of the state-apparatus and its members. (Ch. XXX,
Chavannes, III, pp. 583, 585ff. Cf. Balazs, 'Beiträge,'
1932, p. 47, on the T'ang period. For the Ming period cf.
Lee, op. cit., p. 379. The history of this struggle, which
continued into modern times, is still unwritten. See 'Wirtschaft und Gesellschaft Chinas,' 1, pp. 708ff.)
88 The wealthy merchants were attacked by taxation of their
profits, seizure of their property, forced loans and movements toward the 'limitation' of land-owning; in addition,
steep extra taxes were put on their luxuries (chariots)
or their dress would be regulated by law, or they would be
regarded as citizens of a lower order. ('Ssu-Ma Ch'ien,'
XXX, Chavannes, III, pp. 575, 588, 541.) These measures
changed in form and intensity but they never ceased until
the end of China's independence. Even in Ming times we
hear of regulations of clothing imposed upon merchants.
The peasants might wear silk clothes, if they were able to;
merchants were forbidden to do so. (Cf. Lee, op. cit.,
p. 359.)
89 On profits from trade in grain, especially in famine times
(Shang Yang, op. cit., pp. 177ff.; Han Shu, see Lee, op.
cit., pp. 158ff. Kiu T'ang Shu, cf. Balazs, 'Beiträge,'
1932, p. 31. For more recent times, 'Wirtschaft und
Gesellschaft Chinas,' I, pp. 724ff.). On profits from salt
and iron since the Han period, F. Hirth, Notes on the Early
History of the Salt Monopoly in China, 'JNCBRAS,' 1887,
pp. 58ff.; cf. too Chun-Ming Chang, 'The Genesis and
Meaning,' pp. 8ff. See also 'Wirtschaft und Gesellschaft
Chinas,' I, pp. 719ff.
90 Han (cf. Lee, pp. 159 and 187); T'ang (Balazs, 'Beiträge,'
1931, p. 78); Sung (cf. Lee, pp. 313ff.); and so on until
the present.
91 Whereas historical statements concerning merchants' land are
often obscure (speaking frequently only of 'wealthy families',
etc.), reports on the landed property of officials characterize the entire post-feudal history of China (cf. Lee, op.
cit., passim).
92 Han Wu Ti's reform measures arose out of the financial need
of the state exhausted by wars. Because Wang Mang had

usurped the throne, he and his followers tried to break
the resistance against the usurper, by means of an omnipotent
centralized state. Wang An Shih's reforms fell within the time
of the gravest external, political crisis of the Sung dynasty.
93 'Ssu-Ma Ch'ien,' XXX (Chavannes III, pp. 579ff. Cf. too
Chung-Ming Chang, op. cit., p. 22; Hu Shih, Wang Mang,
p. 228. On the history of the institution see Franke,
'Staatssozialistische Versuche,' pp. 11ff.).
94 Cf. Hu Shih, Wang Mang, p. 228; Franke, 'Staatssozialistische Versuche,' pp. 20ff.
95 F. Hirth, 'Notes on the Early History of the Salt Monopoly
in China,' pp. 55ff.; Franke, op. cit., pp. 6ff.; also
'Wirtschaft und Gesellschaft Chinas,' I, pp. 722ff.
96 An attempt to explain the causes of China's 'Asiatic' stagnation is given in 'Die naturlichen Ursachen der Wirtschaftsgeschichte,' p. 606 and passim.
97 Together with the external political factor, which placed
China in the position of a semi-colony, this fact is decisive
in the most recent economic history of China (cf. Wittfogel,
'Die Grundlagen der Chinesischen Arbeiterbewegung,'
pp. 238ff.).
98 Japan's agriculture, like China's, rests upon irrigation.
But in Japan canals a few miles long have been regarded
as large, and those of ten to thirty miles as extraordinary
(cf. O. Nachod, 'Geschichte von Japan,' II, 2, Leipzig,
1930, p. 987; also Yosoburo Takekoshi, 'The Economic
Aspects of the History of the Civilization of Japan', London,
1930, III, pp. 412ff.). The result of this is the development
of a system of production which, although it has an 'Asiatic'
tinge by reason of irrigation, is essentially of a decentralized
feudal structure. Nachod's demonstration that Japan's ruling
class borrowed much from the Han culture of China, but
not the institution of eunuchs, is both true and sociologically
important (Nachod, II, 1, pp. 286ff.). However, this happened not so much because of any abstract 'healthy feeling'
on the part of leading Japanese as because of the concrete,
social situation in Japan, which offered no place for the
eunuch system of 'oriental' society. The foundations of modern
capitalistic activity were built up much more slowly in
Japanese feudalism than in Europe - mainly because of its
somewhat isolated position and because of the structure of
its productive powers. There is additional similarity to
features of European development in the fact that the land-
owning feudal class began of itself to develop certain branches
of production, especially mining (cf. Takekoshi, I, pp. 334ff.,
407 and 545; II, pp. 177, 486ff.). We also hear of pottery on
a large scale, run by feudal lords. (Takekoshi, I, p. 478;
70 workers in one shop!) The feudal starting-point of
Japanese capitalism has been noticed by several different
authors (cf. Tokuzo Fukuda, 'Die gesellschaftliche und
wirtschaftliche Entwicklung in Japan, Münchener Volks-

wirtschaftliche Studien,' 42nd part, 1900, p. 155; also
G.B. Sansom, 'Japan, A Short Cultural History,' New York,
1931, pp. 512 and 515). These and other authors have pointed
out that at the same time there was developing, through the
initiative of commercial capitalism, a system of industrial
production no longer of a medieval handicraft type, but
definitely capitalist in tendency and structure. Takekoshi
gives a detailed account of this development, which may be
criticized from the methodological viewpoint, but nevertheless remains valuable because of the abundance of its docu-
mental material (op. cit., II, pp. 177, 352, but especially
415; III, pp. 264 and 271). When J.E. Orchard compares
Japan's economic position at the end of the Tokugawa period
with that of England in the sixteenth century ('Japan's
Economic Position,' New York, 1930, p. 71), he strikes at
the heart of the matter, although he is reluctant to draw
the historical conclusions. The development of Japan, re-
tarded though it was, essentially bore a much stronger
resemblance to that of the feudal and semi-feudal economic
pattern of the sixteenth- and seventeenth-century Europe
than to that of China. An understanding of this fact is
essential to scientific comprehension of the discordant
development of Japan and China in the last decades.

THE THEORY OF ORIENTAL SOCIETY*

Karl Wittfogel

1 THE PROBLEM

Theoretical concern with the complex of China and its oriental background needs no detailed justification in the year 1938. The world of historical practice has become small. The intimate connection between the Eastern and Western segments of human society is shown most directly by the Sino-Japanese War and its repercussions on the remaining centers of historical activity. The scientific analysis of China, which for ten or fifteen years might have seemed either romantic or at all events peripheral, now returns towards the center and becomes a demand of objective necessity.

Ad hoc analyses show themselves quickly as inadequate, particularly if they are not presented as the result of previous thorough preparation. Critical theory teaches that an actual historical phenomenon becomes understandable only if it is taken in its full historical and spatial perspective. The laws of movement within Chinese society only reveal themselves truly if they are developed genetically. The same is true of the neighboring complex of Japan, concerning which we must know whether it is structurally similar to China or different, before we can truly comprehend the peculiar forms of its historic activity. Thus problem leads to problem. Have we, in the case of China, to deal with a social type like that of the western lands, or have we not? If not, what are its distinguishing features? Are there present, in the structure of this state, basic peculiarities corresponding to that (perhaps present) socio-economic peculiarities, which support or correct the general theory of the state, or show it in a new light? Or yet again: if the analysis is to be socio-genetic, at what level is such an analysis to start? The great French historians of the early nineteenth century conceived history as the movements of socially distinguished strata (classes). Critical theory, not satisfied with this, posed the question of the structural principle of all such social formations and conflicts.

From the envisioning of such a basic analysis of history to its
realization is a long way. Numerous attempts show at least the
extraordinary difficulty of the problem, if they do not indeed
demonstrate something more negative: its insolubility. Perhaps,
when one postulates the methods of production during an epoch
as the first problem of analysis, one has set himself too much
to do. Should one perhaps consider history, as did those great
Frenchmen, mainly as social history - that is, a history of social
movements with introductory remarks on their economic background and with careful simultaneous analysis of political and
social phenomena, whose origin may be considered sufficiently
explained thereby? Comparative consideration of the scientific
historiography of the past decades shows that, apart from the
crude forms of a mechanical Geschichtsökonomismus ('economic
interpretation of history'), scientific analysis has by preference
used just such a socio-historical method.

The investigation of a complex like the Chinese, so 'new'
from the theoretical standpoint and to which a meaningful tradition has not yet been attached, might help to clarify the basic
methodological problem. More: if a basic consideration of oriental
and occidental social development discloses specific basic movement - patterns of the 'East' (whatever that may be), then
possibly a new point of departure for the methodology of investigating 'Western' developments may result. And finally: if material
analysis of the Eastern world finds out by the best known
historical steps that it is not a separately progressive but rather
a stationary social complex, there arise out of such an insight
some essential conclusions for our conception of social history
in general. This conception must then obviously include and explain not only developments but also a historically extremely
significant variation: stagnation.

2 EXPERIMENT IN ONE FUNDAMENTAL ANALYSIS

*Artificial irrigation as a specifying factor in agricultural
production*
The social phenomena which we are studying develop at a stage
of history below that of capitalistic industry and above that of
primitive agrarianism. The latter, in its various manifestations,
forms the point of departure. But this point of departure must
already have been passed before a specific form of work and
life can arise.

The agricultural process of production sets in motion, on the
objective level where material analysis commences, two elements
in combination: land and water. Differences in the type and
productivity of the land, the soil, determine to a large extent
the settled social stage of the resultant labor-process; they
lead, however, only to variations, not to a fundamentally divergent pattern of the labor-process. This effect proceeds from the
second of the chief factors in the means of work - from water.

Water gives to the soil and to the plants contained therein as
in a vessel, the moisture without which no transfer of nutritive
elements into the plant organism and no metabolism of this
organism can occur. It also carries with it dissolved organic
and, above all, inorganic substances which are required for
the nourishment of the plants. The type of climate here comes
into play. Apart from the factor of temperature, which affects
various stages of agrarian development in various ways, differ-
entiating and limiting them, the factor of rain, caused by the
movements of the wind, is decisive. Rain-water may moisten the
soil in a timely and quantitatively sufficient manner, or it may
not. In the former case the agricultural labor-process can proceed without additional irrigation, while in all other cases
agriculture is either completely impossible or can only succeed
by means of the planned application of water through socially
working men - i.e. by means of artificial irrigation.

Several basic possibilities thus present themselves. The
specifying factor in all cases other than the first is the presence
or absence of river- or of ground-water - i.e. of rain-water which
'originates' outside the area in question and then enters this
area. (See Table 1.)

Table 1 offers the specifying types of the water situation in
a manner newly defined according to our categories. Of course
these facts have not been exactly represented in the maps of
physical or economic geography, and yet the agrarian-economic
categories which we are trying to analyze can be recognized
without difficulty in their spatial arrangements on good maps of
climate and vegetation. It is indeed to be borne in mind that
the factors we have treated indicate merely the possibility of a
corresponding nomadic or agricultural development, not yet its
reality. In order for a primitive social organism to proceed towards agriculture or cattle-raising, certain natural pre-conditions
(apart from the activity of the people and the means of work at
their disposal) are necessary - above all, in socially simple
conditions, the physical existence of a reproductive flora and
fauna. Failure to fulfill these conditions either leads, as in the
case of Australia, to stagnation on the level of food-gathering,
or, as in the case of America, prepares a one-sided agricultural,
not pastoral, development. Socially active man creates this form
of his history, just like any other form, yet the actuality of his
history on each achieved phase of his social production is determined by the pattern of the natural productive forces that can
be realized at the moment.

Our table makes one thing immediately clear: just as food-
gathering societies originate and exist under very different
conditions, so do the societies that cultivate plants and animals.
Apart from extreme forms (the desert C b and the super-abundant
variant of A 1 a), the water-situations fall into three chief types:
the case A 2 with its marginal forms 1 b and 3 a and b (rain-
agriculture); the case B 1 a, 2 a, 3 a, and C a (irrigation agriculture); and the case B 1 b, 2 b, and 3 b (nomadism), whereby

Table 1

Variations of the water-situation			Rain for specific agricultural types		Rivers, ground-water, etc.		Tendency to—
			Timeliness	Sufficiency	Present	The task:	
A	1	a	+	++	a+	a 1 Drainage, Protection / a 2 —	Patterns of water-control
		b			b—	b —	
	2		+	+	Insignificant		Rainfall agriculture. in case of temporary nomadism, transition to rainfall agriculture plus cattle-raising.
		a	.	+	a—		
		b	+	+—	b+	Transition to B 1 a	
B	1	a	+	— But enough for pasture	a+	Supplementary irrigation	Irrigation agriculture
		b			b—		Nomadism
	2	a	—	+	a+	Irrigation for the sale of insurance	Irrigation agriculture
		b			b—		Nomadism
	3	a	—	— But enough for pasture	a—+	Irrigation to make agriculture possible	Irrigation agriculture
		b			b—		Nomadism
C		a	—	—— No rain at all	a+	Irrigation to make agriculture possible	Irrigation agriculture
		b			b—	Desert	

+, sufficient or present in considerable quantity; ++, present in excess; —, present only in slight, insufficient quantity; ——, not present at all.

the last two types remarkably enough grow in the main out of
similar situations, which will only be varied by a substantial
factor (presence or absence of river- or ground-water). The
relations of the theoretical categories reflect adequately the
relations of the real conditions. The two forms of production are
found indeed regularly as the two sides of a related natural
basis. Specific social features in the relations between these
two production organisms, the mechanism of the economy in the
border territories, invasions, the so-called nomad dynasties,
the phenomenon of the Great Wall, receive their scientific explanation here from the structure of their productive forces. The
great nomadic societies developed in Africa and Asia on the
borders of agrarian societies with irrigated agriculture, on
which they inflicted from outside effective sociological, military,
and political disturbing factors. On the other hand, the nomad
areas (B 1 b, 2 b, and 3 b) also border on agrarian lands of
Type A 2 and 3, and on more primitive and early feudal agrarian
organisms (A 1 without facilities for water control), as in Central
Africa. In each case a truly scientific analysis must take into
account the already-established categories for the clarification
of the entire socio-economic mechanism.

If the 'B . . . b' -variant (nomadism) lies outside the sphere
of agricultural production, then the water-situation within it
is split into two basic types: A and B . . . a plus C a, i.e.,
into areas where agriculture may be carried on by means of the
quantity of water 'naturally' furnished by rain, and areas where
this is not the case. In the second case it is sometimes possible
to find agriculture in the lower forms (more or less A 3 b) set
up at first in the hinterland of the still untamable rivers. Agriculture only attains higher productivity when its practitioners
manage to make good the water-deficit by furnishing supplementary water: i.e. by artificial irrigation. The new agrarian
technique thereby achieved may be in the beginning an act of
necessity, which makes agriculturally poor areas for the first
time yield any crop at all. But very soon the new productive
method is generalized and now concerned not only with the
creation and insurance of a single harvest, but also with the
intensification of this harvest and often with its multiplication.
Socially working man has found access to a new 'nature-machine',
whose appropriate application in some circumstances may lead
to a garden-like intensification of agriculture such as areas
given to rainfall agriculture do not know - at least, not to any
considerable degree.

The differentiation of productive powers on the side of nature
develops necessarily, as soon as Man brings it about concretely
in his work-process, a corresponding differentiation of the socially
conditioned means of work, as well as of animal and human working
forces. The more intensive the work-process becomes in the
matter of irrigation, the smaller becomes the soil-surface necessary to the reproduction of the immediate producers and the less
rewarding becomes the use of work-animals. On the other hand,

the more the system of rainfall agriculture extends, the more
necessary become effective working tools in order to mobilize
the powers of the soil, and the more desirable becomes the use
of work-animals (oxen, horses) able to pull these effective
implements over the wide fields. In areas of irrigation agriculture, suitably primitive tools will suffice, with a whole arsenal
of very refined installations for irrigation often supplementing
them. In regions of rainfall agriculture, a more highly developed
plow and plow-animals come completely into the foreground,
while facilities for irrigation are naturally lacking.

The peculiarity of irrigation-economy expresses itself indeed
not only in the special formation of tools and such items. Here,
since the success and failure of production depend in the highest
degree upon the carefulness of the laborer, a specific type of
labor force turns out to be essentially unsuitable: the full
slave, without property or family. We have already treated this
point in earlier systematic analysis, and further studies in the
field of oriental social economy have confirmed us in this view.
There were, to be sure, house-slaves in abundance in the zone
of irrigation agriculture, but the slave played an altogether
peripheral role in the process of agricultural production. In so
far as he entered it at all, he appeared always, as in handicrafts, in circumstances which mitigated his status as a slave
and offered him real incentives to qualified activity. The different
position of slaves in late Rome and the 'Orient' is thus explained
by its material basis. And at the same time the peculiar development of late Rome is similarly explained. As the slave system
decayed in consequence of the lack of opportunity to import
unfree persons from the border territories, there resulted a
reaction on the natural economy of Western Rome - i.e. on the
sector of the state whose agrarian economy was centered about
slavery. Eastern Rome, which likewise was affected, managed
to maintain itself more or less on a higher economic stage by
the production of simple merchandise, and by towns and trade
capitalism. For the chief areas of the East - Egypt, Syria and
Asia Minor - rely essentially on artificial irrigation. The immediate
producers were not as a rule slaves, but peasants of the most
diverse types.

*The general pattern of waterworks as a second specifying
factor*
The presence or absence of artificial irrigation lends a very
definite coloring to the productive powers of agriculture. Yet
the qualitative factor alone is not enough. The irrigation water
may be of purely local origin. Then the reservoirs may be
either individually or locally managed, and no further need for
supplementary labor arises. But perhaps the water must be
tamed on a larger scale. Streams are to be dyked, reservoirs
are to be built, canals are to be dug. Then the great task of
waterworks arises, with which at this stage neither the techniques of individuals nor those of local groups are able to cope.

In this case the task of water-regulation must be carried out
socially, either through a state already established by some
other means or through separate groups which unite and become independent through these and other such tasks and so,
attaining economic and political power, set themselves up as
states.

Another factor enters into this: time keeping. The farming
year is throughout determined by the rhythm of the seasons,
and a knowledge of this rhythm is everywhere important. But
only when the beginning and ending of the rains, the rise and
fall of the rivers, becomes vitally important, does the need for
a relatively precise calendar arise as an immanent problem of
the production-process. In general, where a leading group
concentrates in its hands the control of the waters, we find this
group and the state which stands behind it as well, in direct
or indirect control of astronomy. Thus two new types of productive forces are found which give to the state functions which
in other agrarian societies it did not have to fulfill. In the basic
strata of an extensive rainfall agriculture, there are only certain
military, juridical, and religious functions which are carried out
by special ruling groups. The state growing out of this includes
a multitude of individual landlords, who finally set up a feudal
and hierarchical order, but only in a loose and federal manner.
No central economic function gives the state such a decisive
preponderance that it can suppress the rebellious feudal lords
and make itself absolute in its position. Even if, at the beginning
of capitalistic development, the absolute monarch raises himself
above the bourgeoisie and the feudal lords, playing the one
against the other, he still cannot completely sweep away the old
lords of the manor and divert the surplus of the peasant production which hitherto flowed into the granaries of the feudal lords,
into his own treasury.

It is quite otherwise in an irrigation society, in which the
control of the waters is carried on socially - i.e. at this stage,
by the state. The bureaucratic-priestly-military hierarchy which
appears as the state is either from the beginning the sole master
of the peasants' fate - in other words, of the peasants' production
and its fruits - or it becomes so in the end, with the growing
significance of its double economic function (waterworks plus
astronomy), by conquering, expropriating, transforming, replacing a feudal landed aristocracy which possibly existed in the
economic hinterland of the rivers.

We bring together the results of our considerations so far in
Table 2, which indicates how the various types of productive-
power structure lead to different patterns of social-political
type. We add Variant III (the society with a slave economy)
to Type II (feudalism); they only show higher features than I
and II in one aspect, that of local work-organization, while
otherwise they show an extremely low development of material
and personal productive powers. The other two types are of
universal scope; the society with a slave economy, on the other

Table 2

Powers of production				Methods of production			
Agriculture				**I**	**II a**	**II b**	**III**
The work Medium	nature		Fruitfulness of the soil	+	+	+	+
			Water for irrigation (no water except rain)	+	+	−	−
	Conditions set up by society		Apparatus for irrigation	+	+	−	−
			Tools and work-animals	−	+	+	+
Labor	Qualification		Eventually in two stages	+	+	+	−
	Organization	Local		+	+	+	+
		Territorial		+	−	−	−
Astronomy				(+)	−	−	−
Essential social conditions				Village-bound or free peasants. Absolute sovereign and bureaucracy (scribes, priests, administrative officials, officers)	Serfs, feudal lords		Slaves, slave-holders
Character of the state				Centralized absolute bureaucratic state	De-centralized feudal state		Slave-holding state
General designation				'The Orient'	'Oriental' Feudalism	'True' Feudalism	Antiquity

hand, has found but one full expression, in the Graeco-Roman
world. They rely upon rain-agriculture as does the feudal
agrarian order, out of whose seed-forms they probably grew in
particular historical circumstances. A perhaps unique situation
in the Mediterranean permitted the concentration of a patriarchal
agrarian predatory state, to which the enfeebled condition of
the oriental states offered a rich booty. As soon as the state
had reached the limits possible to its communications and
military techniques, the productive structure, which was only
able to reproduce itself by the continual addition of cheap slave
labor from the border regions, quickly decayed. The extraordinary world-historical significance which the culture of this
epoch has gained for the history of mankind entitles us, in
spite of the uniqueness of Type III, to place it beside Forms I
and II, by whose side it always goes as a theoretically interesting variant.

A specific totality of essential productive powers is gathered
together in a specific manner of production, which corresponds
to a specific social condition and an appropriate political order.
We designate Type I as oriental society, not simply because
it appeared exclusively in the Orient, but because it appeared
there in its most powerful form and because the word 'Orient'
recalls specific circumstances of soil and, above all, climate,
which were indeed of decisive significance for the genesis of
this social-economic formation.

Variant II a, which developed most spectacularly in Japan,
may be counted, by reason of its basis in irrigation, as an
orientally colored form of the feudal society, a fact which can
be expressed terminologically by the designation of oriental
feudalism. Following historical traditions, we designate the
means of production in Type I as Asiatic means of production,
the circumstances of production growing out of them as oriental
society, and the state which expresses this society as oriental
absolutism or - in order to emphasize the peculiar strength of
this type of state - as oriental despotism.

With the establishment of these terms we have concluded the
first section of our analysis, which lays bare the essential
structure of the social organism we are considering. The phenomenon shows a number of features which do not appear to accord
with the basic structure. The continuation of the analysis has
to determine whether these apparent contradictions are only of
a subjective kind or whether they express objective contradictions which may be scientifically explained.

The problem of social and state order
1   *The simple form* The first question that should be answered
concerns the character of the ruling class. Economically we see
the same stratum of the population which dominates either over
the totality or at all events over the decisive mass of the most
important media of production (land and water). The mass of the
national surplus product accordingly falls into their hands. Is

there such a class in oriental society, and what is it like?
In order to clarify the basic relations, we must start with a
purely oriental society. Apart from the early forms of Indian
society, the Inca state offers a particularly clear picture for
this. The peasantry, organized into clans (Ayllus), reproduced
their own existence by means of collectively regulated agriculture. The surplus production of this very strictly controlled
commandeered labor went to the state, which applied it both
to the reproduction of the material machinery of the state and
to the maintenance of the court, the administrative officers,
priests, and the military – i.e. officialdom in its diverse categories. The situation here is completely transparent. The
sovereign and his bureaucracy exercise material control over
the totality of the cultivated lands, and consequently the surplus
production engendered by these lands falls to the ruling class
organized as the state. A peculiar form of the relationship
between ruling class and state-bureaucracy-plus-court emerges:
they are identical. The natural revenue of the land-bound
peasantry's labor is paid directly to the ruling bureaucracy.
The revenue here is like a tax. A primitive class-arrangement
of a classical transparency obtains. The kin group persists
and cultivates the soil in partnership, just as the old Indian
community, which has partially continued into modern times,
knew collective ways of tilling the communal fields.
2 *The developed oriental society* As soon as one has acquainted
oneself with the specific peculiarity of the oriental state (the
ruling class = the sovereign and the bureaucracy), the problem
seems to offer no essential difficulty. In fact, things are not
that simple. In few areas of social science is the condition of
theory so confused as here. Certain phenomena of the oriental
world, which in their visible form during the eighteenth and
nineteenth centuries and in the accounts of oriental historical
science appeared just as importantly meaningful as the picture
we have just drawn, are the cause of this. Private property
has extended to the countryside. The village community has
decayed in many cases, and new social classes have arisen;
apart from the private land-owners of bureaucratic origin,
above all the little and big merchants who as profit-makers and
in part even as land-owners bring completely new features into
the picture which at first was so simple. Can theory explain
this change, or is not oriental society rather dissolved by it
and transformed into another kind of social formation?
A great part of the prevailing confusion can be explained by
the fact that the original conception of oriental society occurred
in a period – before and shortly after the middle of the nineteenth century – when neither the social history of Egypt and
Babylon nor that of India and China had been handled by modern
means. Meanwhile, in the last decade, the documents of the
great early Asiatic cultures have been deciphered in great
quantity and applied to the unveiling of every land's social
history. Especially great steps have lately been made in research

into the social history of China. China's 1000-year-old historiography transmitted to the present age a uniquely rich and
coherent material for history, which now will be made available
with the methods of critical analysis to the social scientists of
China and, one may hope, to those of the rest of the world. If
one adds the results of the researches carried on in the last
decade to the realm of theoretical consideration, it is shown
that through the new facts the conception of oriental society
is not shaken or destroyed, but that it is, on the contrary,
deepened, strengthened, and made more specific.

The disintegration of the simple oriental society may take place
from two sides, but basically the dual phenomenon may be reduced to one basic cause. The powers of production grow. Metal
implements for agriculture (above all, iron agricultural tools),
as well as the development of the technique of irrigation and,
in addition, in a certain degree the application of draft animals,
make the results of tillage much more productive and, at the
same time, make individual farming more profitable than the old
forms of communal cultivation. Artisanship and also certain
branches of industry grow at the same time, together with the
greater industrial needs of agriculture and the state-economy.
The 'free' individual peasant who buys and sells on the market,
arises, so that in the end everything, including his individual
land-property, can become merchandise.

On the other side, and in varying degrees, the merchant
arises as the agent of the growing circulation. Trade - and some
industry - and even money-lending capital is to develop out of
this, to some degree in Egypt, to a high degree in Mesopotamia,
and quite highly in India and especially China. They attract to
themselves a part of the revenue in the form of profit. Private
property in land carries with it a further consequence: If the
farmer is free to sell his land, then others are free to acquire it.
Large private landholdings appear, occasionally in the hands of
the new commercial capitalists, often in the hands of the members
of the old upper class, who can obtain the ownership of land not
only by purchase but also by gifts from the ruler or by retaining service-land. The social-economic order has become much
more flexible. But are its bases thereby abolished?

In Pharaonic Egypt, the influence of commercial capital appears
to have been particularly slight. Also in Babylon trade and
money-lending remained in the hands of the ruling classes,
namely those of the priesthood, the 'temple'. The phenomenon of
a land-owning merchant class appears much more extremely in
China. In all cases, however, the mass of the free peasants
remains a 'public sector' paying taxes and service to the state
and its bureaucracy. The key economic positions remain therefore in the hands of the office-holding bureaucracy which controls
the greatest part of the surplus work and produce of the farmer.
Whatever the new legal forms may be - there may be many - the
bureaucracy remains by far the greatest land-owner, bringing
together in its own hands the greatest proportion of the surplus

production and the forced labor. The levies on trade- and
money-lending-capital - which by the way may take very diverse
forms - and on the private great land-owners, who pay less
taxes as they grow more powerful, weaken the central power,
complicate and contradict the action of the oriental despotism,
though they do not abolish its specific character.

At the end of the development, we therefore find dissolved
and still-undissolved village communities, partly unified in a
social complex as in India. We find the power of disposition over
the farm work and farm production in the hands of the state
bureaucracy, in the hands of private officials and officers, of
priests (temples), and also in certain circumstances of great
merchants. The simple oriental society has, in consequence of
the development of the basic powers of production, transformed
itself into the developed oriental society.

3   *Military and civil functions of the dominant bureaucracy*
Heretofore we have described the upper classes in the simple
and the developed oriental society as if they were at the same
time military and civil. This description is not wrong, but it
needs more exact definition. We see before us such societies
as Babylon and China, where apparently administrative officials
and priests or officials with priestly functions stand at the peak
of the social pyramid, while in India the priesthood had to share
its power with the warrior caste. In Assyria, the military element even steps clearly into the foreground. Does not this change
in the type of the ruling class indicate at the same time a difference in the social-economic substance?

The difference we have noticed is indeed significant, but not
in the way an outsider might expect. An officer finds himself
economically on the same level as a priest or an administrative
official. Both social categories live on revenue in kind or money,
which the state has taken from the peasants in the form of taxes,
especially the land-tax, and which is distributed, via the state-
treasury, to its bureaucracy. The officer, therefore, although
professionally resembling the feudal knight is economically far
from him, but close to the civil official. His relationship to the
mass of the farm population is, in the case of the Orient, indirect, impersonal and temporary, while in the case of the
knights it bears a direct, personal, concrete, and permanent
character.

The difference in the two groups of officials, therefore, does
not lie in their economic positions, but in the significance of
their day-to-day functions. This is conditioned in a far-reaching
way by the international situation, in which a given oriental
complex exists. If the complex in question is surrounded by
powerful states, then the external military functions are of first
importance beside the internal political repressive and economic
tasks (police, justice, public works). The civil bureaucracy
does not disappear. This would be impossible, since it forms an
integrating part of the social organism. But in its activity in
some circumstances it will not only be supplemented but even

overshadowed by the functions of the military bureaucracy, in
the event that the latter must fulfill protective tasks of great
importance and duration. In Assyria this was apparently almost
always the case, while at the same time the typically 'oriental'
features (waterworks, administration, etc.) were relatively
limited. In India several great waterworks-complexes were
centered around a number of isolated river-systems, so that the
antagonistic political conditions and the corresponding large-
scale military tasks always arose anew. A complete victory of
the military or secular elements, which found a potent weapon
in Buddha's anti-Brahman teaching, thus became impossible.
Even in the realm of Asoka, Buddhism, which had reached the
height of its power, could still not fully displace the ideas of
the Brahman priestly caste. Finally Buddhism was completely
defeated, and an interesting balance of power between the priesthood and the warrior caste, the Brahmans and the Kshatriyas,
came about - a social balance which the functional duality between
the two upper groups of the Indian oriental upper class plainly
expressed.

China's social structure accommodates itself without difficulty
to the principles we have indicated. A variant of situation
A 1 a 1 (too much summer-water in the periodically rising loess-
rivers, insufficient rain at the right time) enforced the adoption
of local irrigation, but above all water-control measures on the
grand scale. In the time of Confucius (about 500 BC) the management of waterworks had already taken on the character of a long-
perfected, legendarily revealed, heroic accomplishment.

The first epoch of unified empire (221 BC to AD 220) left
many military tasks undone. It established the civil element and
its particular ideology, Confucianism, in a solid and program-
matical manner. Only Central and South China were actually not
assimilated. The push into the southern areas disrupted the
temporarily unified state order. In the Period of Disruption
(221-589) we find military accomplishments again attaining high
social prestige. The T'ang era (618-906) mourned the decline of
the knightly customs. What had happened? After the building
of the Grand Canal under the Sui dynasty (589-618), which, by
means of an artificial Nile, combined China's two great stream-
valleys in an economic-political unity, after the establishment of
the examination system, after the reorganization of all-important
branches of political life, the civilian element ruled supreme in
the reunified country. The literary oriental officialdom (and
Confucianism) pushed the warrior officials (and Buddhism) into
the background. Conqueror nations such as the Mongols and the
Manchus were able to proclaim once more temporarily the supremacy
of the military virtues. In vain. The world 'within the four
oceans', Greater China, had grown together into a rather loose
but dynamic unity. Even in the shadow of the conqueror peoples
the Chinese representatives of oriental bureaucracy, the
Confucian officials, occupied the supreme positions in the Chinese
hierarchy of social prestige.

4   *Agrarian and state crises in oriental society* In a simple
oriental society agrarian crises are mainly the result of too much
or too little water. Large-scale catastrophes of flood or drought,
endangering the physical existence of great populated regions,
drive the victims of flood and hunger to mass emigration and
uproar. The crisis in such a case still bears an immediate natural
character. At this stage the subjective factor of a good or bad
government – maintenance or disrepair of the waterworks –
plays an especially great role for the economy of the country.

The mechanism of crisis becomes very involved if oriental
society with its productive powers develops its property-based
conflicts. This can occur essentially in two ways, according to
the two basic forms of the developed oriental society. The agricultural production may be closely concentrated about the life-
giving irrigation system; or this system may bear a loose, un-
even, dispersed character. The second type is very clearly
represented by China, which in this sense may serve as an
example of a specific variant of oriental society (as its B-form).
Inner crises occur in both cases, but in case B they take on
especially violent forms.

The development of productive powers allows generally an
extension of the basic complex over a larger region. With the
extension of the dominant economic area, the ruling bureaucracy
grows in number and function. In case the original area retains
its political superiority, it can successfully prevent the growth
of outside rival groups. In Pharaonic Egypt the merchant class
is insignificant. India's trade develops more strongly than the
Egyptian, but even here the bureaucratic central-power blocks
each threatening development of the merchants and money-
lenders. The weakening results accordingly, not from the outside,
but from an internal conflict of the ruling bureaucracy itself.

The origin of the internal conflict can be of two kinds. The
machinery of the state may more or less effectively prevent the
steady acquisition of peasant lands by centrifugal elements of
the bureaucracy; but, since the means of communication often
are imperfectly developed, it cannot prevent – in consequence
of the great territorial extension of the developed oriental
state – the relative independence of its territorial organs (satraps,
governors, tax-controllers, and district officials). With the
increasing appropriation of the locally collected land tax by the
local bureaucracy, the power of the local officialdom increases.
The consequence is either a shrinkage in income of the central
power or, if this income remains constant, an intensified oppression of the peasant masses, from which the additional local
appropriations must be extracted by added local pressure. In
the first case, the possibility of financing public works is
materially damaged. In the second case the excessive abuse of
the peasantry by the 'grinding bureaucracy' – excessive from
the standpoint of the traditional average – strikes vitally at the
personal foundation of the public works. In both cases with
the weakening of public works the productivity of agriculture

shrinks and thereby the very foundation of political power
within the state in question. The tendency towards disorder
and uprising grows. In case the neighboring nomadic societies
have then reached a point of relative concentration and high
aggressive striking power, the decadent irrigation state can be
easily overrun by nomadic conquerors. If the outside cause is
lacking, then the cataclysm follows as soon as the internal disintegration is far enough advanced. Either way, however, it is
an inner crisis-mechanism which loosens up the oriental social
structure and prepares it for civil war or for defeat from the
outside.

A variation of the just-described cases is the concentration
of land in the hands of individual members of the ruling bureaucracy, who exert excessive military, priestly, or administrative
power. The temple may get much land and pay little or no taxes.
High local officials acquire land as special gifts or as regular
pay, and this assignment, considered as temporary, may develop
into a permanent possession in consequence of economic-political
conditions. As the income of the central power weakens, it is
unable to fulfill its specific oriental task in the process of agricultural production. At the end of the Old Kingdom in Egypt,
before the invasion of the Hyksos, before the ascent of the
Ammon priesthood, and before the coming of Alexander the
Great, the central power seems to have been weakened in this
fashion.

These cases show a great many varieties, but they all belong
to the same basic type. Every time the crisis develops out of
the conflicts of oriental bureaucracy, which permits beside itself either no or only peripheral economic elements. The B Type,
China, in consequence of the essentially looser form of the basic
irrigation system, has prepared a favorable soil not only for the
training of centrifugal elements in the ruling bureaucracy but
also for the growth of influential non-bureaucratic elements.
A higher mobilization of land-ownership was technically possible,
was economically desirable from the standpoint of the tax-levying
bureaucracy, and was therefore officially established. The function of a commercial bourgeoisie could only be exercised partially
and imperfectly by the state bureaucracy. The cycle therefore
had to assume a particularly complicated form.

The 'public sector' (formed by the free peasants) at the beginning of a new epoch surpasses the 'private sector' (great land-
holders and tenants), since the collapse of the old regime destroyed countless peasants and many landlords and laid waste
large stretches of farming land. A farm policy which distributes
land, seed, and tools and which keep taxes low brings the land-
less elements back to the villages. The peasantry grows. The
crop grows. The taxes grow. The power of the state grows,
which increases its initial efforts to build canals and dykes.
However, the upswing carries within it the seed of its own downfall. With the growing well-being of the villages and the state,
the profit of the merchants and the income of the officials also

grow. The accumulated liquid capital tends to change into an immobile form, into land. The private sector expands and therewith that portion of the peasantry which is fully taxed. The active or retired land-owning officials and the mighty merchant landlords pay little tax. With growing extension of the private sector the public income decreases. The effectiveness of the state to carry out its economic functions is decreased. It gives less to the farmers but takes more from them. Taxes rise and with them - circulus vitiosus - the tendency of the free farmers to flee from the open into the private sector, i.e. to transfer their land to a private land-owner (possibly also a monastery) and in this way to escape from the tax-collector. At the same time with the decay of the waterworks, the pressure rises even in the private sector. Peasant uprisings, initially local, take on larger dimensions. The internally weakened regime, whose centrifugal elements form a destructive block of interests with the merchant land-owners and eventually even simple land-owners, becomes ever less secure in its power, more cynical in its morals, more cruel in its taxation. An uprising, led by uncompromising elements of the old upper class, or a nomadic invasion, or a combination of both, accomplishes what the weakness of the economic sphere has thoroughly prepared: the fall of the dynasty, on whose ruins the leader of the victorious movement newly establishes himself with a reform program, as the absolute chief of a rejuvenated, though basically unaltered, oriental despotism.

5   *Why no independent development towards industrial capitalism?* The discovery of a crisis-mechanism, which runs in the form of a vicious circle, involves another principle which shows only a particular side of the same process: the agrarian and political crises moved in a vicious circle and not in a steadily increasing spiral. The social-economic system reproduces itself instead of developing. Why?

The Orient has applied a specific type of nature- and culture-conditioned productive power (irrigation-water, facilities for irrigation, intensive manual labor) which played little or no role in the agriculture of the West. Thus in the oriental zone a higher productivity of agricultural work is reached than in the West. This result is based on public works of great magnitude. While in the West managerial and intellectual centers of agrarian order were not needed, in the Orient they become a conditio sine qua non for the simple reproduction of the agricultural life. The political center in the Orient becomes economically more important and politically more powerful because - in contrast to the feudal world - in one way or another it has a direct bearing on the activities of the peasants. The oriental state and its representatives thus arrogate to themselves the bulk of the land revenue, whereas Western (and Japanese) absolutism conceded a great part of the land rent to the feudal lords.

In the West the transition to capitalism occurred in cities, which knew how to make themselves independent by a chain of more or

less political aggressive movements. The surplus from handicraft and commercial activity remained in the hands of the
bourgeois class, under whose control it grew. The accumulation
of capital was thus possible politically as well as economically;
the surplus was not taxed away, confiscated, or taken away as
a pseudo-loan. The industrial productive powers could be
developed as well, without negatively touching the agricultural
hinterland, in whose urban centers the new development took
place; and the absolute central power had to encourage industrial
progress all the more eagerly, since here, in contrast to the
feudally controlled villages, it expected additional income and
power for the court and its administration.

Quite different in the 'Orient'. The city centers here dominated
the economic reproduction either of decisive parts or of the whole
agricultural life. Anyone who, for the purpose of gaining bourgeois freedom, drove the central power from the cities, upset
the nerve centers of the agricultural hinterland. Obviously then,
the position of the oriental administrative city is a different one,
not only in the sense of the immediate production techniques.
It is so likewise in regard to the distribution of surplus products.
Occupying the social-economic center and either destroying or
decisively limiting the existence of feudal revenue-getters,
oriental absolutism gains supreme power also over the city's
production. The wretched artisans of the Orient and the proud
and independent artisans of the West are the product of different developments, in the course of which one side is kept at a
bare minimum by taxes, etc., while the other side learns to
defend its interests in powerful guild cities.

The development of Western industrial capitalism accordingly
occurred because there a peculiarly decentralized structure of
the agricultural productive power permitted - economically,
sociologically and politically - the commencement of capital-
accumulation, while the centralized structure of the highly productive oriental agrarian order worked in the opposite direction,
namely toward the reproduction of the existing order, toward
its stagnation.

Despite any cyclic collapses, Asiatic society, in so far as it
was not physically destroyed from outside, could therefore re-
establish itself in principle after the worst disaster. The classical
type of a society which tenaciously reproduces itself, a stationary society, is created.

NOTE

* From M. Fried, ed., 'Readings in Anthropology,' New York,
Crowell, 1968, vol. 2, pp. 180-98.

WITTFOGEL ON THE ASIATIC MODE OF PRODUCTION*

Gunter Lewin

To evaluate correctly Wittfogel's views on the Asiatic mode of production, one must know which factors, in his opinion, have a controlling influence on the character of a particular society.

Among the factors that can influence the development of the human community, Marx includes 'external, climatic, geographic, physical, etc. conditions' (1857-8a, p. 472) which the primitive community ('Stammgemeinschaft') finds present when it settles down, as well as its 'tribal character'. These special conditions only influence, but do not determine the particular features of social development, as Wittfogel would have it in the first section of his work (1932, pp. 469-72, 594, 730). It is true that he occasionally quotes opposing views by Marx and Engels; for example, the following passage from 'The German Ideology': '[Feuerbach] does not see that the sensuous world around him is not a thing given direct from all eternity, remaining ever the same, but the product of industry and of the state of society' (Marx, and Engels, 1976, vol. V, p. 39). But this does not prevent Wittfogel from further developing his pseudo-Marxist ideas. For further on, he writes (1932, p. 483):

Not every transformation of natural substance into something of value for man leads to the release of new natural forces. . . . But work alone is not enough. Labour must preserve the possibility of a corresponding novel application. Where this is not the case the productive process is marking time.

Wittfogel furthermore states (ibid., p. 713) that necessity and outward usefulness determine the character of co-operative labour. From the need grow the activity and the will to evolve ever new productive forces. 'Outward utility', the *particular characteristic of the natural surroundings present* [emphasis G. Lewin], decides the outcome of this activity. This is also true for the capitalist epoch. Of course, co-operative activity always moves continuously in a specific direction; but this direction results from the objective or in the last analysis, the natural basis, of its function. . . . So the release of the primary or secondary properties of wool,

iron, wood, coal, falling water, electricity, petroleum and
rubber has each time altered the direction of man's cooperative
labour.

This perspective is quite obviously absurd since it depends
on the 'activity of man working in co-operation', whether or
not 'primary or secondary properties of natural materials' are
'realized'. Even before the appearance of man, water, for example,
had the objective capacity of quenching thirst, watering fields,
driving mills, and much more. Only when man 'realizes' these
productive forces, can they contribute to furthering human
development. The manner in which man acts on nature, whose
chemical and physical properties remain the same throughout
history, is decided by the development of productive forces
and the relations of production and their interrelationship.

Marx and Engels never spoke of any 'decisive' role of natural
conditions for human social development, a form of geographic
determinism. One must obviously take historical differences
into account. The more primitive the stage of human development,
the greater the influence of the existing natural conditions on
man and his social organization. Nevertheless, even in the production process of the communal mode of production, the action
of man on existing nature and the resultant change in the
natural surroundings are decisive.

C. Parain demonstrates very effectively, with the example
of Quémé Valley culture in Dahomey, how similar natural conditions in no way lead to similar forms of social organization. In
terms of nature, this valley offers irrigation possibilities not
unlike those of the Nile Valley. Instead of an irrigation economy,
here we find 'a striking example of nonevolution, whereas a
similar natural environment in Egypt produced the most fertile
development thanks to the creation of a centralized and despotic
power' (1966, p. 43).

In Wittfogel's view what then determines the development of
human society? We have already seen that for Wittfogel the
'external, climatic, geographic, physical, etc. conditions' of
the natural environment control the nature of the human
community living within it; this determining influence persists
almost perpetually and determines the fate of a people throughout their existence. Parain's example of the Quémé Valley
demonstrating a case of devolution in social development proves
the error of Wittfogel's position. As we shall see, Wittfogel
conceptualizes this determinism as operative even in capitalist
society, indeed, even in socialist society.

Using the example of the development of French industry
since the revolution of 1789, Wittfogel tries to prove that the
causes of France's general economic backwardness compared
with Great Britain and Germany are to be found in its inadequate
stocks of raw materials, its lack of coal and iron, i.e. in its
natural resources. With a totally unjustifiable reference to Marx,
Wittfogel declares that 'without an analysis of these resources
and without deriving the development from these resources in

these countries, one cannot scientifically demonstrate the
applicability of theoretical principles to the historical process
of industrialization' (1932, p. 471). The untenability of this
perspective is especially clear if we look at the example of
Japan, which has become one of the leading industrial countries,
but which lacks any significant raw material resources. China,
on the other hand, a country rich in raw materials, lagged
behind both economically and politically until 1949.

In his consideration of the development of France, Wittfogel
clearly confuses two concepts: social order and economic structure. No doubt the latter is to a certain extent influenced by
natural factors. Thus, the German Democratic Republic cannot
produce sufficient quantities of pit-coal and produces no
coffee at all. It must therefore concentrate its economy on areas
which offer the best possibilities. This fact, however, has
nothing whatever to do with the social order. On the other
hand, a country can only make use of certain natural resources
to the extent which its social development permits. Man can free
himself from the restrictions of 'natural resources' the higher
he climbs the ladder of social development. Automation, the
synthetic production of basic materials, the mastery of space,
are problems which can only finally be solved by classless
society. There are inevitably barriers to solving them within
a class society.

Wittfogel refuses to contemplate these sorts of ideas and therefore renounces the possibility of discerning the direction of
human progress and deriving any practical application from it
for himself.

While Marx correctly emphasized the effect of human labour
on dead nature, Wittfogel makes human labour absolutely
dependent on the originally existing natural resources. This
false interpretation of the materialist view of history is the
starting point for all his observations on economic history and
an interpretation which he has retained up to now, with a few
minor alterations to which we will turn later. In his argument,
natural conditions play the decisive role. Despite the fact that
he occasionally admits that 'nature' is not immutable, but subject to alteration by man, nevertheless this realization has had
no influence upon his basic viewpoint. He persistently maintains that the artificial irrigation essential to the economy,
given certain natural conditions, gives 'Asiatic' types of human
society a static character, with minor and very exceptional
variations. A notable feature of Wittfogel, the fanatical anti-
communist, is that he is still essentially able to build upon the
work of Wittfogel, the one-time Marxist.

Such incorrect premises inevitably have had a negative influence on the value of his work in Chinese economic history. He
prefaces his research into this subject (1931, p. 7) with an
allusion to the Marxist classics:

According to a proposition by the founders of Marxism that
we have carried out, 'all historical writing', *a fortiori* all

economic and social history writing, must set out from the
'natural bases and their modification in the course of history
through the action of men' [Wittfogel's source is *The German
Ideology* (Marx and Engels, 1975, vol. V, p. 31).] Now, of
course, this proposition in its general form is far too abstract
to be of any practical use. Therefore even Marx and Engels
made it more precise. According to them, the 'natural resources'
of all history fall into two sets. First is 'the physical constitution of man himself', a concept which is especially aimed at
the racial factor [G. Lewin], as observations from other
quarters have demonstrated. Second, however, that formula
means *those natural conditions first found by man: the geo-
logical, oro-hydrographical, climatic and other conditions!*
[For Wittfogel's source, see above - Lewin.] A presentation
of *Chinese* socio-economic development, in so far as it lays
claim to methodical correctness according to the *Marx*ist
school, must also take its starting point from both these
named factors.
We have quoted this passage not just because it sheds light on
Wittfogel's views, but also because we wished to use it to demonstrate his methods when using quotations from classical authors.
    The complete reference from Marx from which Wittfogel quotes
reads as follows (Marx and Engels, 1976, vol. V, p. 31):
    The first premise of all human history is, of course, the
    existence of living human individuals. Thus the first fact to
    be established is the physical organization of these individuals
    and their consequent relation to the rest of nature. Of course,
    we cannot here go either into *the actual physical nature of
    man, or into the natural conditions in which man finds himself -
    the geological, oro-hydrographical, climatic and so on. All
    historical writing must set out from these natural bases and
    their modification in the course of history through the action
    of men.*
    Men can be distinguished from animals by consciousness, by
    religion or anything else you like. They themselves begin to
    distinguish themselves from animals as soon as they begin
    to *produce* [emphasis Marx] their means of subsistence, a
    step conditioned by their physical organization. By producing
    their means of subsistence men are indirectly producing their
    material life.
Wittfogel, by omitting the second part of the quotation, the
decisive role of human productive activity, shifts the emphasis
to the meaning he desires.
    Marx and Engels here neither draw upon the concept of 'race',
nor do they claim that natural conditions determine the fate of
human society. On the contrary, here they emphasize the
significance of material production. They establish that humans
only distinguish themselves from animals by their ability to
produce their means of subsistance and that they are in a position
to do this by virtue of their 'physical organization', i.e. by
their physical and psychological development. Wittfogel, with

the assistance of classical quotations arbitrarily taken out of
context and mixed with his own commentary, knew how to give
the impression that his opinions agreed with those of Marx and
Engels. Obviously, however, this is not the case.
Let us return to Wittfogel's statements. After explaining that
'any socio-economic examination of these [Asiatic - Lewin] agrarian
societies, which does not adopt the essential Marxist concept of
the Asiatic mode of production . . . is doomed to fail' (1931,
p. viii), he attempts to impose the significance of natural conditions on the Asiatic mode of production as well. He puts up the
following two prerequisites for the evolution of specifically
'Asiatic' forms of society:
1   Agriculture is based on artificial irrigation.
2   An enlargement of irrigation works is achieved communally,
requiring the intervention of government in the form of 'public
works', i.e. it cannot be carried out with volunteers (1931,
p. 87). The two factors that dominate Wittfogel's thinking are
the special natural conditions demanding extensive irrigation
systems, and the necessary intervention by a despotic government to achieve them. Both factors are alien to Marxist theory.
Since Wittfogel, even in his early work, clung grimly to this
way of thinking, research into the conditions of production
always took second place for him. Therefore he never worked
toward a genuine class analysis of the society he examined and
came nowhere near a Marxist analysis of the different stages
through which Chinese society had passed. Even allowing for
the fact that he intended to follow up 'Wirtschaft und Gesellschaft
Chinas' with a second volume about conditions of production,
he is still to be reproached for not having gone deeper into
this subject in his works of this period. The significance of
production is always secondary to 'natural factors' for him. This
leads him to approach the Chinese question in a totally ahistorical
way, even when he occasionally makes the attempt to periodize
the stages of development of Chinese society. To a great extent,
he disregards the laws of movement of social development and
is therefore also unable to explain which evolutionary principles
underlie his phases of Chinese social development. This inevitably takes him into the antithesis of Marxism.

BIBLIOGRAPHY

Marx, K., 1973, 'Grundrisse: Foundations of the Critique of
    Political Economy', Harmondsworth, Penguin, 1973.
Marx, K. and Engels, F., 1975, 'Collected Works' vol. 5, London,
    Lawrence & Wishart.
Parain, C., 1966, Protohistoire méditerranéenne et le mode de
    production asiatique, 'La Pensée', 127, pp. 24-43.
Wittfogel, K., 1932, Die naturlischen ursachen der Wirtschafts-
    geschichte, 'Archiv für Sozialwissenschaft und Sozialpolitik',
    67, pp. 466-91, 579-609, 711-31.

Wittfogel, K., 1931, 'Wirtschaft und Gesellschaft Chinas',
Part I: 'Produktivkräfte, Produktions und Zirkulationsprozess',
Leipzig.

NOTE

* From Von der asiatischen Produktionsweise zur 'hydraulic
society': der Werdegang eines Renegaten, 'Jahrbuch für
Wirtschaftsgeschichte' (Berlin), 4, 1967, pp. 205-8. Translated by J. Gordon-Kerr.

WITTFOGEL'S 'ORIENTAL DESPOTISM'*

Arnold Toynbee

This is a queer book** by a fine scholar. Professor Wittfogel, in his preliminary acknowledgments, tells us, as every author should do, what was the genesis of his book. His belief in human values caused him (to his great personal credit) to be interned in Hitler's concentration camps. 'My final thoughts,' he records, 'go to those who, like myself, were passing through that inferno of total terror.' Some of his fellow victims asked him, 'if ever opportunity offered, to explain to all who would listen the inhumanity of totalitarian rule in any form. Over the years and more than I can express, these men have inspired my search for a deeper understanding of the nature of total power.'

Here indeed is a strong incentive to research and writing. But what result would one expect Professor Wittfogel's experience to produce? Surely a full-dress study of the National Socialist regime in Germany, and this for at least three reasons. First, Professor Wittfogel has had direct experience of this regime; and even the most penetrating and imaginative scholar understands better something that he has been through himself than something that he has been able to study from documents only - as Professor Wittfogel has studied the regime of the Liao (Khitan) dynasty in medieval China. Secondly, the Nazi regime is the example of total power that is most easily comprehensible to Western writers and readers, because Hitler and his National Socialist movement are entirely home-grown products of our Western civilization. There is nothing 'oriental' or 'hydraulic' about Nazi Germany (I will explain in a moment what Professor Wittfogel means by his rather quaint jargon). A third reason for beginning with the Nazi regime in the contemporary Western world if one is going to make a serious study of total power is that it is the best (i.e. the worst) example of this atrocity so far known to us. Search all the surviving records of human behaviour all over the globe since the beginning of recorded history, and you will find nothing to equal this. The Nazis' cold-blooded murder of millions of people, their tortures, their espionage,

their devilish policy of inciting children to denounce their
parents - these leave the Assyrians and the Aztecs dead beat.
So one would expect Professor Wittfogel to begin with the
Nazis, and to take them as the standard for his comparative
study. But not a bit of it. After his illuminating reference in
his acknowledgments, he hardly mentions the Nazis again.
And, when he does mention total power in our Western world or
in the Graeco-Roman world, it is always in general terms and
nearly always in the form of an apologia for it: it was not really
so bad after all, or at any rate it has never lasted very long.
The wielders of total power whom Professor Wittfogel is stalking
in this book are not our Western Nazis; they are the Russian
communists; and I have a most uncomfortable suspicion that the
book - beneath its load of authentic and massive learning - is
really a political book and not a scientific one. At any rate,
what it all leads up to is a common-form contemporary Western
indictment of the communist regime in the Soviet Union. Of
course, what is trite may nevertheless be true, and, from the
standpoint of the minority in the Western world that believes
in, and tries to practise, what we consider to be genuine
democracy, Russian communism is about the next-best target
to German National Socialism for a democratic Western attack on
totalitarianism. Then why does not Professor Wittfogel attack
the Russian communist regime direct, on the basis of its, to our
minds, glaring intrinsic demerits? The way is open, but he does
not take it. Instead, he gets at the Russian communists by the
roundabout route of an attack on what he calls 'oriental or
Asiatic hydraulic agromanagerial despotism'.
Here I have three bones to pick with him. First, he is doing
mankind a serious disservice in trying to resuscitate the propaganda myth - invented by European Greeks in the fifth century
BC - of the antithesis between a good Europe and a bad Asia.
Second, he is flying in the face of the considered opinion of his
colleagues the Egyptologists, Assyriologists, and Indologists in
trying to apply to the magically vicious soil of Asia - though
not to the magically virtuous soil of Europe - the Marxian thesis
that the means of production rigidly determine all other elements
of social life. Third, he is barking up the wrong tree: for,
when one makes even a cursory comparative study of total
power, one finds that the worst cases of it have occurred, not
in 'hydraulic' societies, but in societies in which agriculture
gets its water from the rainfall.
In my own association of ideas, the word 'hydraulic' suggests
drills, brakes, and elevators, but Professor Wittfogel uses it,
rather oddly, to mean 'irrigational'. 'Hydraulic' societies, in his
parlance, are societies in which agriculture gets its water from
the artificial tapping of lake water or river water by irrigation
canals.
But, to get back to my three quarrels with Professor Wittfogel
and, first, his attempt to revive the propaganda myth of good
Europe versus bad Asia. We know all about this smear's origin

and history. It was invented by European Greeks in the fifth
century BC in their very natural animosity against the Persian
Empire, which had made an aggressive, though unsuccessful,
attempt to conquer them. Alexander the Great was brought up
on this myth by Aristotle; but, when he made his own successful conquest of the Persian Empire, he discovered, to his
astonishment, that the myth was a lie. As soon as he met the
Persians, he recognized in them the qualities of character which
had enabled them to give peace and unity to Western Asia for
200 years. His response was to take the Persians into partnership with the Macedonians and the Greeks in the management of
the world. But for his premature death, Alexander would have
carried this policy through, and that would have been the end
of the lying myth. Unhappily, his insight and his policy died
with him, and the myth lived on, unscotched, till it was borrowed
from the Greeks, in recent times, by the modern Western conquerors of the world.

This myth of a magic contrast between European virtue and
Asiatic vice is the ancient Greek and modern Western counterpart of the medieval Byzantine Greek propaganda myth that
the Western Christians were misguided barbarians who were
bound to come to grief very soon. This myth was borrowed from
the Byzantines by the nineteenth-century Russian Slavophils,
and from the Russian Slavophils by the Russian communists. We
know how much it irritates us to be told that Russia alone is
holy and orthodox (whether in Christian or in communist terms)
and that we Westerners are erring and doomed schismatics or
capitalists. It is just as irritating for the peoples who happen to
live within the conventional boundaries of 'Asia' to be told by
us that, as an inescapable result of their geographical location,
they are bound to languish under 'oriental' despotism of the
'hydraulic agromanagerial' type. Away with both these propaganda myths! Both are lies, and both are heinous offences
against the human race, because both of them are engines deliberately designed to stoke up those tribal animosities by which
mankind has made itself so miserable.

I will now take up my third point before my second. I contend
as against Professor Wittfogel, that all the worst cases of total
power and an atrocious abuse of it have occurred in regions where
agriculture depends on rainfall, not in regions where it depends
on irrigation. National Socialist Germany: there was nothing
'hydraulic' about her. Tsarist Russia and her successor communist Russia: there is nothing 'hydraulic' about her either.
Professor Wittfogel cannot deny that in Russia agriculture
depends not on irrigation but on rainfall; but he still claims
the Muscovite form of Russian despotism for his 'hydraulic agromanagerial' type on the ground that the Muscovite despots may
have learnt their trade, not from themselves or from the Byzantines, but from their Tatar overlords, and that the Tatars of the
Golden Horde, at the western end of the great Eurasian steppe,
may have learnt the tricks of the 'hydraulic agromanagerial'

trade from brother Tatars who had ridden off in the opposite direction to conquer 'hydraulic agromanagerial' China. Professor Wittfogel cites no serious evidence in support of this surely very far-fetched piece of speculation; yet it is one of the key theses of his book.

Professor Wittfogel himself bears witness that, though the cultivators in 'hydraulic' oriental societies have no say in the government of their community, they are not slaves either. But plantation slavery has played a big part in mankind's economic history. Where has it been rife? In Roman Italy in the last two centuries BC, and in the West Indies and the Americas since their conquest by West Europeans in modern times. The only Asiatic case that I can call to mind is the cultivation of the Tigris-Euphrates delta by African (Zanj) slave-labour in the ninth century of the Christian Era.

Finally, there is Professor Wittfogel's thesis that in China, India, Iraq, and Egypt, though not in Lombardy or the Netherlands, artificial water-control on a large scale has produced a uniform type of 'agromanagerial' despotism. If the late Professor Frankfort were still alive, he would have a lance to break with Professor Wittfogel over that. The differences between the 'Asiatic' societies whose agriculture is based on irrigation are much greater than their resemblances. I believe there is nothing more in Professor Wittfogel's thesis than the obvious truth that large-scale enterprises cannot be carried out without a unified and effective high command. This is certainly true of large-scale waterworks, but it is also true of armies and ships and industrial enterprises.

Here I will leave Professor Wittfogel to be dealt with by the Assyriologists and the Egyptologists. I seldom write a condemnatory review and dislike writing one of this book - not least because I much admire Professor Wittfogel's pioneer work in the economic, social, and political interpretation of Chinese dynastic histories. But his present book is, in my opinion, something of an aberration and still more of a menace. So I have said frankly what I think about it.

NOTES

* From 'American Political Science Review', 52, 1958, pp. 195-8.
** Karl A. Wittfogel, 'Oriental Despotism: a Comparative Study of Total Power', Oxford University Press/Yale University Press, 1957.

REPLY TO ARNOLD TOYNBEE*

Karl Wittfogel

Professor Arnold Toynbee has written what he himself considers a 'condemnatory' review of my book, 'Oriental Despotism'. I find this entirely consistent since in my study I rejected his approach to the history of man. On this score, and despite his derogatory language, no answer would be required. But his review contains inaccuracies which so completely obscure both the intent and substance of my analysis that a response is in order.

While setting the record straight, I shall try to indicate (which Toynbee did not do, and from his point of view perhaps could not do) certain basic differences that separate his approach and mine. These differences go far to explain why he disregards many of my key ideas altogether and argues against others so as to make them almost unrecognizable.

His way of handling my thesis that large-scale government-managed works of irrigation and flood control played a crucial role in the establishment of specific types of total power ('hydraulic', 'Asiatic', or 'oriental' despotism) exemplifies his method. Branding earlier versions of the concept of oriental despotism as lying myths and labelling my term, hydraulic, as 'quaint', without mentioning my reasons for introducing it, he faces the core thesis only in the next to last paragraph of his review, and then only to dismiss it with the assertion that it presents nothing but an 'obvious truth' - the idea that a unified and effective high command is necessary for all large-scale enterprises, military and industrial as well as hydraulic.

This statement disregards the crux of my argument, which sharply distinguishes between authoritarian leadership (as the broader phenomenon: 'no society is without its authoritarian segments') and total power (a specific type of authoritarianism). By committing the logical fallacy of confusing these two categories he eliminates the need for investigating the specific qualities of total power which is the main object of my study.

Bad faith? He does not hesitate to impute unworthy motives to the supporters of the concept of oriental or hydraulic despotism:

they fuel 'engines deliberately designed to stoke up those tribal animosities by which mankind has made itself so miserable'. I see no reason to reply on this level. Rather I submit that he would have reproduced my analysis of total power and hydraulic society if it had been meaningful to him. Despite his earlier stress on societies as the decisive targets of historical inquiry, he has failed to indicate my views on political power, economic organization, and social control – and their specific interrelations – apparently because the larger institutional categories which I introduce have no deep reality for him.

His response to my analysis of Greek and Roman conditions confirms this. He accuses me of referring to total power in the Graeco-Roman world (and in 'our Western world') 'always in general terms and nearly always in the form of an apologia'. What are the facts? I describe certain specific orientally despotic features that were present in early Greece and Rome; I note that in the wake of Alexander's conquests the Hellenistic rulers gradually destroyed the political independence of their own co-nationals in Asia and the Greek homeland; and I discuss the role of the Hellenistic variant of total power in the orientalization of Rome. Indeed, the development of Eastern statecraft in the Roman Republic and Empire is so important in my historical review of the major representatives of oriental despotism that Toynbee's comments are truly surprising. No less surprising are his comments on my discussion of total power in the post-Roman West. I indicate, specifically and institutionally, significant features of Western despotism: the rise of centralized armies, census-taking, public works, and the use of torture. And when I mention the institutional and temporal limitations of Western despotism, I do so, not with generalities, but with reference to the societal features that caused these limitations. If he considers my specific explanations wrong, he should say so, but to assert that such explanations are lacking is manifestly incorrect.

He is equally remiss when he claims that I have applied to Asia 'the Marxian thesis that the means of production rigidly determine all other elements of social life', and that I have postulated 'a uniform type of "agromanagerial" despotism'. Invoking the authority of the late Professor Frankfort he seemingly contradicts me: 'The differences between the "Asiatic" societies where agriculture is based on irrigation are much greater than their resemblances.'

Examination of my book will reveal that my view is very different from the one ascribed to me. Far from accepting a rigid economic determinism, I show the dependence of the economic on the ecological factor, and the dependence of the latter on cultural conditions which in open historical situations offer a variety of choices. Causality yes, determinism no – that is my position, which seems to be quite clear, since not a few of the many serious reviewers of 'Oriental Despotism' have cited it, often adding that it refutes the Marxist concept of economic determinism.

And far from depicting oriental society as uniform, I devote over two hundred pages to its multiformity: simple, semi-complex and complex; core (compact and loose) and marginal oriental societies.

Moreover, I do not doubt that in Frankfort's special field (art religion, and the religious aspects of kingship) the civilizations of the old Orient differed profoundly. This, however, does not affect my societal analysis. In 'Oriental Despotism' I distinguish the essential features of a given society (the key institutions of power, property, and social relations) from the innumerable elements of technology, customs, art, and belief which are not specific to the conditions of power, status, and property. The position enables me to identify basic similarities in the power structure of a group of societies as well as basic dissimilarities in their religion, literature, and art. It enables me to point out cultural similarities between hydraulic China and feudal Japan and between orientally despotic Byzantium and proto-feudal or marginally feudal Kievan Russia. It enables me to explain why, despite the strong cultural impact of Byzantium on Russia, it was the Mongol conquest that imposed an 'oriental' despotic order which persisted throughout the tsarist period and which, while not making the victory of the Bolsheviks necessary, greatly facilitated their coming to power.

In one of his few attempts at historical argument, Toynbee attacks my assumption that the despotic order which the Mongols established in Russia was rooted in the great Eastern monarchies, especially China. He calls this a 'very far-fetched piece of speculation' unsupported by any 'serious evidence'. What kind of evidence is required here? If he had given any attention to my concept of a marginal oriental society as a configuration in which the government employs the organizational and acquisitive methods of oriental despotic statecraft with little or no dependence on a hydraulic economy (a concept which, by the way, contradicts any theory of economic determinism, rigid or otherwise), he would have understood that I presented substantial evidence for 'Asiatic' consequences of the Mongol conquest of Russia - consequences that have been recognized by such earlier Russian historians as Karamsin, Kovalevsky, and Kliuchevsky, and by such recent historians as Sumner, Florinsky, and Vernadsky.

Toynbee's request for 'serious evidence' on the Chinese background of the beginnings of Mongol rule in Russia is almost embarrassing. In 1234 the Mongol Great Khan, having completed the defeat of the Chin dynasty, became the official ruler of the vast realm of North China. In 1237, this monarch ordered his subordinate, Batu, to conquer the Russian lands and organize them in accordance with the political and administrative ideas then held in the eastern metropolis. To ask for evidence on these elementary facts is like asking for evidence that after the completion of the reconquista in Spain the conquest of America proceeded in accordance with orders from Madrid.

It is possible that Toynbee's lack of concern for such details
is only a special manifestation of his lack of concern for the
ultimate problem that underlies my comparative study of single-
centered (as contrasted with multicentered) societies, of a state
stronger than society, of total power in its most elaborate
historical form: the problem of the institutional conditions that
permit, or preclude, the growth of political and individual
freedom. His review shows him profoundly disinterested in this
problem. In fact, his dissociation from this sphere of inquiry
is so passionate that he heaps abuse on those who differ from
him. He calls the classical Greeks, who found the monarchies
of Asia lacking in the free institutions they cherished, promoters
of a 'lie' and 'smear'. (Among the guilty he mentions Aristotle.)
He also calls those who in recent times took up the issue of
oriental despotism 'modern Western conquerors of the world',
although he must have been aware, or have been reminded by my
study, that among them were such pioneers of rational institutional inquiry as Montesquieu, Adam Smith, James Mill, Richard
Jones, and John Stuart Mill.

Since Toynbee condemns all previous attempts to analyse
oriental despotism, he could hardly be expected to applaud mine.
However, his attack on me has a special twist. He reprimands me
for dealing with this system and with communist totalitarianism
instead of with German Fascism, as was suggested by my account
of the 'genesis' of my book. Arguing that National Socialism is
the worst known system of total power, and also the one that
is 'most easily comprehensible to Western writers and readers',
he feels that for these reasons and on the basis of my personal
experiences in Hitler's concentration camps, I should have made
'a full dress study' of Nazi Germany rather than of oriental
despotism and communist totalitarianism.

The brutalities of National Socialism are monstrous indeed.
But since, according to Toynbee, that variant of total power
is the easiest to understand, he should rather welcome my
dealing with other types that are less well understood and that
include the USSR, for he himself considers Russian communism
'about the next-best target to German National Socialism for a
democratic Western attack on totalitarianism'. While he does not
specify this rather lame indictment, it does imply that, after the
collapse of National Socialism, the Soviet regime is the worst
totalitarian system functioning today.

Thus he disregards his own premises when he reaches the
conclusion that I wrote the wrong book. And he does not improve
his position when he supports this conclusion by referring to
my account of the genesis of 'Oriental Despotism'. Any reader
who takes the trouble to open the book will find that the development of my ideas is discussed, not in the Acknowledgments, as
he says, but in the Introduction, where I tell in some detail
that I began my study of Chinese and Asiatic society a full
decade before Hitler seized power and that this subject remained
my central interest for more than a quarter of a century.

Luck had it that at a certain point in my inquiry my experiences with National Socialism deepened my insight into such phenomena as total terror, total submission, and total loneliness. Luck had it that my growing awareness of Russia's orientally despotic past led me to see Russian - and eventually Chinese - communism in a new light. In my study I suggest that clarification of the earlier forms of total power will facilitate the comparative analysis of all forms of modern totalitarianism, and I deplore that, in this respect, Italian and German Fascism have not been given sufficient attention. If my efforts help us to analyse and develop our great heritage of human freedom, and if they help us to understand and combat more effectively the total negation of freedom, then I will gladly bear Toynbee's wrath. He calls my book a 'menace'. A menace to what? To the worst form of total power? I hope it is.

NOTE

* From 'American Political Science Review', 52, 1958, pp. 502-6.

WITTFOGEL'S 'ORIENTAL DESPOTISM'*

George P. Murdock

This is a truly great book, one of the major contributions to
the science of man in our time. Its importance to anthropology
in the area of comparative political institutions parallels that of
Tylor's 'Primitive Culture' in the field of comparative religion,
and may conceivably even outrank that of the entire corpus of
theoretical literature in political science.

The author, an eminent professor of Chinese history, has fully
assimilated the comparative perspective of cultural anthropology,
in which he has clearly been aided by his wife, Esther Goldfrank,
a mature and productive member of our profession. Starting with
an immense erudition in Chinese history and society, Wittfogel
has for years been extending his researches in the relationships
between economy, property, and political institutions to the
other agrarian societies of eastern and southern Asia: to the
ancient and medieval states of southwestern Asia, northeastern
Africa, and southern Europe; to the Aztec, Inca, and Maya
civilizations of the New World; and to the medieval and modern
political systems of western Europe and Russia. He has likewise
examined the government and economy of a number of less complex societies, notably the Asiatic pastoralists, the Chagga and
Suk of East Africa, the Pueblo Indians, and the Polynesian
Hawaiians. The book presents, with great vigor and clarity,
the theoretical conclusions from these Herculean comparative
researches, together with a mass of supportive illustration.

The author upholds and demonstrates a multilinear interpretation of political development, within a limited range of typological possibilities. His theoretical orientation is thus close to
that of Julian Steward and the latter's associates, and is
thoroughly consistent with parallel positions in other subject
areas of anthropology, e.g. those of Lowie, Eggan, and the
reviewer in the field of social organization. Its conclusions are
devastatingly at variance with those of all unilinear political
evolutionists from Morgan to Stalin.

The author begins with the recognition that a distinctive type
of political system, absolutistic and bureaucratic in nature, tends

to arise in arid or semi-arid regions which make the transition
from hunting and gathering to agriculture. The construction and
management of large-scale irrigation installations must be undertaken by the states since they lie beyond the capacity of any
other organized group at that level of social development. These
functions require centralized planning and supervision and
absolute power of enforcement, which inevitably initiate trends
toward the suppression or fractionation of private property,
the absorption and integration of the ecclesiastical and military
organizations, conspicuous consumption by the wielders of
power, unrestricted exploitation of the masses, terroristic techniques of control, and ultimate stagnation after a brief initial
period of creativity.

Wittfogel acknowledges his strong intellectual indebtedness
to the classical economists like Adam Smith, James and John
Stuart Mill, and Richard Jones who first noted and characterized
'oriental despotism', and to Karl Marx who further elaborated
their definitions, but he goes much farther than his predecessors
in clarifying the origins, delimiting the distinctive features,
and indicating the widespread geographical distribution of this
major type of political system. He also delineates the variations
it can assume under different conditions of environment, economy,
and cultural contact, and the manner in which states of 'hydraulic'
type can extend their managerial and bureaucratic controls to
other kinds of public works, such as navigation and drainage
canals, massive defense installations, and even modern industrial
plants, always with comparable social consequences. Far more
adequately than his predecessors. Wittfogel contrasts with
oriental despotism, where the state is stronger than society and
men outside the ruling apparatus are essentially slaves of the
state, several alternative types of political organization, especially the feudalism of medieval Europe and Japan and the modern
democracies of the Western world. Both are 'multicentered' in
the sense that the state is effectively checked and restrained by
other strong and competing organizations, such as the church,
craft and merchant guilds, and the private owners of land and
industrial capital. Such political systems not only offer vastly
greater protection to the individual but also provide a basis for
adaptive and progressive social change instead of slow stagnation.

Unless the reviewer greatly overestimates the intellectual
curiosity of his professional colleagues, nearly all of them will
read this book. To report upon it in greater detail would thus
serve only to dull the keen edge of their appetites. Nor will
he indicate his own points of disagreement or disappointment,
which are fairly numerous though relatively inconsequential
when viewed in perspective. He hopes that all readers will find
the volume a whetstone on which to sharpen their critical facul-
ties, and that at least some of them will be stimulated to expand
and refine its conclusions with reference to ethnographic evidence which even the most broadly oriented historian could not
be expected to command.

Although this review must naturally emphasize the scholarly
qualities of 'Oriental Despotism', it cannot ignore entirely the
potential political impact of the volume. It provides for the
first time a solid theoretical framework on which the 'free world'
might base a direct positive assault on the foundations of communism and Fascism to replace the unorganized rearguard defense
which it has presented for some decades to the ideational attacks
of its enemies. The political and social systems of Soviet Russia,
communist China, and recent Nazi Germany fall clearly into the
pattern of bureaucratic Asiatic despotisms. Far from representing
the emergence of new postcapitalistic configurations, they are
obvious reversions to a common pro-capitalistic political typology
and differ from the older societies analyzed by Wittfogel chiefly
in their even more complete bureaucratic control of agriculture,
commerce, and industry, in their even more ruthless subjection
of the individual, in their even more refined techniques of
terror, and in the even more cynical hypocrisy of their rulers.
That the latter are fully aware of the parallel and its validity is
revealed by the tragi-comic story (ch. 9) of the deletion or dis-
guising of references to Asiatic society in recent Russian editions
of Marx's works and by the frantic efforts of communist apologists
to substitute a wholly fictitious theory of Asiatic 'feudalism'.
    Wittfogel presents a broad comparative basis for confidence
that democratic multicentered societies will ultimately emerge
triumphant over their totalitarian enemies IF - and this conjunc-
tion must be capitalized - they realize their enormous inherent
strength, adapt their actions to this knowledge, and avoid the
dangerous pitfalls into which they have been stumbling much
too frequently. With respect to internal policy one concludes
we must preserve at all costs our civil liberties and academic
freedom, our independent judiciary, our free press, our private
property in land and in business and industrial enterprises, our
independent religious sects, our labor unions, our farmers' and
veterans' organizations. Though we may criticize and attempt
to correct their excesses, we must cherish and safeguard our
'pressure groups' for they constitute our chief bulwark against
the corruption of total political power. In foreign relations we
should subject the self-defeating Truman-Eisenhower policies
to thoroughgoing revision. It would probably be safer, for
example, to give material aid directly to the Khruschevs and
Maos of the modern world than to the Nehrus, Nassers, and
Sukarnos. The former would merely bolster corrupt systems that
are already exhibiting unmistakable signs of internal strains,
whereas the latter, in situations where a choice of alternatives
still remains, may actually help those who are steeped in the
traditions of oriental despotism, and uncritical of them, to tip
the scales in the undesired direction. If our author is right, as
communist apologists admit by the very character of their denials,
then the only sound policy of the Western democracies, foreign
as well as domestic, is to support selectively all tendencies
toward multicentered political organization throughout the world

and to refrain, except in dire emergencies, from giving aid
to any overt or covert exponent of oriental despotism.

NOTE

* From 'American Anthropologist,' 59, 1957, pp. 545-7.

'ORIENTAL DESPOTISM': POLITICAL WEAPON OR SOCIOLOGICAL
CONCEPT ?*

Wolfram Eberhard

A good hundred years ago Karl Marx introduced into his system
the concept of 'Asiatic society' as a special form of societal
organization, based upon a low level of 'civilization' and large-
scale irrigation-works, occurring in the desert areas between
the Sahara and the high plateaux of Asia ('New York Daily
Tribune,' 25 June 1853). Marx never attempted to study the
factual basis of his theory; it remained crude and imprecise
and seems to have served mainly as an excuse to explain why
he did not include non-Western countries in his basic theory.
The theory gained importance in Lenin's day when communism
began to spread into Russian Central Asia and later when there
were hopes that China would become communistic (1927): if
these countries were structurally different from Western countries, a different program of action would have to be developed
to convert them; moreover, there was raised the possibility of
a different final product.
    Wittfogel entered this discussion some thirty-five years ago
and attempted from then on to refine Marx's thoughts and to
develop them into a real theory. His work came to a provisional
climax in 1938 with the publication of 'The Theory of Oriental
Society'. In the present book he devotes a whole chapter to the
changes in the theory made by Marx, Engels, Lenin, and Stalin,
and shows how the changes were not the result of deeper insight or the collection of better data, but of changes in the
political line. It is, however, unfortunate that the author puts
into the same category with these politicians such scholars as
G. Childe who have brought to light new information and accordingly have changed their theories for reasons other than a
mere change in the party line. Wittfogel's history of the theory
would also have been more valuable had he reported on the discussions among scholars in China, Japan and India, and the
occurrence of the theory at the present time within the Soviet
orbit. It does not seem to be dead there (p. 411), as evidenced
by J. Prusek ('Archiv Orientalmi,' Prague, 22, 1954, pp. 1-2)

who includes in his characterization of Chinese society the same traits that Wittfogel mentions as typical traits of 'oriental despotism': large-scale irrigation-works and extensive road construction necessitating centralized government and a bureaucracy whose members due to their power accumulate wealth and thus become landlords.

The present form of the theory as presented by Wittfogel exhibits a number of differences from the 1938 form, although a special study would be needed to gain full clarity because the author discusses only in footnotes and not always clearly his present attitude toward earlier opinions. The main point seems to be that he is now against using the term 'feudal' for any period of Chinese society because he now accepts the narrow definition of feudalism instead of the former wide, Marxian definition. He has dropped his concept of a special 'oriental feudalism' and speaks only once of 'quasi-feudalism' in the loosely dependent periphery of oriental despotic states (p. 310). On the other hand, he still upholds the theory of a 'cycle of crises' in oriental societies (p. 171 n) - a concept parallel to the Marxist crisis theory in Western societies - but he does not give any detailed discussion of this important feature (only brief general remarks on retrogression, as on p. 417, seem to refer to this theory). While in the 1938 theory the natural factors (rain versus irrigation agriculture) were regarded as of fundamental importance - because they supposedly induce definite modes of production which in turn produce specific forms of society, thus making possible a classification of societies and of their stages of development - the 1957 theory centers around the concept of 'agromanagerial despotism' which is regarded as an essential and also as the only specific trait of oriental society (p. 414). The other 'essential but not specific traits' are: large-scale government-managed irrigation-works, corvée labor, and serfdom. These essential traits occur also in other societies, but Wittfogel states that in an oriental society they become specific through their dimension and/or their special configuration.

Oriental society can be 'tribal' (p. 142), primary or secondary (p. 414), compact or loose (p. 166), marginal (p. 173), and submarginal (p. 195). It is one of a minimum of five societal 'conformations' (p. 419); it surpasses all other stratified pre-industrial societies in duration, extent, and the number of persons dominated (p. 418). It is in character stationary (p. 445). Among the oriental societies are: Hawaii, the Suk in East Africa, the Chagga in Africa, the Zuni in New Mexico, the Roman Empire, Byzantium (although their hydraulic activity consisted mainly in bringing drinking water into towns), the Arab Empire, Moorish Spain, Iran, ancient Peru, and Mexico (these two are, therefore, even now 'imperfect democracies' - p. 432), India, and Russia during and after the Mongols. Oriental despotism spread 'into the woodlands of Germany' and only later did Western Europe return to a non-hydraulic form of society (p. 212). But Japan,

Holland, Belgium, and Northern Italy, which all have been using
extensive irrigation systems, are not hydraulic societies, although once the Etruscans supposedly created in Northern
Italy a 'sub-marginal' hydraulic society (p. 197). Hydraulic
society spreads by diffusion into areas which do not have irrigation. The use of the concept of diffusion in this manner is
rather surprising. On the other hand, in spite of strong Chinese
culture influence upon Japan, and in spite of the existence of
irrigation in Japan, the decisive 'trends were unable to shape
Japan' (p. 198).

At this point, several remarks may be made: 1   What is the
usefulness for modern research of a theoretical model which is
so crude that it encompasses dozens of societies which differ
from one another in thousands of important traits and which are
regarded as stationary in character over thousands of years?
2   If 'agromanagerial despotism' is introduced as one of the
essential and the only specific trait of oriental society, large-
scale government-managed irrigation works cannot be called
another independent essential feature: it is already contained
in the first term (cf. p. 8) which consists of three, in reality
non-specific, elements. 3   The term 'despotism' is not clearly
defined. What, for example, is the difference between 'occidental
despotism' (p. 299) and absolutism? This point, as does the
previous one, refers to the loose use of terminology, which has
been criticized by other reviewers. 4   There are 'early, non-
state-centered hydraulic societies' (p. 416) which, therefore,
did not have large-scale government-managed irrigation works,
and yet are called oriental societies. How can we determine
when a society still can be called an oriental society in spite
of the absence of one or more essential traits? How can we
determine whether the dimensions of essential traits in a
society having only essential traits but not the specific trait
are such that the former become specific?

Objectionable from the methodological point of view is Wittfogel's
failure to distinguish between institutional and functional data,
as well as his incomplete evaluation of their interaction in time
and space. For example, Wittfogel describes how in despotic
societies almost all details of the daily lives of subjects are
regulated by the government. As proof, he mentions either
some law or some actual occurrence. In the first case, he
specifies neither the period when this law was in force, nor the
geographic or societal sector in which it applies, nor the degree
to which it was enforced. If he would examine all United States
legal codes, he would find that they contain innumerable laws
concerning many intimate aspects of our private lives; on the
basis of these laws, our society could be called 'despotic'. But
the existence of a statute does not mean that it is enforced.
Few nations ever had the Roman concept of law. In the second
case, he takes his example from one definite historical period,
without checking whether the example really indicates a general
usage or not, or whether the usage was limited to a specific

moment in time or was typical for a long period.
Because the basic concepts are too unclear, they are too
inclusive. For example, anyone who has lived in irrigated areas
knows that there are many different forms of irrigation: by
well, by tank, by terraced fields, by canals above the level of
the fields, and canals below the field level; irrigation by bringing
water, by draining water away, by a combination of both, etc.
All these may occur as individual enterprise, and/or on a family,
clan, tribe, community, or state basis. Can such different forms
all result in the same type of society? Are the Egyptian Pharaoh,
the Russian Tsar, and the Chinese emperor really the same type
of despot? Are the Chinese mandarins, recruited from a number
of upper-class families in spite of an open examination system
(p. 351), the same 'managerial bureaucrats' as the slave-officials
who served some Near Eastern despots?

The perhaps most crucial test of the theory is whether it does
justice to the data. Unfortunately, the book suffers from a lack
of that which Wittfogel claims Toynbee has too much (p. 371),
namely 'attention to detail'. For instance, it is incorrect to call
the Chinese emperor a 'despot'. There are as yet no monographs
which have studied the position of the Chinese emperor over
the last 2,000 years, but it is known already that the emperor's
role underwent various changes: for example, he was not always
regarded as the only man responsible for the conduct of government; he was, in some periods, manipulated by factions of
powerful families; in many periods he was bound to follow laws
and ethical principles which ne had not made and could not
change; there are other periods in which the emperor seems
to have been an absolute ruler. It is not true that imperial
China 'tolerated the predominance of privately owned land over
a long period of time' (p. 303). Japanese scholars have outlined
the development of private land-ownership in China over the
centuries in all detail. The government has not 'in this
case . . . restricted the owner's proprietary position by . . .
directives as to what crops should be grown' (p. 303): such
interference was typical of some periods only, and not always
of those when the power of the emperor was strongest. The
government also did not restrict the land-owners' position by
'a fragmenting law of inheritance' (p. 303). This law of equal
inheritance of sons was an old customary law in China and in
many other cultures. Primogeniture is typical of only very few
societies and over historically short periods. Equal inheritance
does not lead to much fragmentation if the average mortality is
high and life expectancy low, as can be proved by the study of
genealogies and land registers. It is not true that despotic
power was total (p. 101) in oriental societies such as China or
India: there were organizations which in certain periods of
history were nuclei of power and which the 'despot' had to take
into consideration – for instance the secret societies, the
Buddhist church, and federations of land-owning gentry families
in China, or the castes in India. Artisans were organized in

China and India. In China, they became free in exactly those
centuries in which 'despotism' was supposedly strongest. It
makes no sense to speak of the absence of privately owned land
in the Islamic Near East without discussing the whole Islamic
concept of property and the concept of religious conquest; the
situation is much more complex than in other countries. It is not
true that the Chinese were interested only in corvée while the
Japanese were interested in revenue (p. 199): Sung China replaced corvée by taxation. Almost all data adduced by the author
could be discussed in a similar way and as a result, in my
opinion, different generalizations would have to be made.

As Wittfogel's theory stands today, it may perhaps be used by
somebody as a political weapon (p. 10), but I can hardly believe
that it will be used as a tool in sociological analysis.

NOTE

* From 'American Sociological Review,' 23, 1958, pp. 446-8.

WITTFOGEL'S 'ORIENTAL DESPOTISM': A SOVIET REVIEW*

I.A. Levada

In the 1920s and the beginning of the 1930s a German historian
and sociologist named Karl Wittfogel published several works on
the history of China. At that time he considered himself a
Marxist though even then it was evident that, despite plentiful quotations from Marx and Engels in his writings, he did not
reach a truly correct materialistic understanding of history.
In particular, he propagated the idea of a supraclass state in
the Orient.

In the middle of the 1930s Wittfogel broke with the progressive
movement. Nowadays, Wittfogel, as professor of Chinese history**
at Washington University and chairman of the Chinese History
Project at Columbia University, apparently cannot forgive himself for at one time 'employing Marx's own socio-economic
criteria' (p. 6). In the first paragraphs of the book under review, published simultaneously in the United States and England,
he overtly declares that he already adopted anti-Soviet positions
in the 1930s and therefore revised his relationship to Marxism.
Wittfogel seeks to emphasize his hostility to the socialist
revolution on almost every page.

The publisher, advertising Wittfogel's book, characterizes
it as the 'first thorough analysis of Eastern despotism', and
assures the reader that it gives an ' "entirely new evaluation"
of such basic institutions as private property, class and bureaucracy . . . new explanation . . . of communism and basic
elements of Marxist-Leninist doctrine'. Even the author himself
is far from modest in assessing his work. In his words, the
communists 'are wrong when they hold us incapable of producing
big and structured ideas' (pp. 9-10). After recalling the scientific merits of Newton and Montesquieu, Adam Smith and Darwin,
he transparently hints that his work is called to fill a 'theoretical
vacuum' and to give the bourgeois world a 'system' which could
serve as a weapon in the struggle with Marxism.

** Retired in 1962 (translator's note).

Now with what weapon does Wittfogel threaten Marxism?
At first sight his book gives the impression of an extraordinary
broadly conceived and firmly grounded work. If Wittfogel's
earlier works were based on data from the history of China,
'Oriental Despotism' abounds in references to studies of India,
Sumer, Egypt, Hawaii, Mexico, Peru, Byzantium, Russia, and
other countries. The author pretends that he has ascertained the
laws characterizing social life in all the countries of the East,
and in some cases even of the whole world. However, the seeming broadscope of the formulation in fact permits Wittfogel to
avoid the concrete analysis of social development and gives a
semblance of authenticity to his purely speculative constructions.
The basis of Wittfogel's argument is not new. Roughly, it can
be reduced to the following: in the countries of the East, agriculture, which is based on artificial irrigation, plays a decisive
role. The necessity of administration (management) for the
creation of centralized irrigation systems leads to the formation
of the despotic state which also builds vast aqueducts, roads,
colossal defensive constructions, temples, palaces and tombs,
registers time, conducts a census of population, etc.
The idea about the relationship between the despotic organization of society in the ancient East with agriculture, based on
irrigation, is by no means original to Wittfogel. Similar ideas
were advanced by Adam Smith, J.S. Mill, and R. Jones. In his
articles for the 'New York Daily Tribune', in 'Die Grundrisse
der Kritik der Politischen Oekonomie', 'Capital', and a number
of letters, Karl Marx gave the principal basis for a scientific
analysis of this specific feature of the East. He pointed out
that a natural economy of small communities, combining agriculture with domestic craft, was the foundation of the social
structure in the countries of the East. These communities,
however, appear as hereditary holders subordinated to the
supreme or the only proprietor - the state (monarch).(1) In
contrast to the ancient world, where a collectivity of formally
free citizens enjoying equal rights was ranged against the
slaves, a 'general slavery' ('allgemeine Sklaverei') was characteristic of the East.(2) In this way, the key to the solution of the
'riddles' of the East was seen by Marx in the particular forms
of ownership and exploitation, i.e. economic relations, formed
in correspondence with the conditions of production and exchange. After a number of discussions in Soviet science,(3) a
firm view emerged that the analysis of ancient Eastern societies,
in the light of Marx's remarks, does not substantiate any conclusion about the existence of a special 'Asiatic' social-economic
formation. One should speak only of the specificity of forms of
slave-holding, or feudal societies of the East. It is to this
specificity that Soviet orientalists direct their attention.(4)
Wittfogel fiercely assaults the Marxist methodology. Leaving
aside economic relationships, he tries, first of all, to infer the
state form of 'oriental despotism' from particularities of geographic environment. Agriculture in given natural conditions

requires irrigation; construction of dikes and canals requires
concerted labour; team-work requires leaders (p. 26). From
these incontrovertible truths Wittfogel reaches controversial
conclusions: the necessity to obey authority leads allegedly
to the emergence of despotism by the sovereign and the general
enslavement of the population. In order to be more convincing
he invents a new terminology: 'hydraulic society', 'hydraulic
despotism', 'hydraulic state', etc.

However, the 'hydraulic' variety of 'geographic materialism'
does not represent the main characteristic of Wittfogel's
methodology. Geographic environment, he asserts, provides
the possibility for the emergence of 'hydraulic' society and
despotism, which 'materializes' only under certain conditions.
The latter are found by Wittfogel in the . . . free will of people.
Against the 'economic determinism' of the materialistic conception
of history Wittfogel advances the thesis that history gives people
'a genuine choice' (p. 17).

From his point of view, the primitive tribes of hunters,
gatherers, or cattle-breeders faced a dilemma: either to move to
'hydraulic' agriculture and to succumb to despotism, or to maintain the low level of their economy and with it 'freedom of
personality'. He devotes sentimental lines to those primitive
peoples 'who endured lean years and long periods of famine
without making the crucial changeover to agriculture . . .
demonstrate the immense attraction of non-material values, when
increased material security can be attained only at the price
of political, economic and cultural submission' (p. 17).

In other words Wittfogel's starting point is that people,
firstly, always possessed and valued a certain (namely bourgeois)
concept of the individual and his/her rights, and, secondly,
always weighed the historical consequences of their activities.
Both premises are equally wrong. The very concept of the
individual and his/her rights could appear only after the long
development of a private economy broke the primitive unity of
man and collectivity. This concept has an historical character.
Forms and terms of the transition to higher forms of social
organization may vary according to different concrete circumstances, including the conscious activities of people. (It was
Marxist-Leninist theory that revealed the significance of the
purposeful activity of the progressive forces in society, provided
with a clear theory of social process, and therefore able to
anticipate and consider both the near and remote consequences
of their actions.) However, the attempt to pretend that primitive
tribes were able to evaluate the social consequences of irrigation
cannot be judged other than completely absurd.

The method of historical voluntarism with which, as we can
see, Wittfogel approaches problems of the Orient, leads to the
substitution of the study of the objective laws of the history of
society by its subjectivistic interpretation, i.e. that in which
the ideologues of the bourgeoisie are now interested. In Wittfogel's
conception, the idea of 'freedom of choice' plays a double role:

with its help the peoples of the Orient who 'chose' despotism
stand against the 'primordially free' peoples of the West (more
exactly - Western Europe, the United States, and Japan).
Moreover, this idea is contraposed to the Marxist thesis of the
inevitable transition of the whole of mankind to communism.
Wittfogel calls his method of study 'institutional analysis', which
first of all means the examination of the political institutions,
i.e. the superstructure, which is pictured as a determining force
in the social structure. Having related not only Marxism but
also classical bourgeois political economy to a 'propebased
sociology' (based on property), he sets against them a thesis
allocating the determining role to the state and only a 'secondary,
if serious' role to property (p. 404).

Wittfogel certainly knows Marx's classical definition of the
notions of base (infrastructure) and superstructure as well as
the relations between them; and the Marxist definition of property
(ownership) is also known to him. Having set himself the goal
of 'refuting' the Marxist teaching on the role of economic relations,
Wittfogel resorts to an apparent substitution of concepts. Thus
he reduces property to a legal form, derived from the state.
Property is defined by him as 'the individual's recognized (!)
right to dispose of a particular object'. It presupposes the existence of a guaranteeing government and therefore 'being a legal
and social institution, property is a political phenomenon'
(p. 228).

Given such a definition, the 'conclusion' is already anticipated.
The voluntaristic manner in which Wittfogel interprets this
category of political economy can be seen even from the fact
that he adds proletarians* to the category of owners, as 'owners
of the . . . capacity to work' (!) (p. 49). Wittfogel does not
see in ownership an objective economic relationship, an historical
social form of appropriation of material goods by people. Therefore he is unable to pose those real questions of the history of
the East which need investigation, such as: the problem of
divided property (the combination of the supreme property of
the state with the ownership of the community), (5) the problem
of the forms of property of the direct producer and the forms
of his personal dependence, etc. He only mentions some external,
legal expressions of these relationships: instability of the position
of individual private proprietors in the despotic state (threat
of confiscation; a hereditary law, excluding primogeniture and
making obligatory the division of property).

However, these issues are dealt with extremely superficially
by the author. Thus, instead of a serious analysis of rights of
inheritance in the East - a question which is closely connected
with the explanation of the role of patriarchal-clan survivals -
Wittfogel is content with simple reference to the wish of the

---

* Wittfogel does not mention 'proletarians' in this context (translator's note).

despot to undermine the influence of the proprietors. In the
chapter devoted especially to the forms of property in the
'hydraulic society', he gives a list of various forms of private
property that can be distinguished only in the East, according
to Wittfogel. Beginning with the established fact that in a number
of Eastern countries private owners were in the majority controlled, in one or other way, by the state, and did not constitute
an 'independent' force competing with the state, he jumps from
the problem of the political position of traders (in the assessment of which Wittfogel makes wild assumptions) and reaches
the conclusion that property, in any form, plays an unimportant
role in the East.

Speaking about the despotic states in the East, Wittfogel tries
first of all to prove that their foundation is not a certain form
of property but a form of 'management', or organization, of
the economy. This statement, directed against Marxist-Leninist
teaching about the state in general, and, as we shall see below,
against the socialist state in particular, Wittfogel tries to support
with reference to Engels. This concerns the well-known passage
in 'Anti-Dühring' where Engels discusses the transformation of
functionaries in primitive communities from the 'servants of
society' into 'masters', representatives of the state.(6) This
is not the first time this statement has been used by adversaries
and falsifiers of Marxism to prove the correctness of their argument about the supraclass 'organizational' character of the state.
However, Engels underlines ('critics' of Marxism usually hush
this up) that in the passage he consciously left aside the question about how such a transformation concretely happened and
'how different ruling persons ultimately united into the ruling
class'.(7) As Engels showed, this could happen only on the
basis of the development of private property and the separation
of the propertied groups from the whole mass of community
members. When Wittfogel departs from the conception of the
'managerial' (or otherwise 'apparatus-like') state, he recognizes
only two classes in 'hydraulic society' - the rulers and the
ruled: 'The men of the apparatus state are a ruling class in
the most unequivocal sense of the term; and the rest of the
population constitutes the second major class, the ruled' (p. 303).
Such is the 'new sociology of classes' which he desires to substitute for Marxist teaching. Let us remark first of all that
Wittfogel speaks not about an exploiting class but about a
'ruling class', which is, of course, far from being the same
(the merchants in conditions of feudalism are an exploiting but
not ruling class). According to Wittfogel, peasantry, merchants,
and craftsmen do not form classes as they are 'not organized
politically'; on the other hand, the whole 'apparatus', i.e.
bureaucrats and army - from scribe and soldier up to the
emperor - appear as a 'ruling class'.(8) In the 'hydraulic
society', declares Wittfogel, there are no economic classes, but
only 'bureaucratic' ones, i.e. groups appearing thanks to the
state: landlords, capitalists, and 'gentry' (p. 4). Thus two

different questions are mixed up: who carries on the administration and in whose interest is it done? The administration appears as an end in itself. The 'managerial' class with all its force strives only towards the enlargement of its power.

Wittfogel argues that he discovered two 'laws' determining the activity of despotic power: the 'cumulative tendency of unchecked power' (p. 106) and the 'law of diminishing administrative returns' (p. 109). If we translate these terms into common language, it appears that the argument is about the despot's goal to maximize his supremacy, although the enlargement of the sphere of administrative control is beneficial for 'bureaucracy' only up to a certain level, beyond which this control becomes disadvantageous.

Wittfogel alleges that from the Marxist point of view it is impossible to explain such phenomena as, for example, bureaucracy and shen-shih in ancient China. In his words, 'a property-based sociology' makes it necessary to classify the imperial clerk, given the level of his income, with commoners rather than with the ruling class (p. 304). However, Marxism does not consider the 'level of income' as the only criteria for determining the social position of people, but stresses rather the place of people in the system of economic relationships. Therefore it does not see in a bureaucracy a special class, but a stratum, which performs the functions of administration in the interests of the economically ruling exploiting class and thus serves these interests. Even if this bureaucracy, in its embryonic form, emerged as an organ for fulfilling social functions, with the split of the society into classes it becomes a tool in the hand of exploiters and remains such while the class-antagonistic society remains. After the liquidation of private property in the means of production, i.e. in socialism, the functions of administration are not a monopoly of any special group but a matter of all the working masses. Consequently, a 'bureaucratic class' never existed in the past and does not exist now.

Equally wrong is Wittfogel's assertion that 'oriental' ('hydraulic') land-ownership differs completely from European feudalism as a result of its 'bureaucratic character' (p. 275). First of all, many of the concrete ways in which landed property arose in the countries of the East (grants, commendations of peasants, direct seizure of peasant lands), and its subordinate position in the relation to the paramount property of the state (of the sovereign) are also known in feudal Europe. But irrespective of how the feudal monopoly of land arose, its realization creates the source of benefit for the landlord, who lives from the surplus product of the producers, the latter relegated to serfdom, i.e. he becomes the feudal lord.(9)

The scheme constructed by Wittfogel of the 'extraeconomic' classes of 'hydraulic society' contradicts the real data of class struggle, which fills the whole history of the East, betokening first and foremost a confrontation of economic interests. In order to find a way out, Wittfogel simply rejects class struggle

in the East. He contends that there were 'many social antagonisms but little class struggle'* (p. 327). According to him,
social conflict becomes class conflict only when the participants
are a recognizable and representative fraction of a class (p. 328).
Consequently, the spontaneous struggle of peasantry in the
feudal society in the East, lasting for centuries, is not class
struggle, because Wittfogel, as he wrote, 'does not recognize'
the peasantry as a class. The first reason for the 'absence' of
class struggle in the feudal East is, according to Wittfogel, the
despotic state, 'paralysing' this struggle (p. 328). To prove
his point he introduces the following fact: in 1746 in Fukien the
tenants attempted to unite in a struggle for a reduction of the
rents. The government discovered the conspiracy and executed
its organizers. Wittfogel declares that such measures of the
government (which only proves that the feudal state defends
the interests of landlords) 'precluded mass political action [class
struggle] as a legitimate [!] form of social protest' (p. 328).
At the same time he concludes that the 'class struggle is a luxury
of multicentered and open societies' of the West (p. 329).

As we can see, Wittfogel is faithful to his 'method'. Having
decided to 'liquidate' class struggle in the East, he announces
that a claim to such a title is manifested only in a 'legal' struggle,
and, of course, does not discover it. However, there was never
any 'legitimate' or 'organized' class struggle in pre-capitalistic
Europe either. Even in the most 'democratic' capitalism only
reformist, i.e. accessory forms of the struggle of proletariat,
are 'recognizable'.

One can hardly wonder that having 'done away with' the
problems of the social structure of the countries of the East
in a similar way, Wittfogel sees the only foundation and moving
force of their existence in the state which creates and determines
everything. Yet about this very state he is unable to say more
than it is 'despotic', 'total', 'terroristic', 'all-suppressing', etc.
One of the chapters is called 'A state stronger than society'.
As Wittfogel himself admits, this formulation was taken from
P. Miliukov.(10)

The 'total' character of oriental despotism is found by Wittfogel
in the absence of 'independent'** groups (clan, military, propertied
ones), which would be able to vie with the state (p. 49). Spheres
of pitiful 'beggar's' freedom (in family, village, community, craft
guilds) occur in 'hydraulic society' only to the extent that
despotic interference in everyday life is limited by a consideration of the sovereign's advantages (the already mentioned law of
diminishing administrative returns). The power of the despot
does not encounter either 'constitution'(!) or 'societal' limits

---

* Title of para. F - translator's note.
** In Russian original 'politicheski nezavisimye' (politically independent) does not stand in the reviewed book: independent and
semi-independent (translator's note).

nor do laws of nature and custom limit his arbitrary rule.(11)
   Through such reasoning the voluntarism of Wittfogel's methodology appears clearly. He does not try to recognize inevitable
and lawful social relationships behind the 'arbitrary' activity
of individuals, the sovereigns, who, in fact, were of course
far from being so 'total' as he tries to present them. For example,
a certain independence of traders and craftsmen in the despotic
state has objective economic grounds in the very character of
the basis of feudal society.(12) Arbitrary rule by the monarch
in relation to one or other representative of a class cannot mean
arbitrariness towards this class as a whole. A Chinese emperor
could, of course, within his discretion execute any official,
any landlord or trader, and confiscate his property. On the
other hand, no despot could by any means destroy the respective classes or strata, because he himself depended on them,
constructed his rule on them. Wittfogel does not want to recognize precisely these necessary ties determining the activity of
any despot in the slave-holding and feudal societies of the East.
Instead of explaining the real conditions and real role of the
despotic state in the East as a form of political development,
inevitable at a certain stage - which is one of the urgent tasks
of science - he concerns himself with the sentimental 'exposure'
of the horrors of despotism with its terror and servility, the
isolation of the individual, his loss of rights, etc. The idea
that this political form was in a sense necessary and progressive
is rejected by him.
   One should remark that Wittfogel tries to turn the whole attack
on despotism against the socialist system. He constantly looks
for parallels, and even identifies 'oriental despotism' and
socialism. As a 'theoretical' support for such analogies he uses
his invented and ahistorical category, the 'managerial state'.
We have already shown the falsity of this concept. Slave-holding
and feudal states were exploitative and not simply 'managerial',
and thus the socialist state differs fundamentally from them. As
Engels pointed out, the first act of the state, serving the whole
society after the socialist revolution, is to bring the means of
production into public ownership.(13) As soon as this is achieved,
the management of the economy is directed to serve the interests,
not of individual persons, monopolies or exploiters, but the
interests of the whole society.
   A centralized economy gives birth to centralized extra-economic
coercion (despotism) only when it is imposed upon a predominantly
petty, dispersed economy.(14) This was the case in ancient
China or in Egypt, where the direct producers, farmers, were
bound for most of the year in a petty economy, on which their
lives depended. Irrigation, and other large state works, even
if objectively useful for the mass of agriculturists, did not appear
to them to be urgently necessary.
   Therefore, large-scale co-operation depended upon coercion.
This meant - in the given conditions - that co-operation required the help of actions of the despotic state (as the result

of its ownership of the individual producers). In developed capitalism, extra-economic coercion is no longer in principle required, because the economic necessity to work appears irresistible to the proletarian who is completely alienated from the means of production. In socialism, economic necessity no longer stands in opposition to the interests of the worker, so that co-operation becomes a conscious and therefore free form of social relations.

Wittfogel certainly knows these elementary Marxist truths. Because he is unable to disprove them, he tries simply to avoid them and, with the help of his 'institutional analysis', to use his perverted interpretation of the social structure of the ancient East to attack modern socialism. And not only socialism! As is known, the state sector achieved considerable development in the countries that liberated themselves from colonial oppression. Attempts are made to plan the economy in the interests of strengthening national independence. Given this fact Wittfogel sounds the alarm. He tries to impress on 'the intellectual and political leaders of non-Communist Asia' the danger of 'administrative politics' and planning (pp. 9 and 442). He considers free private enterprise of the West European and North American type the saviour of Asia and all mankind from the 'totalitarian danger'. The servile character of this idea in relation to imperialism needs no comment.

The false idea of 'hydraulic society' provides Wittfogel with the basis for constructing a 'new' scheme of world-historical development. The central thrust of this scheme is the division of the world into a 'multicentred' West, where private property and initiative flourish eternally and the individual is free, and the 'hydraulic', 'mono-centered' East, with its despotism, constraint on private property, etc. Having introduced the concept of 'hydraulic density' patterns (taking into account the varying role of irrigation in agriculture), he adds the idea of 'core', 'margin' and 'submargin' and 'hydraulic world' (marked pseudo-scientifically with a whole system of indices - Ca_1, Ca_2, Cr_1, M_1, L_2, etc.). He then considers it possible to add, for example, Byzantine and post-Mongolian Russia to the East, as 'margins'. He interprets the history of Russia according to the reactionary 'Eurasian' conception, disseminated by white émigré historians (G. Vernadsky and others). Thus until the Mongolian conquest, Russia was on the 'margin' of the West, and after the Mongols introduced their 'Eastern' order (according to Wittfogel, borrowed by them from China), it became a 'margin' of the East.

This whole absolutely anti-scientific and speculative construction - leaving aside its defamatory character - is used to demarcate an impenetrable boundary between West and East, and to deny the possibility that 'Eastern' institutions may independently emerge in the West and vice versa. Thus, through the influence of the East (the 'orientalization' of Rome in the epoch of Hellenism) Wittfogel tries to explain the Catholic church's attempt to subject secular power, the ceremony of kissing the shoe of the German

emperor, the decree on the general census in medieval England,
etc. At the same time he categorically rejects an independent
origin for feudalism and the bourgeoisie in the 'Orient'. If one
were to speak of feudalism in the East, then the 'fundamental
difference' between East and West would be undermined. Then,
Wittfogel says in horror, 'we run the danger of abandoning the
freedom of historical choice, because we are paralysed by the
fiction of a unilinear and irresistible development' (p. 369).
This overt declaration shows once again that his whole sociological scheme had been constructed in order to reject the idea
of the inevitability of the transition of mankind to communism.

The countries of the East, asserts Wittfogel, can liberate
themselves from the 'hydraulic despotism' only with the help
of the capitalist West; in the East there is only 'diversive'
change.* The fact that in Asia there is no local capitalism at
the moment of the intrusion of Western European capital, and that
therefore the development of bourgeois relations did not exactly
follow the European pattern, he tries to elevate into a principle
that the East is incapable of independent development. True, in
his opinion, Western influence was too weak to effect substantial
changes in the majority of the countries of the East. He sees
a confirmation of this in the considerable role of state capitalism
in these countries. From this he draws the conclusion that
European capitalism is still called upon to play a transforming
role in the East.

This conclusion is false and the development of historical
events refutes it again and again and, with it, the dream of
transforming the East into a citadel of Western capitalism - in
other words, the perpetuation of colonial slavery which is
cherished by ideologues of imperialism. The positive role of
European capitalism in colonies and dependent countries consisted in undermining the foundations of the natural economy,
and in the impetus it gave to a commodity economy in both city
and village.(15) However, imperialist domination has been for
a long time the main brake on the development of the economy
of dependent countries. Escape from the imperialist yoke is
essential for the East. Wittfogel did not notice that peoples
of colonies and semi-colonies, having experienced capitalist
exploitation in its most barbaric forms, are looking for ways of
independent development.

The Chinese people found freedom with the victory of the
socialist revolution. Many a country in Asia, now on the capitalist road, sees an important guarantee of its independence in the
development of a state economy which plays a progressive role.
Why does Wittfogel lament this so bitterly? Because he cares

---

* In the Russian original 'razvitie' = development; Wittfogel distinguishes between development, which is endogenous, and
change, which is externally effected (see Wittfogel, 1957,
p. 420) (translator's note).

about the interests of Western imperialism. It should be pointed out that while he regrets the weakness of a 'middle class' (i.e. private-economy bourgeoisie) in the East, liberating itself from colonial supremacy, Wittfogel considers Turkey the one 'bright spot'; it is true that Turkey lacks a 'developed' bourgeoisie, and uses state-capitalist forms, but she is none the less 'closely linked with the West' in her political system and orientation (p. 431).* This is the catch! 249

It is obvious that Wittfogel treats the socialist revolution with utmost hostility. While he characterized the February revolution in Russia as 'anti-totalitarian' and a result of the 'influence of the West', he pictures the victory of the Great October Socialist revolution and the construction of socialism in the USSR as a return to the 'hydraulic' system on the basis of modern industry (p. 441).** Wittfogel makes a similar evaluation of the Chinese revolution. He is exasperated by the 'short-sightedness' of bourgeois investigators such as J. Fairbank who, though they do not show affection for a socialist order, recognize its progressiveness.

The attempt to judge history in terms of the 'individual', and to reduce history to a confrontation of 'totalitarianism' and 'anti-totalitarianism', is characteristic of many contemporary ideologues of the bourgeoisie who do not wish to see that this yardstick is, first of all, historically conditioned and, secondly, determined in the final instance by the development of productive forces. Marx had already destroyed similar arguments advanced by Sismondi and his school. (16) The practice of socialist construction in the Soviet Union, the People's Republic of China, and other socialist countries serves as a persuasive confirmation of the fact that a true emancipation and flourishing of the individual is possible only on the basis of a progressing socialist economy.

Wittfogel devotes a whole chapter of his book to Marxist views about the 'Asiatic' mode of production. Gleaning isolated individual statements from Marx, Engels, and Lenin on the particularities of the East (and simply phrases where the word 'Asiatic' was used in one or other connection), he tries to create an impression that the classics of Marxism recognized the existence of a special, 'Asiatic' order in the East (including Russia) in all periods. However, they were not 'consistent' as they did not identify the ruling class with the 'apparatus'. He cites passages in the classics of Marxism that emphasize the common traits of class order in both West and East in order to accuse them of a 'retreat'*** from the 'Asiatic theory', and

---

*The passage about Turkey is to be found on pp. 434-5 (translator's note).
** Wittfogel wrote: 'The October revolution . . . gave birth to an industry-based system of general (state) slavery' (translator's note).
*** Both 'retrogression' and 'retreat' appear on pp. 369-98 (translator's note).

'unconscientiousness'. Summarizing the discussion by Soviet historians of the 'Asiatic' mode of production, he argues that the Marxists as it were 'renounced' the 'Asiatic concept' for purely political reasons, because they feared dangerous implications for Marxism. (He considers his own voluntaristic scheme such a conclusion, but, of course, it has no relation to the statements of the classics of Marxism on the 'Asiatic' mode of production.)

In order to prove the impartiality of his 'theories', Wittfogel tries to ascribe to Marx a rejection of party spirit in science. Marx contraposed the scholarly honesty of Ricardo, objectively standing on the positions of capitalism, to the conscious falsification of Malthus, who spoke as an apologist of reaction.(17) Marx's evaluation of Malthus as a man whose scientific conclusions were 'fabricated' from the hindsight of the ruling classes in general, and reactionary elements in these ruling classes in particular,(18) should be recalled when it comes to characterizing Wittfogel and his latest work.

In his introduction Wittfogel says that 'Oriental Despotism' is a result of investigations which took him half of his life. This is the result - Wittfogel's renunciation of scholarship. Rejection of Marxist teaching about the nature of the state was an intellectual prerequisite of this fall. Having disavowed a class analysis of social structure, Wittfogel progressed down the inclined plane of reactionary sociology, which eventually led him into the role of a scholar-apologist of capital and colonialism. This is an instructive lesson for contemporary revisionists, defending ideas alien to Marxism about the supraclass state.

NOTES

* From 'Sovetskoe kitaevedenie', 1, no. 3, 1958, pp. 189-97. Translated by Peter Skalnik. Professor Adam Kuper kindly revised this translation.

1 See K. Marx, 'Forms preceding capitalist production', in Marx, 1857-8a, pp. 472-3.

2 Ibid., pp. 495-6.

3 See 'Diskussia ob aziatskom sposobe proizvodstva' (Discussion about the Asiatic mode of production), Moscow-Leningrad, 1931; V.V. Struve, 'Problema zarozhdeniia, razvitia i razlozheniia rabovladelcheskikh obshchestv drevnego Vostoka' (Problem of origin, development and disintegration of slaveholding societies of the ancient East), in 'Izvestiia gosudarstvennoi akademii istorii material' noi kul'tury', 77, Moscow-Leningrad, 1934, and other works.

4 Among the most recent we can mention: A.I. Tiumenev, 'Gosudarstvennoe khoziaistvo drevnego Shumera' (State economy of ancient Sumer), Moscow-Leningrad, 1956; I.M. Diakonov, 'Istoria Midii' (History of Midia), Moscow-Leningrad, 1956.

5   As N.V. Kolganov supposes, in these cases it is right to distinguish between 'sobstvennost' and 'vladenie' (i.e. property and possession - perhaps ownership - translator's note). (See N.V. Kilganov, 'Sobstvennost'v sotsialisticheskom obshchestve' (Property in socialist society), Moscow, 1963, pp. 12-15.)

6   See F. Engels, 'Anti-Dühring' (Russian edition), Moscow, 1957, pp. 167-8 (1939).

7   Ibid., p. 168

8   See F. Engels, 'Origin of the Family, Private Property, and the State', 1972 (Russian edition: 'Proiskhozhdenie sem'i, chastnoi sobstvennosti i gosudarstva', Moscow, Gospolitizdat, 1953, p. 170).

9   In the same way the property of a capitalist, originally accumulated, e.g. via sea piracy or by embezzlement of state, becomes capitalist independent of its origin, because it served as a means of exploiting the hired labour force.

10 See P.N. Miliukov, 'Ocherki po istorii russkoi kil'tury' (Essays on the History of Russian Culture), part I, St Petersburg, 1896, p. 115.

11 The latter is 'proven' by reasoning that the victim of despotism does not feel better because he or she is executed on the basis of certain customs.

12 See P. Grinevich, K voprosam o istorii kitaiskogo feodalizma (Problems of Chinese feudalism), 'Problemy Kitaia', 1935, no. 14.

13 See F. Engels, 'Anti-Düring' (Russian edition), Moscow, 1957, pp. 264-5 (1939).

14 See B.F. Porshnev, 'Ocherki politicheskoi ekonomii feodalizma' (Essays on the political economy of feudalism), Moscow, 1956, pp. 38-40.

15 See Mao Tse-tung, 'Selected Works' (Russian edition: 'Mao Tsze-dun, Izbrannye proizvedenia'), vol. 3, p. 143.

16 K. Marx, 'Theories of Surplus Value' ('Teorii probavochnoi stoimosti'), part II, Moscow, Gospolitizdat, 1957, p. 111.

17 Ibid.

18 Ibid., p. 113.

WITTFOGEL'S IRRIGATION HYPOTHESIS*

Julian H. Steward

1 A MATTER OF PROBLEMS AND INTERESTS

A scholar's contributions to science should be judged more by
the stimulus he gives to research - by the nature of the problems
he raises and the interests he creates - than by the enduring
qualities of his provisional hypotheses. Karl Wittfogel's hypothesis
concerning the role of irrigation in the development of early
civilizations was first formulated during the 1930s (Wittfogel,
1935, 1938, 1939-40, 1946), when most students of the development of culture were still basically oriented toward descriptive
and historical rather than explanatory analyses. Anthropology
and history were mainly concerned with cultural differences.
Any formulations of cultural development which recognized cross
cultural similarities, and especially any which postulated causal
processes which might account for these similarities, were
thought to be impossible and almost certainly erroneous.

There was, however, an undercurrent of interest in identify-
ing cross-cultural regularities or developmental similarities, but
in America this was largely a tenuous heritage of earlier theories
of cultural evolution (White, 1959). Even after the stimulus of
the centennial of Darwin's 'Origin of the Species' in 1959,
evolutionists were still largely concerned with the question of
general stages of cultural evolution (e.g. Sahlins and Service,
1960) rather than with the determination of the specific causes
of different kinds of cultural development or evolution (Steward,
1960; Steward and Shimkin, 1961). Interest in cultural causality,
however, has continued to grow even though comparatively little
research is guided by interest in the still nebulous theory of
cultural 'evolution'.

Wittfogel's contribution to recent trends originated in his extensive research on China, which had created vast hydraulic
works including irrigation systems, drainage canals, and internal
routes of water transportation. He developed the hypothesis that
the early civilized states of both the eastern and western hemispheres were integrated by the managerial controls required to

construct and maintain the irrigation - and more broadly hydraulic - systems. As water was brought to arid lands, food production and population increased and became the basis for class-structured states and the achievements of civilization. While historians of culture were emphasizing differences between civilization, Wittfogel was postulating a single basic factor that brought all of the civilizations into being.

The thirty years since Wittfogel's first publications, especially the two decades since the Second World War, have produced a vast amount of field research which has thrown doubt upon the universal applicability of the irrigation hypothesis. It is clear that in many instances irrigation had been ascribed excessive importance and that in others, its development seems to have been the result rather than the cause of the growth of states. Much of this research has obviously been directly stimulated by Wittfogel's theory, and it is safe to say that even Wittfogel's most vigorous critics have advanced our understanding of the role of irrigation precisely because their interest had been directed to the subject and they had a theory which could be tested. Inadequacies of the irrigation hypothesis leave us with the challenge of finding alternative hypotheses to explain the growth of early states.

The basic problem is broader than irrigation and states. It is one of seeking explanatory formulations for the development of any culture. My own interest in this was first published in 1936 (Steward, 1936) and dealt with primitive bands. Later (Steward, 1949), Wittfogel's irrigation hypothesis stimulated me to ascertain whether apparent cross-cultural regularities in the development sequence of early civilizations could not be extended farther back in time and made to include features of culture other than those concerned with the emergence of states. This 'trial formulation' included Meso-America, Peru, Egypt, Mesopotamia, and China, and provisionally postulated similar developmental eras which were designated Hunting and Gathering, Incipient Agriculture, Formative (of States), Regional Florescent States, and Empires and Conquests. Research since 1949 has required many revisions of these sequences and many modifications of the assumed role of irrigation (see Steward, 1955a).

In the criticisms of Wittfogel's and my formulations, three trends are discernible. First, there are some who have delved so deeply into the details of single areas that they not only repudiate any cross-cultural formulation but apparently hold an a priori conviction that general causes do not exist. Some even seem to assume on philosophical grounds that the creative operations of the human mind are not reducible to causal under-standings. Second, an opposite extreme is represented by the more systematic evolutionists who are content to place all these early civilizations in a stage called the 'agricultural' or 'urban revolution'. There is a third, or intermediate, position which is empirical in its attention to the details of individual cases and which applies the insights of each case to other cases to be

tested and revised. That factors and processes of culture change
must be abstracted from the hundreds of minutiae in each instance in no way invalidates the methodology.
This third view, unfortunately, is often obscured by confusion
in its application. In a recent symposium on the prehistory of
Latin America (Meggers and Evans, 1963), experts on special
areas from Mexico to Argentina described their areas in some
detail, but the period diagnostics ranged from presence or
absence of agriculture or pottery to presumed development of
states and cities. Cross-cultural comparability was badly obscured
in most of the articles. The increasing knowledge of early history
and cultural developments in the Near East lacks systematic
presentation owing to the diversity of interests, as illustrated
by the essays in 'City Invincible' (Kraeling and Adams, 1960).
The classical orientalists tend to pay so much attention to
humanistic achievements that no two subareas seem comparable.
Even general concepts, such as 'urban revolution' and 'civilization', lose validity in the assumption that a civilization had to
have writing and, since cities were the 'containers of civilization', Jericho is in doubt while the Inca Empire, despite certain
huge urban centers, is completely disqualified.
The sociologists and anthropologists, especially the dirt
archeologists, have generally given us more insight into what is
significant cross-culturally in the Near East. The sociologists
who contributed to 'City Invincible' rightly pointed out that
the apparent absence of cities in Dynastic Egypt did not
necessarily disqualify it from the stage of 'urban revolution',
since no one had defined a city. The diagnostics of a city plague
students of the New World for the same reason. Whether Classical
Mayan ceremonial centers should be called 'cities' has not yet
been decided. The concept of urbanization will remain vague
until it is recognized that there are many kinds of cities, each
with one or more special functions. This requires an empirical
comparative study, which must have a starting point and some
provisional hypotheses.
We are really on more solid ground in comparisons of societies
during the initial phases of agriculture, probably because irrigation was of minor importance, communities were small, writing
was absent, and there were no state institutions to confuse
the issues. Many studies of the New World have traced the
origins of plant cultivation and reported on the earliest farm
communities. Braidwood (1964) has raised interesting questions
about the degree of incipiency of agriculture and possible
differences in the natural environments where farming first
sustained permanent villages. In each case the sequence started
at some point when dependency upon domesticated crops began
to supersede reliance upon wild foods. There have been no
genuine civilizations which lacked agriculture, or agriculture
and a partially symbiotic animal husbandry dependency.
My first essay (1949) attempted to trace the sequence of socio-
political types and of cultural achievements from what we must

now consider a vague and ill-defined era of primary village-farm communities (Braidwood, 1964), through later eras, and to define each era in terms of its achievements and culminations. In all early civilizations the sequence of these cultural features was extraordinarily similar. Virtually all basic plants were domesticated before the higher civilizations developed. The increased populations and stability of communities based on better food supply entailed amalgamation of formerly independent villages into larger sociopolitical units. The trend also entailed social differentiation into classes and specialized occupations as well as roles.

Whereas the era of incipient or early agriculturalists had pottery, permanent houses, and other simple technologies, ensuing developmental eras witnessed remarkable parallels in other achievements. Metallurgy began quite early with gold and culminated in the age of empire in the New World with bronze and in the Old World with bronze and later iron. Specialization of role freed certain groups of individuals or classes, perhaps priests, so that they developed astronomy, mathematics, and writing. Writing was not common to all civilizations, and systems of writing differed, but it appeared, if at all, in about the same developmental era in both hemispheres. Similarly, the wheel was invented at about the same stage, although it was only a toy in America and was later abandoned.

The development of monumental architecture, elaborate priest-hoods, state religions, and humanistic expressions of state political and religious concepts in art, literature, architectural style, music, and other media were understandably manifestations of state development and of classes of specialists. The sequence of cultural achievements of all kinds was surprisingly similar in each early civilization, and there was little disjunction. The Inca had no writing, the New World abandoned the wheel, and the stylistic and symbolic expressions of state development were largely unique in each case. In a functional sense, however, the particular styles of art and architecture, the various religious ideologies, and other humanistic extensions of state development were similar in sanctioning the basic institutions.

This brief recapitulation of similarities in the development of early civilizations oversimplifies the picture. It will be unacceptable to the cultural relativists who, in the name of empiricism, must push examination of minutiae to the point that each case seems unique and any valid cross-cultural understandings are ruled out. As a broad, cross-cultural hypothesis, however, it is concerned with basic factors and processes that underlie cultural growth. The value of such a generalization is that it provides a foil or a target for criticism which, devastating as it may be, will hopefully advance understandings a step further.

2 THE IRRIGATION FACTOR

The importance of water to any civilization is becoming painfully
evident today. If the value of irrigation to pre-industrial
civilizations has been overemphasized as the all-important factor
in creating the state - or Wittfogel's 'oriental absolute state' -
the hypothesis, like many others, is first presented in bold
strokes and later requires modification. The crucial issue of
whether large-scale irrigation created the state or was created
by a developed state is an academic issue to the extent that
neither could exist without the other in extremely arid lands.
More concretely, however, the issue must be broken down into
special considerations of soils, terrain, topography, climate,
and other factors. The principal criticism of Wittfogel's hypothesis, like that of so many pioneering hypotheses, is that 'irrigation' is far too broad a rubric to have precise heuristic value.
One must ask in each case such questions as 'how arid is arid?'
The rainless Peruvian coast will support no vegetation without
irrigation except along the stream borders. How adequate is
rainfall? In some areas it barely supports plant life. In many
areas, even the eastern United States, there are drought years
when irrigation would be a tremendous asset. What is the nature
of stream flow? In some areas rivers deposit alluvium, overflow
and change their channels to the extent that they can be readily
tapped by short ditches or canals. In others streams are deeply
entrenched and require major construction of dams and complicated systems of canals. What features of social structure may
affect maximum use of land, whether irrigated or not? Warfare
may expose dispersed settlements - and their irrigation works -
to disastrous raids. The nature of the state may involve
specialized production for trade with its attending hazards. What
capital, advanced technology, and control of manpower can a
state put into maximum development of irrigation? There is
incredible expansion of irrigation today, based upon water
storage in huge dams and construction of long canals. How much
land is available for cultivation? This may be related on the one
hand to topography and water flow and on the other to terracing
of hills where flat land is scarce, as in the Philippines, Peru,
Japan, and elsewhere. How much water is needed by the crops?
Rice requires far more water than some of the grains.

Empirically, revisions of Wittfogel's hypothesis are clearly
required by much recent research. It has been shown by Adams
(1965) that in the Diyala Plains of lower Mesopotamia significant
enlargement of irrigation systems was a sequel to, not a precondition or antecedent of, state development and planning and
of urban growth. Meso-America, despite much recent search,
has not yielded convincing evidence of large irrigation systems
prior to the Spanish Conquest. The prehistory of Peru, however, reveals a fairly concomitant growth of population, states,
and almost certainly irrigation, except that population seems to
have dropped off just prior to the Inca Empire. China's early

well-and-ditch irrigation probably did not require much supra-
community coordination, but its great irrigation and navigational
canals of later periods were clearly state-controlled. The cycles
of population rise and fall which were closely associated with
China's dynasties may or may not imply periodic neglect and
reconstruction of irrigation.

The obvious difficulty of correlating irrigation works, land
use, population, and community and urban development is in
dating irrigation systems and determining their extent at any
given period. Any irrigation system would require an enormous
amount of excavation and, since canals were undoubtedly used
over long periods - often re-excavated - they are difficult to
date (Woodbury, 1960). Undoubtedly, the most comprehensive
study of this problem is Adams's reconstruction of some 6,000
years of the Diyala Plains of Iraq (1965).

In all the world areas classified as arid and semiarid, crop
production is increased through irrigation. In all cases, moreover, irrigation requires some kind of supervision. The problem,
then, is whether a society has centralized authority which
developed apart from needs of food production and which took
over the managerial functions of irrigation, or whether irrigation itself created a managerial class. Suggestions are provided
by some of the simpler cases.

The most elementary irrigation known was that of the Paiute
of eastern California, who lacked domesticated plants (Steward,
1930). These Indians, who lived along the eastern escarpment of
the Sierra Nevada mountains of California, diverted streams by
means of small dams and ditches which were carried a mile or so
in order to further the growth of wild seeds. This practice was
evidently the logical sequitur of their observation of the great
natural growth of these plants along the stream margins and
marshy terrain as compared with the extremely arid areas
between streams. The small, localized bands of these Paiute
had no chiefs or headmen, and 'irrigation bosses' therefore had
to be accorded power to get the job done.

In the arid Southwest, the Indians irrigated in varying degrees
and for different crops. The Hopi, who were among the most
developed and culturally complex, had adapted maize, beans,
and squash to the meager rainfall of the high Colorado Plateau.
Some of their maize was exceptionally deep-rooted. Occasionally,
the Hopi built dams in the arroyos to catch the periodic rainfall, but these were local enterprises and they involved no
changes in the sociopolitical organization of the Indians. It is
interesting that the Hopi developed a new kind of irrigation in
post-Spanish times. Onions and a few other garden crops adopted
from Europeans were planted near small springs which irrigated a few hundred square feet of land.

The prehistoric Hohokam peoples of the Salt-Gila River valleys
of hot, arid southern Arizona, however, had constructed irrigation systems wherein the canals and ditches totaled several
hundred miles and served many communities. These canals

brought large areas of desert under cultivation. This magnitude
of planning and labor seems to imply some centralized authority,
and recent archeological research suggests that a state-like
sociopolitical system somewhat resembling that of Meso-America,
but diminished in scale, was involved. Whether the state
developed first and created the large irrigation system or resulted
from it cannot be known. If the Hohokam were migrants from
much farther south, the state may have preceded the irrigation.

There are other examples of very impressive irrigation works,
such as the mountain terracing in the Philippine Islands, which
resembled that of the Central Andes in land use but not in size.
In the Philippines, however, fairly small 'tribes', communities,
or clusters of peoples who were not at all organized into states
seem to have managed their hydraulic problems by informal
understandings. The Andean irrigation systems, on the other
hand, were vastly more extensive, and they must have resulted
from large-scale planning over a long period. The canals extend
far back into the mountains, collect water from a vast highland
area of considerable rainfall, and distribute it within valleys.
In some cases they even combine the waters of several valleys.

A contrast of two irrigation areas in Japan is illustrated by the
findings of Toshinao Yoneyama, who worked on my recent study
of native communities. In the Nara Basin, a fairly small stream
irrigates the land of 24 hamlets which occupy only two or three
square miles and have a total population of about 3,000 persons.
Maintenance of ditches and distribution of water are entirely a
matter of local concern. Interestingly, however, Kaminosho,
the last hamlet downstream, has first rights to the water and
enforces the cooperation of other communities through certain
ditch-cleaning ceremonies each year. Possibly Kaminosho was the
first settlement in the area and later hamlets remained subordinate to it. The other community is Kurikoma in the mountainous
area of the northern Tohoku district. It had no access to irrigation water until several centuries ago when an overlord put his
people to work building a dam and ditches. The regions differ
in that streams in the Nara Basin flow sluggishly from the
mountains over alluvial fans, and irrigation requires minimal
dams and ditches, whereas in the Kurikoma area the river is
deeply entrenched in gravels as it comes out of the mountains
and requires extensive work on dams and ditches.

The importance of topography in Peru is unparalleled. In the
great highlands or puna, potatoes and quinoa were grown with
rainfall up to an altitude of 15,000 feet, but the population seems
always to have been dispersed with few cities prior to the growth
of Tiahuanaco and later of the Inca capital at Cuzco. Rainfall
decreases with altitude, and the coast is virtually without rain.
Except for limited flood-plain farming along the immediate margins
of the rivers, the farm productivity in the lower altitudes could
not have compared with that of later periods. Unlike aggrading
rivers, such as those in the Nara Basin, the Hohokam area, and
the lower Tigris-Euphrates system, Andean streams are deeply

entrenched in mountain valleys and gorges. The earliest irrigation seemingly expanded until whole valleys were included in single systems. Later, several adjoining valleys might be integrated. Construction of these irrigation systems in Peru clearly required central managerial controls, whether by priests, warriors, civil rulers, or special engineers.

The applicability of the irrigation hypothesis in its extreme form to lower Mesopotamia has clearly been discredited by the work of Robert Adams. The larger area has many microvariations in landscape and climate. The earliest known farming was on the 'hilly flanks', where rainfall was adequate for plant cultivation and where the first permanent farm villages are found. In the lower valley, however, as in the Diyala Plains of Iraq, the rivers aggrade to the extent that they overflow their banks, change courses, and create swamps. Under these conditions cultivation could be carried on through construction of fairly short lateral canals and drainage ditches which did not require highly centralized state management. Between 4000 BC and the beginning of the Christian era, there were some fluctuations in population which evidently represented sociocultural factors, such as militarism, but there were no urban centers and the population never exceeded 100,000, in contrast to several later periods when cities of several hundred thousands and a total population of more than 800,000 were reached. The many factors other than use of water and land indicated by Adams's research (1965) will be discussed subsequently.

3 ADDITIONAL FACTORS

While it is clear that managerial controls of irrigation or hydraulic works alone were not everywhere the principal factor underlying the growth of early civilizations, it is equally certain that irrigation of some form increased food productivity in all these areas. No matter how small the irrigation system, cooperative effort in its construction and maintenance and in distribution of water had to be coordinated. Many local factors of physical geography caused variations, ranging from the need for small amounts of water to supplement rainfall to complete dependence upon irrigation, and they presented engineering problems ranging from drainage of swamps caused by river overflow to damming of deeply entrenched streams.

In every case some forms of managerial controls were needed. In each culture, however, there were apparently early developments of theocratic and perhaps other controls and later of militaristic authority. The question, then, is how these ruling classes came into being, whether they assumed control of irrigation, and whether the expansion of irrigation systems, increase of population, appearance of cities, and development of civilization were enhanced by the irrigation factor. To the extent that a society depended upon irrigation water and/or lands to which water could be diverted, the role of sociocultural institutions and hydraulic controls must have been intricately interrelated. The identification of the critical factors, however, requires far more research into the nature and antiquity of irrigation on the one hand and far better understanding of the early social and political structures on the other. Several basic propositions, however, seem tenable.

First, the agricultural revolution in different parts of the world - the domestication of what were to become the plant staples - occurred in areas where rainfall farming, flood-plain cultivation, or small-scale irrigation sufficed. At some later time large-scale irrigation (and/or drainage) increased production.

Second, a precondition of early states - that is, of regionally integrated sociopolitical units consisting of many villages - was sufficient farm yield to support permanent communities and to relieve a portion of the population from work in the fields. As the nonfarming population increased, a variety of specialists in construction, crafts, and religious and political functions developed.

Third, amalgamation of formerly independent communities into states required integrating factors. Assessment of possible factors must take into account state functions and enterprises other than managerial controls of irrigation.

Initial state integration around theocratic controls has been postulated for several areas. In the western hemisphere warfare for conquest, slavery, tribute, or territories seems to have been absent during the culmination of pre-imperial states. The early Maya and highland Mexican cultural culmination prior to about the eighth century AD (the so-called Classical Periods) probably was based in part on a religion in which victims were captured for religious sacrifice, but they seemed to have lacked true nationalistic warfare. A similar culmination apparently occurred in the Central Andes, and perhaps also in the Mississippi Valley Temple Mound cultures.

Just why thousands of people submitted to theocratic controls is not clear. It has been suggested that the early Chavinoid cultures of Peru had qualities of a messianic cult. Eric Wolf has speculated that Teotihuacan, like historic Mecca, combined the functions of trade and religion. In Mesopotamia state integration seems to have been peaceful, while warfare was beginning in the hilly flanks. The case for nonmilitaristic, theocratic integration in pre-Dynastic Egypt and China is strong but not wholly convincing.

In all these instances it is obvious that eventually a populous state that was internally specialized and socially stratified required centralized controls. Allocation of labor, distribution of goods, adjudication of disputes, and performance of religious rites could no longer be carried out by the kin group or the isolated village. A civilian, or civil-religious, government had to develop.

Control of hydraulic works passed from the implicit authority
within a small cluster of villages to the explicit authority of the
state. In some cases this transfer was necessarily early; in others
it came late. Little farming could have been carried out on the
coastal deserts and the steep mountain slopes of intermediate
altitudes of Peru unless a central authority mobilized the people
of many localities to construct the necessary dams and ditches.
Whether state organization had already existed in Peru is probably unimportant. The growth of the state and of agricultural
production are aspects of a single process. The Peruvians,
however, increased their public works after the Andes were
unified under one huge empire. At the same time, population
possibly declined, which might be interpreted to mean that
farm production had previously reached the limits of environmental potentials and productive technology. By contrast, the
low flat valley of the Tigris-Euphrates was easily watered by
short ditches from the river. Indeed, its swampy nature had
made drainage as much a problem as irrigation. The irrigation
maximum in this area came much later, when authoritarian
and militaristic governments built enormous canals and created
cities in areas previously uninhabited.

Other world centers probably represent intermediate situations. Where farming had to expand into arid areas, centralized
control over the waterworks were required as local conditions
imposed various difficulties. Expanded farming concomitantly
increased the population, strengthened state controls, and
permitted further state expansion.

With the fairly primitive technology and engineering skills
of these early civilizations, however, a limit had to be reached.
In some cases strong authoritarian states realized maximum productivity. In others, state expansion was accomplished by
military conquest rather than by peaceful, internal development. Thus all early civilizations sooner or later reached an era
of conquests or successions of dynasties.

As comparative studies are pursued, and an empirical, case-
by-case approach is utilized, each study illuminates others. It
now begins to appear that the factors and processes in any
cultural development are fairly limited in number and are rather
similar cross-culturally. The differences lie more in their
combinations and relative importances than in any total uniqueness of given cases.

Irrigation is not a single factor, but the varieties of irrigation and the effects of each variety differ in ways which have
scarcely been touched by research. The other factors in the
context of state and urban development, especially in prehistoric periods, are just beginning to enter the realm of
speculation. A strong religious factor is evident everywhere,
but its varieties and influence have not been clearly assessed.
Militarism is widely evidenced by capture of sacrificial victims,
but the time when wars of conquest became a factor that integrated independent states is speculative. The nature of land-

ownership, production, and control and trade of commodities
is virtually unknown except in eras of written history.
In short, instead of 'throwing out the baby with the irrigation
water', the need is to recognize the particular combinations of
factors, *including the kind of irrigation*, which operated in each
case. Wittfogel's hypothesis challenges the disbelievers to produce alternative explanations which are more than accounts of
the uniqueness of individual cases.

BIBLIOGRAPHY

Adams, R.M., 1962 Agriculture and urban life in early south-
western Iran, 'Science', vol. 136, pp. 109-22.
Adams, R.M., 1965 'Land Behind Baghdad: A History of Settle-
ment of the Diyla Plains', University of Chicago Press.
Braidwood, R.J., 1964 More complex regularities? in Manners,
R.A. (ed.), 'Process and Pattern in Culture', Chicago, Aldine,
pp. 411-17.
Kraeling, C.H. and Adams, R.M. (eds), 1960 'City Invincible'.
University of Chicago Press.
Kroeber, A.L., 1948 'Anthropology', revised ed., New York,
Harcourt Brace.
Meggers, B.J. and Evans, C. (eds), 1963 'Aboriginal Cultural
Development in Latin America: An Interpretative Review',
'Smithsonian Misc. Coll'., vol. 146, no. 1.
Sahlins, M.D. and Service, E.R. (eds), 1960 'Evolution and
Culture', University of Michigan Press.
Steward, J., 1930 Irrigation without agriculture, 'Papers', Michi-
gan Acad. Sci., vol. XII, pp. 149-56.
Steward, J.H., 1936 The economic and social basis of primitive
bands, in 'Essays in Honor of Alfred Louis Kroeber', Berkeley,
University of California Press, pp. 311-50.
Steward, J.H., 1949 Culture causality and law: a trial formula-
tion of early civilizations, 'American Anthropologist', vol. 51,
pp. 1-27.
Steward, J.H., 1955a (ed.), 'Irrigation Civilizations: A Compara-
tive Study', Washington D.C. Pan American Union.
Steward, J.H., 1960 Evolutionary principles and social types,
in Tax, S. and Callender, C. (eds), 'Evolution after Darwin',
University of Chicago Press, pp. 169-86.
Steward, J.H., and Shimkin, D.B., 1961 Some mechanisms of socio-
cultural evolution in Evolution and man's progress, 'Daedalus', vol.
90, no. 3, 'Proc. Amer. Acad. Arts and Sciences', pp. 477-97.
White, L.B., 1959 'The Evolution of Culture', New York, McGraw-
Hill.
Willey, G.R., 1953 Prehistoric Settlement Patterns in the Viru
Valley, Peru, 'Bur. Amer. Ethnol. Bull'. 155.
Wittfogel, K.A., 1935 The foundations and stages of Chinese
economic history, 'Zeitschrift für Socialforschung', vol. 4,
pp. 26-60, Paris.

Wittfogel, K.A., 1938 Die Theorie der orientalischen Gesellschaft, 'Zeitschrift für Socialforschung', vol. 7, nos. 1-2, Paris.

Wittfogel, K.A., 1939-40 The society of prehistoric China, in 'Studies in Philosophy and Social Science', Institute of Social Research, vol. 8 (1939), pp. 138-86, New York, 1940.

Wittfogel, K.A., 1946a General Introduction (to 'History of Chinese Society', 'Liao', by Karl A. Wittfogel and Feng Chia-Sheng). 'Amer. Philos. Soc., Trans'., vol. 36, pp. 1-35.

Wittfogel, K.A., 1957 'Oriental Despotism: a Comparative Study of Total Power', Yale University Press.

Wolf, E.R., 1959 'Sons of the Shaking Earth', University of Chicago Press.

Woodbury, R.B., 1960 The Hohokam canals at Pueblo Grande, Arizona, 'American Antiquity', vol. 26, pp. 267-70.

NOTE

* From 'Evolution and Ecology', University of Illinois Press, 1977, pp. 87-99.

HYDRAULIC SOCIETY IN CEYLON*

Edmund Leach

IRRIGATION BASED 'FEUDALISM' IN SINHALA

The engineering constructions of ancient Sinhala were of the
type that Wittfogel indicates should be typical for an hydraulic-
based oriental despotism, and although the traditional reputation
of Sinhalese monarchs corresponds to the pure stereotype of
the oriental despot, nevertheless the actual Sinhalese social
system was not far removed from European feudalism.

I propose now to give some further details of this 'hydraulic-
oriental' feudalism and to suggest that this particular species
of service tenure is peculiarly well adapted to the ecological
requirements of an irrigation-based society. Whereas Wittfogel
implies that hydraulic society and feudalism are mutually in-
consistent polar types of political structure, my own view is
the exact opposite. I suggest that it is the mutual compatability
of hydraulic agriculture and 'feudalism' (in the sense now to be
described) which may help to explain the political success of
many early experiments in hydraulic civilisation.

What actually happened in the Ceylon case was that the king
made grants of land to his provincial governors and to monasteries.
The grant covered both land and population - the ground, the
irrigation works, and the people - and the grant included the
right to levy corvée. Although corvée duty was (and is) known
as 'king's work' it was in fact an obligation due to the grant
holder rather than to the king in person. The evidence for this
is not only in modern practice with regard to the lands of
religious institutions but in the fact that the ancient records
include such contradictory expressions as disāve rajakāriya -
'provincial governor's king's work'.

The resulting delegation of powers produced a 'feudalism'
which differed from the European pattern mainly in that, in
Ceylon, the service due from a fief holder to his lord could
be of various kinds and was not limited only to military service,
or money in lieu thereof.

In Ceylon, as in feudal Europe, the monarch's overriding and

perpetual problem was to devise a means of keeping his feudal
barons under control. The extreme frequency of insurrection
and civil war shows that effective power usually lay with the
local landlords rather than with the crown.

In Knox's* day the king's principal device for maintaining order
was that of compelling his lords to keep one of their close
relatives at court by way of hostage. He also rather frequently
changed the land grants around from one baronial family to
another. Knox cites as evidence of the monarch's tyranny the
fact that the hostages were very frequently executed; it might
also suggest that the great lords were very frequently in
revolt!

Nor was such a state of affairs a peculiarity of the later decadent phase of the Sinhalese state. The great Parakrama Bahu I
himself started life as a provincial governor and attained the
throne by leading a successful rebellion against his cousin the
king. And if we credit the Ceylon chronicles ('Mahavamsa' and
'Culavansa') with any historical merit at all, we must conclude
that the frequency with which kings were overthrown by their
rebellious relatives was at all times strikingly high.

In theory the outer provinces of the realm were ruled by
governors (disāva) appointed by the king; in practice the lord-
ship of the local hereditary baron (vanniyar) was virtually
absolute. In many cases the rank of disava was simply the
titular office of a court official who never went near his domain.
When the customary law of the North Central Province was being
recorded in 1820 it was stated that : 'from ancient time the
vanniyar had been deemed to possess power nearly equal to that
of the disava, but that he is restrained in the exercise of it when
the disava is in the province'.(1)

Knox, who discovered to his cost the extreme difficulty of
travelling in the remoter part of the Kandyan kingdom, attributes
this to the subtlety of an all powerful monarch (1681, p. 70):

The King's policy is to make his country as intricate and
difficult to travel as may be and therefore he forbids the
woods to be felled especially those that divide province from
province and permits no Bridges to be made over his rivers,
nor the paths to be made wider.

A simpler explanation would be that the provincial barons
were beyond control. Certainly these are not the characteristics
we would expect if government were in the hands of a centralised
bureaucracy.

Finally we may note one further way in which, in the later
period at least, the Ceylonese baronial class adopted behaviours
comparable to those of the European feudal nobility.

Wittfogel, generalising far too easily, asserts that the rules
of inheritance in hydraulic society always require sub-division

---

* Robert Knox, 'An Historical Relation of the Island of Ceylon
in the East Indies' (1681), Glasgow, 1911.

of the landed property among the children of the deceased. In
this way, he explains, the despotic monarch is able to allow a
form of private ownership of land among his people, since he
knows that no such holding can ever become very large. Wittfogel contrasts this egalitarian inheritance rule with that of
the European feudal nobility who developed a system of primogeniture.(2)

Now the interesting point is that, in Ceylon, Wittfogel's
generalisation applies perfectly to the ordinary peasantry where
inheritance rules do require division of land to both sons and
daughters. But among the aristocracy a different system prevails - the daughters take their share in the form of a monetary
dower and the sons inherit only when they stay at home and
reside patrilocally. Consequently, among the aristocracy (but
only among the aristocracy), the Ceylonese developed a system
of patrilineal descent associated with the holding of large undivided estates. This is not quite the same as European primogeniture; but it comes very near.

Having considered some of the similarities let us consider the
differences between Ceylonese 'feudalism' and the more familiar
European type.

The most fundamental feature of European feudalism was the
association of land tenure with military service. Everyone from
the top to the bottom of the feudal hierarchy held his land under
the condition that he must, when required, render military
service to his immediate lord. The Ceylon system was also one
of 'service tenure' but with the difference that the services due
to the landlord were of various kinds and not exclusively military.
Knox having noted that the king made grants of 'towns' (i.e.
villages) to his nobles 'with all the fruits and benefits that before
came to the king from them' then continues (1681, p. 69):

In each of these towns there is a smith to make and mend the
tools of them to whom the king hath granted them and a potter
to fit them with earthenware, and a washer to wash their
clothes and other men to supply what there is need of. And
each one of these hath a piece of land for this their service
whether it be to the King or the Lord, but what they do for
the other people they are paid for.

Peasant society in Ceylon is even today divided according to
occupational castes. The farmer caste which is the highest in
status is also the most numerous. Other castes have the occupational duties of washerman, drummer, blacksmith, potter, spirit
medium and so on; formerly there were others - for example,
in the days before motor cars and bicycles, palanquin bearer
was an important functional caste occupation.

The crucial point to note is that, as to their main occupation,
all these castes are peasant cultivators, cultivating irrigated
rice lands according to identical techniques. Where the castes
differ from one another is in the service duties which they
formerly owed to their feudal lords in payment for their land.
Thus the farmers rendered military service or tax in kind but

the lower castes rendered their profession, i.e. the washerman
washed clothes, the blacksmith made tools, the potter made pots.
Today the feudal duty to the overlords has mostly been
abolished yet even so the villagers still feel themselves under an
obligation to carry out their caste services for their neighbours;
and they usually do so without any explicit payment.

For example, in an area I know in northern Ceylon, a community of washerman caste provides the ritual washermen for
about thirty different villages of other castes - usually one
particular washerman household doing the chores for one particular patron village. The washerman receives no piece work
payment but he receives various perquisites from the festivals
at which he attends in the course of his duties. So also in the
case of the drummers and the spirit mediums. Potters and the
blacksmiths, however, are paid for the actual work they do on
a piecework basis.

A very important aspect of this kind of labour organisation
is that it permits the development of quite complex systems of
labour specialisation without any extensive use of a money
medium or of credit banking. In Ceylon, in the 'feudal' period,
land was so distributed among the professionally specialised
castes that the tenants of any particular superior landlord
formed an overlapping set of reciprocal work teams. In effect
each man paid rent for his land in the form of specialised labour
contributed to the work team as a whole. But at the same time,
each man, or rather each sub-caste group (variga), was separately
in client status to the local landlord.

In the dry zone region each variga included some ten to twenty
small single caste villages and was corporately organised with
its own caste court. The variga as a whole was the client of the
local baron and the baron was ex-officio president of the caste
court of each variga within his domain whatever its caste.

Each individual held land privately within his own village -
but the village was the village of a particular variga which was
in effect 'owned by' the local baron or sometimes by the local
temple high priest. Each individual was thus under obligation
to contribute the duties of his caste to his variga lord, i.e. to
the local baron - but the latter transferred this duty, so that
(in practice) variga services became reciprocal. Blacksmiths
and mediums receive service from washermen; washermen and
blacksmiths receive service from mediums and so on.

Perhaps I should add that these service duties which attached
to the lands of each particular variga group were over and
above the rājakāriya - 'king's work' - which was the general
obligation to corvée. This now applies especially to the maintenance of irrigation works, but, formerly no doubt, included all
the monarch's constructional ventures.

Although closely comparable systems of labour organisation
exist in other 'hydraulic regions' of the Indian sub-continent,(3)
there is no explicit evidence, other than that of Knox, that this
kind of 'caste-feudalism' is ancient for Ceylon, but it is certainly

highly probable. The differentiation of service obligations in
terms of caste duties is closely bound up with the requirements
of temple service. Temples, in their modern form, cannot be
maintained without the specialised services of washermen,
musicians, potters, painters and the rest and there is no reason
to imagine that this is in any sense a modern development. If
we couple this with the fact that, from the earliest times, we
have reports of villages being gifted to temples along with their
gamvera - that is the 'dues' of the tenants; that the tenants of
such villages are in some cases specified as carpenters, artisans,
masons, serfs, drummers, washermen, etc.; and that a tenth-
century inscription shows that villages were then administered
not singly but in clusters (dasagam - lit. 'ten villages'); then
we can at least say that it seems likely that a caste organisation
of labour such as I have outlined above generally prevailed in
ancient Sinhala.(4)

Let me recapitulate and bring the argument to a conclusion.
The fact that the earliest large-scale agriculturally based urban
civilisations were so often hydraulic societies is an historical
fact of great interest. Clearly the historical fact poses the
question: are there common features in these societies, economic,
political and so on?

In isolating his concept of 'oriental despotism' Wittfogel has
drawn attention to a number of common features in these societies.
But his analysis is distorted by his anti-totalitarian bias and by
his tendency to very large-scale generalisations. Since most of
the more impressive hydraulic societies existed only in remote
antiquity, generalised statements concerning the details of their
governmental organisation are difficult cither to verify or con-
travert.

Sinhala is an exception. It existed as a 'pure type' of hydraulic
society down to the twelfth century AD. The hydraulic system
itself has survived in attenuated form down to the present day.
Many features of present-day social organisation in the Northern
Dry Zone are commonly regarded as direct survivals from the
classic period of hydraulic civilisation.

Classical Ceylon (Sinhala) was confined to the irrigated dry
zone from roughly the third century BC until the twelfth
century AD.

Down to the end of the eighth century AD the capital was continuously at Anuradhapura and around this city there gradually
developed a vast network of major hydraulic works (in Wittfogel's
sense). This hydraulic system served two functions; it provided
water for the capital and it provided a large region around the
capital with a much more reliable source of irrigation water than
that obtainable from the small-scale village tanks alone.

We may concede that if the whole of this impressive hydraulic
system ever functioned efficiently for any substantial length of
time then it would have needed a bureaucratic organisation to
run it. This is Wittfogel's thesis.

But there is no evidence that this was the case, and there is no

evidence for the existence of an hydraulic bureaucracy which
played a significant role in the politics of the Sinhala State.(5)

The documentary sources go back as far as the fourth century
AD (Dipavamsa) and, though their historical quality is always
dubious and much more attention is paid to ecclesiastical than
to political history, they can hardly be dismissed as irrelevant.
If hydraulic engineering was centrally organised and of central
importance to the Sinhalese State then there would surely be
some indication of this fact even in 'fabulous' history?

But engineering activities, when mentioned at all, are treated
as the arbitrary acts of merit of individual monarchs, and mostly
such work relates to dagoba shrines and pleasure gardens rather
than to agricultural irrigation. The political background is one
of constant civil war and palace murders.

When kings are credited with the construction of a tank they
almost always give the revenues to a religious institution (vihāra).
True, the very last paragraph of the Mahāvamsa (sixth century
AD) credits King Mahasena with the construction of sixteen
named tanks and a canal. But the tanks do not form a single
hydraulic system and several of them have already been mentioned in earlier chapters as presented by previous monarchs to
various temples. None of this suggests centralised control,
or efficient maintenance.

The only direct evidence linking the kingship, as such, with
the irrigation system is that the ritual of accession to the throne
was marked by warlike watersports held at the Tissa tank at
Anuradhapura.

From the eighth century onwards the capital oscillated between
Anuradhapura and Polonaruwa, cities sixty miles apart, and
there were repeated occasions when the country was overrun by
foreign conquerors or split up into small warring states. Despite
these political troubles the irrigation system survived until the
end of the twelfth century and even expanded. But there is
nothing to suggest that this expansion involved any kind of
centralised planning or control. Many of the modern ruins
simply represent replacements of still older ruins dependent upon
the same water supply. At least one highly expert judge holds
that there never was a period when the whole system was all
intact.(6)

One of Wittfogel's theses is that, once the hydraulic bureaucracy has established its centralised total despotism, the despot
will thereafter only advance the needs of agriculture to provide
himself with serf labourers and soldiers. In general, he argues,
the most impressive major irrigation works of hydraulic despotisms are created by the kings primarily to enhance the beauty
of their capitals. This thesis fits well with the Ceylon facts.
But in the Ceylon case, although the major irrigation works
provided food for labourers as well as amenities for palaces,
the hydraulic system was not of crucial economic significance
for the society as a whole. When the central government was
disrupted and the major works fell into disrepair, village life

could carry on quite adequately; for each village still possessed its own small-scale irrigation system which was maintained by the villagers themselves.

The major works collectively represent a colossal investment of labour effort but their construction was haphazard and discontinuous and spread over many centuries. We cannot infer from an inspection of these works the existence of a large labour force under central government control; nor can we make inferences about the size of the population which was fed by irrigation system. Still less can we make inferences about the nature of political authority in the ancient state.

In later times, Knox described the seventeenth-century monarchy as an unqualified despotism, but he makes no reference to the kind of bureaucratic core which, on my reading of Wittfogel, the state should still have possessed if Wittfogel's general thesis were correct.

What was true of Ceylon was not necessarily true elsewhere, but the Ceylon evidence as a whole clearly counts against the general validity of Wittfogel's argument.

Some of the facts of the Ceylon story fit the thesis well; others do not. Let us concede that it is a reasonable proposition to assert that it cannot be a mere accident of history that so many of the ancient major states should have started from a 'hydraulic core'. Some common sociological principle must be at work. But what kind of common factor should we look for? And are we necessarily dealing with a universal social law, or only with a principle which crops up in some cases and not in others?

Wittfogel claims that he has discerned a universal principle - the natural and intrinsic association between the formation of large-scale irrigation systems and the formation of bureaucratic-despotic political structures.

It may be that, in Chinese economic history, development was as he says. But did history repeat itself elsewhere?

The Ceylon data reminds us there is at least one other kind of social factor which is common to a great many hydraulic societies - namely that they are, from a structural point of view, peculiarly static. Wittfogel himself recognises, and indeed emphasises, this stasis but attributes it to the stagnant conservatism implicit in 'monopoly bureaucracy' (1957, pp. 422-3). The real explanation is much simpler.

Under Ceylon dry zone conditions, once a village and its irrigation tank have been constructed, it is there for ever and since the irrigation area must always remain the same size, the population of the village itself can only vary between very narrow limits.

Such a situation of enforced physical immobility lends itself to the development of social arrangements whereby the populations of neighbouring villages stand in fixed socio-economic relations one with another.

Durkheim's analysis of the 'Division of Labour' inclines us to imagine that complex 'organic' systems of labour specialisation

are intimately associated with the existence of a monetary exchange medium, the payment of cash wages, and the free mobility
of labour. But in the 'caste feudalisms' of India and Ceylon,
specialised occupations are professed not by individuals but by
whole groups (sub-castes), and, provided these sub-caste groups
are physically immobile, complex organic structures of specialised
labour division can be maintained from generation to generation
without any intervention of a money medium. The economic con-
tract is a permanent one between settled groups continuously
resident in particular localities.

Could it be that the sociological explanation of why so many of
the ancient societies were 'hydraulic' societies is that, in a
wide variety of circumstances, hydraulic society lends itself
very readily to the development of specialised labour on a non-
monetary basis?

I am not putting forward a new universal law but I suggest
that there are many cases of hydraulic society where the
characteristic pattern is 'Indian' rather than 'Chinese' and, in
which, in Wittfogel's terminology, the authority structure is
'feudal' rather than 'oriental'. I believe that this links up with
the basic difference between the Chinese acceptance of legitimate
authority as manifested in patrilineal kinship and bureaucratic
government as against the Indian reverence for personal
charisma. The Indian type of hydraulic society, of which Sinhala
is an example, is cellular not centralised in structure; localised
groups of technical specialists form a work team centred in a
leader. The major hydraulic works are not created rationally
and systematically but haphazard as pieces of self advertisement
by individual leaders. But once started, such constructions
survive and can be enhanced by later adventurers of the same
type.

The political conditions in 'Indian' hydraulic society may vary
very greatly in the course of a few centuries and for the analysis
of such changing circumstance the labels oriental despotism and
feudalism may both prove helpful. But to turn Marx's unelabor-
ated concept into the demon of our time will get us nowhere at
all.(7)

NOTES

*   From 'Past and Present', no. 15, April 1959, pp. 16–25.
1   Sir John D'Oyly, 'A Sketch of the Constitution of the
    Kandyan Kingdom (Ceylon)', Colombo, 1929, p. 27; cf.
    Ralph Pieris, 'Sinhalese Social Organisation', Colombo, 1956,
    pp. 233–5, 249–50.
2   K. Wittfogel, 1957, pp. 80–2.
3   F. Barth, The System of Social Stratification in Swat, North
    Pakistan, in E. Leach (ed.), 'Aspects of Caste', Cambridge
    Papers in Social Anthropology, no. 11, Cambridge University
    Press, 1969, pp. 113–46. Cf. McKim Marriot (ed.), 'Village
    India', Chicago, 1955.

4 H.W. Cordrington, 'Ancient Land Tenure and Revenue in Ceylon', Colombo, 1929, Chs III and IV.
5 M.B. Ariyapala, 'Society in Medieval Ceylon', Colombo, 1956, Ch. 3; S. Paranavitana, Glimpses of the Political and Social Condition of Medieval Ceylon, in Sir Paul Pieris's 'Presentation Volume', Colombo, 1956. Page 73 mentions a body called the dolos-maha-va-tana which was a government 'irrigation department' of some sort but there is no evidence at all that it played a political role.
6 B.H. Farmer, 'Pioneer Peasant Colonisation in Ceylon', London, 1955, Ch. 2.
7 R.K. Mukherjee, 'The Dynamics of a Rural Society', Berlin, 1957, pp. 69-80. It is of some interest that this author who now writes about India from a Marxist point of view, appears to find no inconsistency between earlier and later formulations on this topic.

IRRIGATION: SOCIOPOLITICAL DYNAMICS AND THE GROWTH OF CIVILIZATION*

Barbara Price

The application of Wittfogel's synthesis to test the case of developments in the Americas constitutes one special instance of the more general problem of broader assessment of its impact upon anthropology in the United States. A number of questions, both epistemological and substantive, are raised, even where this brief treatment is unable to examine all of these in any detail. First there is the problem of translation into the terminology of the distinct but closely related paradigm of cultural materialism (Harris, 1968), probably the most powerful strategy developed in anthropology for the investigation of problems of cultural similarity and difference, stability and change. This paradigm is marked by a strong Darwinian component, referring explanation to the operation of natural selection mechanisms upon differential bioenergetic efficiency under stated conditions. Concomitantly, causality is presumed to act in a systemic fashion, governed by positive and negative feedback loops; this model, predicts essentially incremental, continuous, and nonsaltatory manifestations of change. A more traditional typological model, more commonly used to address these questions, leads to other predictions. The result in turn is profound misunderstanding and misinterpretation of the hydraulic hypothesis, as a function of weakness or irrelevance of the initial research strategy.

Particularly in archeology there seems to be an at least implicit expectation that large-scale irrigation works in a hypothetical Phase I must be succeeded by evidence of large-scale centralized states in a Phase II - that this sequence of observations is necessary and sufficient for verification. Literalism of this sort traduces the original theory, and, moreover, militates against explanation; ultimately too it masks problems of system - of establishing and operationalizing context-dependent functional similarities and differences of 'traits' and the relationships among them. Wittfogel's original formulation, interestingly, avoids the pitfalls into which both his critics, and, at times, his admirers, regularly if often unwittingly tend to stumble.

In this paper, evolutionary and developmental problems raised by the hydraulic hypothesis are treated as incremental and systemic, an approach which may appear obvious in the case of a parameter so essentially quantitative as scale, far less so in the operational distinction of intensification and shift in the mode of production. The present aim is explanatory and its method comparative; in these respects it is consonant with Wittfogel's original formulation, from which it departs primarily in its reliance upon a causality explicitly governed by natural selection. Given the controversy attendant upon this departure, it is noteworthy that the translation into these terms is accomplished with ease and accuracy, and permits analysis from a somewhat distinct, albeit highly convergent, perspective.

We may now consider the nature of the relationship between irrigation and sociopolitical structure. It is here that controversy in contemporary anthropology is most acute. One of the reasons for this fact is the general failure to take into account the interrelated factors of agricultural productivity in general, and population growth. Only in this fashion can the impact of irrigation be evaluated: if irrigation is postulated to have certain effects, then what other techniques or processes might have similar kinds of impact on local ecosystems? We shall in turn relate these considerations to general problems of the interpretation of the evolution of social organization.

Childe (1950) states several characteristics of civilization as a culture type, among them large size, social stratification into distinct classes, and economic specialization. Writers on social organization from Durkheim (1933) to Fried (1967) have used the distinguishing criterion of institutionalized force to define the state: the state exists when some body in the overall structure arrogates to itself all legitimate use of internal and external force. Put another way, the state is founded on relationships of differential power. This returns us immediately to the criterion of social stratification, defined by Fried (1960) as differential access to strategic resources. Economic power is ultimately political power as well; the two are inextricably interwoven. Some resource bases, among them hydroagriculture, are inherently more controllable in this fashion than are others. The relation between what archeologists term 'civilization' and what in political terms is called the state may be debated by some. However, for the present discussion the two terms may be considered equivalent, and will be used interchangeably (Sanders and Price, 1968). In the case of the New World pristine states which developed in Mesoamerica and the Central Andes, our analyses of social and political structure are necessarily based on the material remains these societies have left in the sequence of their development. It is a methodological requisite that we establish a degree of equivalence between archeological remains and the kind of society that produced them. This will be the operational basis of our inferences concerning the evolution of culture (ibid., Ch. 2).

A basic archeological criterion of civilization is that of
architectural monumentality (Childe, 1950). The existence of
large-scale public works of any kind tells us something about
the society that built them: they are the material expression of
a kind of social organization. White (1949, 1959) maintains that
cultural evolution is based on the increase of the total energy
content of the society. Monumental architecture is, in a real
sense, the frozen, permanent indication of the amount of energy
harnessed; it is thus possible to compare societies on this
quantitative basis. It is further possible to consider social
organization of any kind as a sort of flow diagram of the
utilization of energy captured and consumed by a population.
Monumental architecture, because the capital and labor (energy)
investment in its construction is high, is thus a legitimate and
justifiable indication, not only of total energy capacity, but of
the fact that this energy was used in a particular way. It is an
aspect of technology that has very clear sociological requisites
and implications.

Large-scale irrigation systems, if they are functionally similar
to various other kinds of agricultural methods in their effects
on productivity, are functionally similar in other respects to
any other kind of architectural monumentality. They share with
pyramids, temples, fortification walls, palaces, and burial
mounds the need for large quantities of systematically amassed
material; a large, organized and directed labor force; and a
diversity of labor force, from unskilled hewers of wood and
drawers of water, to professional specialists in engineering,
architecture, planning, and administration. Thus, on the level
of process, the existence of large-scale hydraulic works of the
kind described by Wittfogel (1957) can be analyzed as resembling
any monumental civic construction, in that similar kinds and
organization of resources and manpower are requisite for them
all. Where hydraulic works differ from these other examples is
precisely where they resemble chinampas, terraces, pukios, etc.
They represent energy inputs that result in augmented productivity: they are capital investments in a way that temples,
palaces, and fortifications are not. Irrigation systems require
investment of energy to produce more energy, where monumental
pyramids and massive walls can be viewed as taking energy out
of circulation. The latter category of monumentality is thus more
properly considered an effect or product of a certain kind of
society whose techno-environmental causal bases lie elsewhere
(Sanders and Price, 1968, p. 9). The question is the extent to
which canal irrigation can be regarded as a major component of
these techno-environmental causal bases, and can thus be seen
as a determinant of institutional structure.

Irrigation agriculture is not a unitary phenomenon of unique
characteristics. Irrigation in general shares many of the causes
and effects common to other systems of intensive agriculture.
Further, not all systems of canal irrigation are themselves strictly
alike. Many of the characteristics in which irrigation systems

differ among themselves are those which determine the actual
or potential scale or degree of monumentality of the system.
Many of these determinants are basically environmental. The
conditions of construction and use of an artificially controlled
water source and its impact on the size and institutions of a
population are not everywhere the same.

Most obviously, environments vary in the degree to which
control of water supply is necessary to the successful exploitation of the habitat by cultivators: environmental challenges
largely govern the direction and degree of success of cultural
responses. In some environments such as coastal Peru or parts
of the American southwest the total productivity of the region is
sharply and dramatically increased with irrigation; without it,
very little land will produce a reliable harvest. In the Teotihuacan
Valley, the difference, while highly significant, is less drastic
than the situation on the Peruvian coast. Even on the Peruvian
coast, however, carrying capacity does not reduce to zero
without irrigation: cultivation is not the only means of possible
or observed subsistence, and a sizable population can be maintained by marine food collecting. But if population is to expand
beyond the limitations imposed by this mode of life, a shift in
technology and economy to agriculture is necessary to open up
additional niches of the habitat to occupation. And if such a
shift is to provide more than merely an occasional supplement
to the diet - that is, for its expansion of the demographic
ceiling of the valley to be significant - in coastal Peru that
shift must include the technology of water control, since there
is no rainfall. A similar analysis is possible for the shift in
the Mesopotamian Plain whereby cultivators assumed dominance
over riverine collectors - not so much by replacing them, since
the contemporary Marsh Arabs continue to practice a similar
way of life, but by substantially outnumbering them. Such a
change indicates a shift in the subzone of the region which is
most favorable to occupation by man: more people can live by
cultivation in the plain than by collecting wild foods from the
river. For them to have done so would involve water control
since, like the Peruvian coast, the most striking orographic
characteristic of this environment is its near-total lack of
precipitation.

What is, or is not, a limiting factor to demographic expansion will vary from one environment to another, depending upon
physical geography and also upon the size and technological
repertoire of the population. To continue our consideration of
environmental parameters, the quantity, nature, and degree
of permanance of the water source will induce differences among
and between various empirical instances of irrigation cultivators.
For cultivators, normally either land or water limitations will
be the most critical inhibitor of expansion. Any population,
human or animal, can expand only to the extent permitted by
that resource basic to its way of life that is in the shortest
supply. In the Teotihuacan Valley, there is more land irrigable

with existing technology than there is available water. Thus
the irrigation system observed today is at its limit and cannot
in practical terms be further expanded (Millon, 1954; Millon
et al., 1962; Sanders, 1965). By Middle Horizon times in north
coastal Peru (Willey, 1953; Moseley, 1969), the expansion of the
river-fed canal systems was already supporting as many people
as it could. All available water was in use; where there was no
water, the rest of the presumably otherwise cultivable valleys
remained desert, lacking any demographic potential. The result
was a relative isolation of the valleys from each other, in a
fashion parallel to the situation obtaining in the American southwest, though on a larger scale.

In the southwest, the Pueblo Sphere of Kroeber (1939, p. 136)
includes a total area estimated at 44,600 square kilometers, and
a total population of 33,800, for an overall density of 0.75 per
square kilometer. The southwestern population is distributed
in a number of small clusters, some internally quite dense, but
each separated from the others by considerable stretches of
uninhabited desert. Each of these nucleated settlements has
fewer than 5,000 and most under 3,000 inhabitants; densities
of individual groups cited by Kroeber range from 0.21 to 2.71
per square kilometer. The subsistence base of many of these
pueblos is irrigation agriculture, particularly floodwater and
seasonal-spring irrigation. Each is surrounded by unused land
for which no water is available. In some instances, however, it
may be land rather than water that is the critical resource in
the shortest supply. In the southern part of the Basin of
Mexico, for example, the lakeshore plain is narrow, and the
response to this restriction under increasing population pressure
has been the terracing of the adjacent hill slopes and the creation of chinampas as artificial islands in the lakes. While the
chinampas are also permanently irrigated by virtue of their
location, most of the terraces observed today are dry (Sanders,
1957). The spectacular terracing of the Cuzco Basin seems to
represent a similar solution to a similar problem.

The nature of the water source, particularly its seasonality
or permanence, as well as its quantity, will affect the technology
and sociology of its exploitation. In the Teotihuacan Valley,
there is a marked difference between the floodwater canals of
the middle valley and the spring-fed perennial flow into the
lower valley canals. Only rainy-season agriculture is possible
with a floodwater system. If the yields and the security are increased over the levels possible with rainfall alone, the total
harvest per year cannot be expanded through double-cropping.
Thus, the carrying capacity of lower valley land is much higher
than that of the middle valley. Conversely, the absence of an
appropriate permanent water source in the middle valley limits
the use of any perennial irrigation there. The valleys of the
south coast of Peru were similarly limited in their demographic
growth; they are drier than the north coast valleys, and many
of their rivers are either seasonal or are dry before they reach
the Pacific.

The carrying capacity of an environment cannot therefore be assumed as given, or evaluated without reference to the total technology of the population. This is the basic methodological error in the discussion by Meggers (1954) of environmental potential; she compares environments and their relative productivity on the basis of a 1950 technology as a baseline. However, the productivity of environments in comparison with each other has notably shifted through time, as the preceding pages have indicated. Many of these shifts, inferred on the basis of settlement distribution, are directly attributed to changes in technology which modify those factors of an environment affecting production and carrying capacity. For some environments, modifiability is a function of investment of labor; success in this case is a function of the extent to which the output of the land is increased by the additional labor input. In other words, it may be uneconomic to invest labor if the return on the labor is less than the input required. For other environments, the technology for 'reclamation' may simply not exist.

The technology exists today, for example (as it did not in pre-Columbian times), to pump water from the Columbia River drainage in Oregon and Washington to the Napa Valley of California. It is expensive to utilize this technology, and yet economic to do so because other soil and climate factors make it possible for this region to then produce an immense variety of truck garden crops of high market value, often either out of season or uncultivable elsewhere in the country. Concomitantly, an efficient transportation system exists to get this produce quickly to areas in which maximum demand exists. If the Napa Valley, however, used this complex and expensive irrigation technology to produce wheat or cotton, which can be grown as well and more cheaply elsewhere, the cost of irrigation would be too high, and the profits too small. Similarly, if there were no means to deliver the produce to its market, or if there were no demand, again there would be no economic basis for the heavy investment in irrigation.

For still other environments no technology yet exists that is capable of intensifying methods of agricultural production. Lowland tropical forest regions in general constitute such an example. The only way to produce maize in, say, southern Veracruz or Tabasco, is by swidden cultivation. Fallow cycles may be longer or shorter depending on population. Where land is plentiful and people few, there may be no effective cycling at all until all virgin territory has been occupied and used. But there are natural, effectively environmental, limitations (empirically variable) on how short the fallow period can be. In some of the best lands in Tabasco a 1:3 cycle is used, and yields fall off if this is reduced to 1:2. The amount of land to which so short a fallow cycle is applicable is, moreover, limited. In parts of Yucatan the cycle may range from 1:6 to 1:10 or longer. Increased weed competition and thus greater labor required, in combination with yields declining from loss of fertility through leaching, make cultivation for longer than two successive years uneconomic. Along river levees of the Gulf Coast, nearly permanent cultivation may be possible (Coe, 1968), where annual inundation and silting renew fertility; but the total quantity of such land is limited. Intensification of agricultural production by technology in practicable terms in the lowland tropics generally takes the form of production based on tree crops which do not deplete the soil to the extent of annual cultivation of cereals. Given such a specialization, in turn, the region involved necessarily depends upon importation of staples from elsewhere, and upon access to markets for its own produce.

Thus, in only a few settings will irrigation be the key to demographic expansion. Its functions in raising carrying capacity may vary. It may increase productivity in a zone already cultivated by more extensive techniques by increasing yield, security of harvest, or by permitting multi-cropping, the production of new crops, or more productive varieties of existing crops. It may open additional areas for agriculture, areas that prior to irrigation were unused or only marginally productive. In the course of the development of the early New World civilizations, hydraulic systems in parts of these nuclear areas could be expanded until at their maxima they were the economic bases of the large, complex societies discovered by the Spaniards. While many other societies were and are known to practice irrigation agriculture, these for various reasons never attained the scale or degree of complexity of the high civilizations.

Irrigation alone, therefore, is incapable of fully explaining the growth of civilizations. Irrigation cultivation, while freeing a population from some of the limitations on its growth, may itself be subject to limitations - both environmental and technological - in its expansion. We must look to some of the parameters discussed above, and to their functional implications, for an explanation as to why some irrigated systems grow large, to include increasingly larger number of people and extents of territory, while others remain small, simple, and local.

Not all irrigation constitutes hydraulic agriculture of the kind described by Wittfogel. And if Wittfogel (1957) does include a discussion of the American southwest, he strays from his central point and to that extent dilutes his own argument. No one can consider a southwestern pueblo to be anything but an essentially egalitarian tribal group, however dependent on irrigation it may be for subsistence, however high its hydraulic density. Groups of this size and degree of complexity are clearly competent to manage this level of technology without the kinds of sociopolitical institutions described for the Irrigation Civilizations. Their reliance on irrigation has not per se transformed them into Irrigation Civilizations.

Wittfogel's oriental society is characterized by a centralized and highly despotic bureaucracy as the locus of power and entrepreneurship in the society, often incorporating both ecclesiastical and secular arms (understandably, since both

church and state are institutions of social control). The bureaucracy is the group of full-time specialists, maintained from the
surplus of the primary producers; economic, political, and
military power are concentrated in the hands of these specialists.
Wittfogel maintains (1955, 1957) that the power of this group
derives in turn from the requisites of hydraulic agriculture and
the need for centralized direction and control of the hydraulic
works in order to keep them functioning efficiently. The capital
and labor needs for construction and maintenance of large-scale
irrigation systems are such that, first, only the central bureaucracy could undertake these projects successfully: no other
segment of the society could afford them. Second, the competitive
situation, within the society and between societies, is greatly
exacerbated when the basic means of production are artificial;
some central authority is thus necessary to control the use of
force and to adjudicate disputes which could potentially disrupt
the entire system. Third, the cooperative nature of the labor
requisites, on a large scale, suggest the advantages of
centralized control in the amassing of manpower when and where
needed. Fourth, the hydraulic management needs per se give
to a group in control extremely effective sanctions and thus
enormous power to back up its demands. As Childe observes
(1942, p. 90): 'Rain falleth upon the just and the unjust alike,
but irrigating waters reach the fields by channels the community has constructed. And what society has provided, society
can also withdraw from the unjust and confine to the just alone.'
The government, in other words, need not even call out the
army to enforce its decisions: all it has to do is turn off the
water.

If it is agreed that Wittfogel's hypothesis of the causal linkage
between a particular kind of economy and a given form of government is applicable only to large-scale systems, we may ask why
it becomes necessary to investigate how large is large enough,
and the processes by which changes in scale are effected. On
the basis of the foregoing, these developments are most produc-
tively treated as incremental, systemic, and in the instances of
growth, governed by consistent positive feedback loops. Any
question of scale, first of all, is inherently quantitative: analysis
should reveal a predicted continuum of observed results rather
than generate a series of mutually exclusive types. Emphasis on
the explanation of processes of change, second, requires examination of the theoretical and empirical constraints acting upon
particular productive regimes, constraints which impose negative
checks upon growth, and which may differ substantively from
one context to another even while resembling each other in the
effects they produce. Virtually everything in Wittfogel's original
model justifes the legitimacy of this approach.

We have alluded to the question of relative scale and expanda-
bility of irrigation systems above, in order to point out the
techno-environmental parameters that may act to impede or to
permit growth. The size of the system is in turn a major deter-

minant of the labor and management needs involved. Since
this, in Wittfogel's developmental scheme, is the underlying
dynamic on which the argument linking production to political
structure rests, it merits more detailed examination. Our best
evidence is obtained through the use of the comparative method.
Both ethnographic and archeological data are relevant to what
is essentially a problem of defining the situation in which a
qualitative change is likely to result from the operation of
essentially quantitative processes.

Millon (1962) has questioned the Wittfogel hypothesis of the
relationship between centralized authority and irrigation agriculture. He has observed quite correctly that the dynamics of
water management are, on the basis of comparative ethnography,
far more variable than Wittfogel's theory would indicate. His
example of the Sonjo (Gray, 1963) and the case of Pul Eliya
(Leach, 1961) are small, single-community systems. They thus
resemble both the contemporary southwest and the inferred
situation for a system such as Hierve el Agua. Comparatively
small population groups are involved in all these cases - in the
case of Pul Eliya, only 146 people, irrigating a total of 135 acres.
Under such circumstances, in spite of the reliance upon artificial
water control, kinship and sodality ties are sufficient to integrate
the society, and to control both water supply and distribution
and the labor requisite for maintenance of the system. Where the
source of water is both local and limited in quantity, the question of disputes between different communities competing for its
control is less likely to arise. And where quantity limits expansion, kinship and other essentially egalitarian ties are sufficient
to control intracommunity competition and conflict. Millon cites
a Balinese example in which parts of several settlements are
involved in irrigation cooperatives, but the numbers include
only 5,500-7,500 people each - again, evidently below the level
at which centralization of authority becomes an effective solution to the problems of management and control.

Irrigation systems in general do have requisites different from
most other types of agriculture in that, no matter how small
their scale, cooperative rather than individual effort is generally
necessary to use and maintain them effectively: the productive
unit is therefore larger than the single household. Such co-
operation of course need not be amicable; indeed it is fraught
with conflict. However, the need for cooperative labor means,
in effect, that the amount of total output of the system per unit
of labor expended will be greater if a number of workers pool
their efforts than it would be if the same number of workers
each operated independently, each expending the same amount
of effort. Expansion of the system will be efficient and economic
only when additional labor investment will add more than the
value of its own input to the total output of energy, by bringing
more land under cultivation or by adding to the yields from land
in production. When putting additional labor into the land does
not result in this kind of increased output, the agricultural

situation, no matter what its technology, may, following Geertz
(1963), be referred to as 'involuted'.

If cooperative enterprise is a virtual necessity for irrigated
land at even the single-community level, this is not necessarily
the case for other kinds of productive systems. In the several
varieties of swidden, while farmers may cooperate in one or
several of the necessary operations, the labor is essentially
individual. Three farmers clearing three fields together involves
the same input-output ratio as each of three farmers separately
clearing each of three fields; it is merely additive rather than
multiplicative. Pot-irrigation, while an intensive technique in
that it maximizes production per hectare by means of labor
investment (Flannery et al., 1967), is similarly an individual-
household enterprise; cooperation of larger groups does not
result in increased output. Both the enormous yields and the
enormous labor inputs of contemporary chinampa agriculture in
the southern Valley of Mexico similarly involve an essentially
individual rather than cooperative patterning. Terracing is, in
a sense, an intermediate case. The labor investment to produce
a crop in terraced agriculture not simultaneously involving
floodwater or permanent irrigation is again basically individual.
But terraces usually occur in groups rather larger than the
isolated holding of an individual family. Terraces, like irrigation
canals, require continuous upkeep to maintain them in good
repair. While chinampas also require such attention, a single
farmer's eroded chinampa menaces the security of no one but
himself. A single disintegrating terrace, however, can threaten
the productivity of all the holdings located below it. Thus, like
small irrigation systems operating at the single community level,
community-level cooperation is generally advantageous in terrace
cultivation.

When either geographical or technological factors, or the
combination of both, inhibit the expansion of an irrigation system
and its demographic ceiling, the system will stabilize at that
level. While the economic advantages of cooperative labor in its
management will still obtain, the total labor force will remain
sufficiently small so that centralized institutions of control will
not be necessary. Nor could they be supported: there would
not be enough work for such specialists to do to justify their
upkeep. Size and elaborateness of any institution of social control
will depend directly on the size of the society; the number of
chiefs, so to speak, depends on the number of Indians.

It is difficult and perhaps ultimately somewhat arbitrary to
attempt precise definition of the point along the continuum where
quantitative changes may be analyzed as qualitative ones. We
may observe that a kind of variable, critical-mass phenomenon
is evidently involved, in which the absolute size and density of
population must be viewed not by itself but in relationship to
parameters such as the degree of environmental circumscription
(Carneiro, 1961), the degree of environmental diversity, the
overall technological level of the population and thus the extent

to which differential utilization of different sectors of the habitat may be both possible and economically efficient. Specialization and symbiosis (Sanders, 1956) may thus themselves be viewed as adaptive developments, techno-economic and sociopolitical means of raising the demographic ceiling of a total region and thereby responding to conditions of population pressure (Sanders, 1968). However, if, as we have done, we stress the applicability of Wittfogel's hypothesis only to large-scale hydraulic and social systems, at least an attempt must be made, despite these cautions, to define the range within which a 'small' system becomes for the purposes of analysis a 'large' one.

Again we turn to the comparative method for assistance. There is clearly a difference of degree that operates as a difference in kind if we compare, say, Hierve el Agua with the Chicama-Moche transvalley system. All change proceeds quantitatively just as all process is incremental; the poles of the continuum are clear, but the intermediate ground far less so. Comparative data, both ethnographic and archeological, may perhaps help to clarify and order this middle range. If we are to speak of 'large' systems, such analysis will be critical. Armillas (1948), for example, considers the hydraulic systems of Late Postclassic Central Mexico to have been essentially 'cantonal'. The multiplicity of individually limited water sources, and their localized distribution, effectively barred the development of a unified and centralized basinwide single system. Limitations of geography thus restricted the size and distribution both of population dependent on a single system, and the possibility of merging several smaller systems into a single larger and more highly centralized one. Adams (1966) similarly notes the late inception of a pan-Mesopotamian single, huge irrigation network - in the Iron Age. It is our view that such super-systems are far larger than either necessary or sufficient for explanation of the sociological concomitants and consequents of large-scale irrigation. Clearly, techno-environmental parameters limit the possibility or probability of this degree of expansion; in Mesoamerica it was not feasible, while in Mesopotamia this did ultimately occur, but as the end-product of long, individual sequences of localized growth. While Wittfogel's formulations are clearly applicable to such super-systems, these dynamics can be regarded as operative at levels well below this point.

The contemporary Teotihuacan system, 'small' in Millon's sense (1962), serves all or parts of sixteen villages. Overall, the contemporary Aztec and Classic populations were similar in size (Sanders, 1965), though varying in composition and distribution. The settlement pattern of the Classic, described in a previous section of this paper, implies that some 85,000-100,000 people at least - the population of the city of Teotihuacan itself - were ultimately wholly or partly dependent on the productivity of this system. Furthermore, this same evidence of settlement patterns strongly implies the conclusion that the regulation of the system involved a high degree of centralized control. First,

Teotihuacan is the largest single settlement in the entire Basin
of Mexico, and very probably in all of Mesoamerica at the time;
its pan-Mesoamerican repercussions were immense and far-
reaching in nearly all aspects of culture (Sanders and Price,
1968; L.A. Parsons, 1969, Ch. 5; Parsons and Price, 1970).
Second, Teotihuacan, while not the only Classic period settlement in Teotihuacan Valley, does seem to have virtually depopu-
lated the lower valley - the permanently irrigated sector of its
immediate sustaining area; its presence evidently inhibited
significant population expansion in the Texcoco Plain area as
well, though this zone, as previously indicated, more probably
constituted a part of the outfield aspect of the economy of
Classic primary production.
      Sanders and Price have suggested (1968, pp. 195-6) a parallel
with the post-conquest Spanish policy of congregacion, or enforced nucleation of population. As in the post-Hispanic instance,
imposition of a settlement policy and its enforcement involves
the wielding of considerable power. The Classic period lower
valley settlement pattern seems largely uneconomic and anti-
ecological. It is generally more efficient for farmers to live on
or near their holdings, and particularly so when they have no
access to any mode of transport more efficient than feet. Yet,
the few rural villages in the lower Valley of Teotihuacan contemporary with the height of the city seem to have exploited
the upper piedmont in which they are located, and to have had
no access at all to the irrigated valley floor. The explanation
may lie in the inferred imposition of a congregation, which would
in turn imply strong central authority. The settlement pattern
looks as though access to the prized lower valley lands may
have actually been contingent on city residence, where the
central authority could exert considerable socioeconomic control
over the population, and could in short have behaved very much
like an oriental despotism. The sanctions involved would have
been extremely potent and not difficult to apply. The overall
degree of local centralization, on the basis of the settlement
pattern, thus seems strikingly higher than that observed at
present in Teotihuacan Valley.
      For the Viru Valley, we have previously cited Willey's estimate
(1953) that a population maximum of 25,000 was reached by Late
Gallinazo times, and the suggestion of Moseley (1969) that the
irrigation system had also reached its maximal capacity then. It
is also in Late Gallinazo times that truly monumental architecture
appears in Viru, particularly so at the largest site of the period.
On this evidence, Willey postulates the existence of a local state
based on control of a unified, valley-wide irrigation system. In
the Moche Valley, the Huaca del Sol and Huaca de la Luna date
from approximately this period or somewhat later, at the peak
of Moche military and political expansion. That expansion, however, indicates that a parallel process to that of local state
formation in Viru was probably contemporary with or somewhat
earlier than the Viru developments. We shall return subsequently

to the questions of militarism and conquest states, in a different
but related context.

As additional comparative material in this question of scale, we
cite the evidence on size of system from Mesopotamia. Braidwood
and Reed (1957) estimate an average population of some 17,000
for each of the Sumerian city-states by approximately 3000 BC.
Associated with each such unit is considerable monumental civic
architecture, thriving long-distance commerce, and social stratification, which are among Childe's previously mentioned criteria
for the archeological recognition of civilization. Parallel to the
evidence of militarism in north coastal Peru in Moche and immediately pre-Moche times (Willey, 1953; Lanning, 1967), and in
Central Mexico in Teotihuacan times (Sanders and Price, 1968),
the Sumerian city-states were in a constant state of warfare with
each other (Adams, 1966). That the local irrigation systems of
Sumer only much later coalesced into the unified pan-Mesopotamian
system is for the present irrelevant. It does, however, appear
to be the case that each individual local city-state system was
large enough to produce the effects observed, just as with the
individual Viru, Moche, or Teotihuacan systems.

Wittfogel is ultimately discussing a shift in the mode of production and its necessary consequences for political organization.
In doing so he justifies the one parameter - irrigation - as
epistemologically and substantively the most powerful of a number
which coexist in feedback relations. But this critical parameter
may, and initially must, come into being merely as a component
of some preexisting productive regime, in which it acts merely
as an intensifier, perhaps as one of many. In other words, it
represents an investment of additional labor to increase production or reduce risk in an ongoing system; selection pressures
favoring retention of this or any other innovation emanate from
that system, not from what it might someday become. Accordingly
the relation between intensification and shift, like that which
governs changes of scale, is a quantitative one. In a shift the
innovation in question has come to produce the bulk of the
society's calories and to absorb the bulk of the total labor imput.

The present argument addresses a widespread misinterpretation
of the hydraulic hypothesis, that all instances of irrigation agriculture should manifest the political concomitants cited by
Wittfogel, an erroneous deduction based on the implicit assumption
that all instances of the occurrences of irrigation necessarily
entail a shift in the mode of production. By contrast, we concur
with Wittfogel that the probability of this developmental trajectory
is actually quite low. It follows that the more productive research
strategy should investigate the conditions under which this - or
some other - innovation ultimately generates a shift, and how
these differ from others in which this process is aborted. Toward
this end the discussion below will attempt to develop the operational basis necessary to the evaluation of functional similarity
and difference of traits within the context of stated productive
systems.

Cultural evolution is seen ultimately as the result of a series of ecological processes operating in time, to intensify or to neutralize each other. No single practice or trait can be assigned as a priori causal preeminence; but some kinds of causal parameters will be more powerful determinants of development than will others. The most important characteristic of the ecological method lies in its broadly interactive approach, which includes consideration of certain kinds of sociopolitical factors as part of the total ecosystem of a people. Thus, it is not the existence of irrigation agriculture as an entity to which causation can be uncritically attributed that concerns us. We ask instead what its repercussions are throughout the ecosystem in particular cases and, thus, by relying on process and function rather than on form, we may examine other cultural phenomena which may, under specified circumstances, produce similar effects.

Irrigation, for cultivators in arid lands, has its initial impact on the productive cycle itself. So too may other technological practices. Conversely, an artefact or structure may represent some cultural means of water control but lack impact on the productive potential of the environment. Coe's (1968, p. 64) stone drain at San Lorenzo is such an example; so is the system of drains observed at Teotihuacan. These represent energy utilization by a population in exactly the way a colossal stone head or a pyramid does: the materials used must be collected, and energy is expended to give these raw materials their final form. While there are practical advantages to removing the rainwater from one's patio, these do not increase the total energy content of the society. Thus, not all observed technology of water control can be analyzed in the same manner. Processes and functional implications as well as form must be considered; not all water control necessarily represents investment of energy that produces more energy.

Besides their impact on productivity in many instances of their occurrence, artificial waterworks are significant because they represent not only a critical resource on which a greater or lesser percentage of the population depends, but also a critical resource which is controllable with great ease. Irrigation is not unique in this respect. Any factor on which a population is dependent may at some level of demographic growth impose a ceiling on further expansion, but some kinds of resources are more subject to technological and sociological controls than are others. Particularly in some phases of demographic expansion, access to trade routes may function as an equally, or nearly equally, vital resource. Where populations have grown to the point where egalitarian direction of trade relations is no longer sufficient to secure regular and consistent access to goods needed but not locally produced, access to trade institutions may stimulate both a nonegalitarian social structure and some degree of economic centralization. This seems to have been the case in Middle Formative Mesoamerica, as we have previously observed. A similar interpretation is the basis of Steward's (1955) treatment of a Teotihuacan for which, at the time, no substantive evidence existed to support its status as an Irrigation Civilization. The basis of political power may lie in the controllability of the 'lifeline' of the population in the ecosystem in question, whatever that lifeline may be. The impact of irrigation in this respect is extremely powerful, but under certain conditions other parameters may operate, to a greater or lesser extent, to produce similar effects.

The relationship between irrigation and centralization is therefore not considered unique. Both centralization of authority and internal social differentiation may be responses to a number of different empirical factors. While irrigation agriculture is inefficient without cooperative effort, such cooperation may be stimulated, again in certain stated demographic and geographic contexts, by other factors. Access to markets, as cited, may be such a factor, especially in areas of close micro-geographic zoning where intense specialization has been a response to population growth and a solution to the problem of maximizing overall production. The economy of each component of such a symbiotic region (Sanders, 1956) depends on regular access to the produce of all other components. Although swidden agriculture, as another example, is usually regarded as strongly centrifugal, the approach of the demographic ceiling may be accompanied by centralized control of the agricultural cycle and the allocation of farmland. There is some indication that this or a related process may have been operative in Late Classic Maya society. The point is that environmental circumscription which deters expansion may be sociological as well as strictly geographical. Under such conditions, the land itself becomes controllable, far more so than is usually characteristic of swidden systems. In circumstances where increasing numbers depend on a resource available in limited supply, that resource becomes more controllable sociologically than it would otherwise be; sanctions may be applied easily and made to stick.

The sociocultural effects of irrigation are, as we have noted, very powerful in all these respects. They will be especially powerful where the combination of geography and productive technology permit, in time, the expansion of small local systems - and small local populations - into large supralocal ones. But irrigation is not unique in this respect. The approach we advocate is an essentially multilinear one: it cannot be assumed that 'the same' trait formally defined will always behave functionally in the same way in all ecosystems. 'The same' trait may be a response to different challenges in different contexts. Similarly, traits that are formally quite different may, depending upon total context, produce quite similar kinds of effects. These are necessarily matters for empirical determination.

BIBLIOGRAPHY

Armillas, Pedro, 1948, A sequence of cultural development in
  Mesoamerica, in Bennett, W.C. (ed.), 'A Reappraisal of
  Peruvian Archaeology', Society for American Archaeology,
  Memoir no. 4, pp. 105-11.
Braidwood, R.J. and Reed, C.A., 1957, The achievement and
  early consequences of food production, in 'Cold Spring Harbor
  Symposia on Quantitative Biology,' 22, pp. 19-31.
Carneiro, Robert, 1961, Slash and burn cultivation among the
  Kui-Kuru and its implications for cultural development in the
  Amazon Basin, 'Anthropologia,' no. 10.
Childe, V. Gordon, 1942, 'What Happened in History,' Penguin
  Books, 1942.
Childe, V. Gordon, 1950, The urban revolution, 'Town Planning
  Review,' vol. 21.
Coe, Michael D., 1968, San Lorenzo and the Olmec civilization,
  in Benson, E.P. (ed.), 'Dumbarton Oaks Conference on the
  Olmec,' Washington.
Durkheim, Emile, 1933, 'The Division of Labor in Society,'
  New York.
Flannery, Kent V. *et al.*, 1967, Farming systems and political
  growth in ancient Oaxaca, 'Science,' vol. 158, no. 3800,
  pp. 445-54.
Fried, Morton H., 1960, On the evolution of social stratification
  and the state, in Diamond, S. (ed.), 'Essays in Honor of
  Paul Radin,' New York.
Fried, Morton H., 1967, 'The Evolution of Political Society,'
  New York.
Geertz, Clifford, 1963, 'Agricultural Involution: the Process
  of Ecological Change in Indonesia,' Berkeley, Calif.
Gray, R.F., 1963, 'The Sonjo of Tanganyika,' London.
Kroeber, Alfred L., 1939, 'Cultural and Natural Areas of
  Native North America,' University of California Publications
  in American Archaeology and Ethnology.
Lanning, Edward P., 1967, 'Peru before the Incas,' Englewood
  Cliffs, N.J.
Leach, E.R., 1961, 'Pul Eliya: a Village in Ceylon,' Cambridge
  University Press.
Meggers, Betty J., 1954, Environmental limitation on the
  development of culture, 'American Anthropologist,' vol. 56,
  no. 5, pp. 801-24.
Millon, René F., 1954, Irrigation at Teotihuacan, 'American
  Antiquity,' vol. 20, no. 2, pp. 177-80.
Millon, René F., Hall, Clara and Diaz, May, 1962, Conflict in
  the modern Teotihuacan irrigation system, 'Comparative
  Studies in Society and History,' vol. 4, no. 4.
Moseley, M. Edward, 1969, Assessing the archaeological signifi-
  cance of mahamaes, 'American Antiquity,' vol. 34, no. 4,
  pp. 485-7.
Parsons, Lee A., 1969, 'Bilbao, Guatemala: an Archaeological

Study of the Pacific Coast Cotzumalhuapa Region,' 2
(Milwaukee Public Museum, Publications in Anthropology,
no. 12).
Parsons, Lee A. and Price, Barbara J., 1970, Mesoamerican
Trade and its Role in the Emergence of Civilization, paper
presented at Burg Wartenstein Symposium no. 47, Austria.
Sanders, William T., 1956, The Central Mexican symbiotic region,
in Willey, G.R. (ed.), 'Prehistoric Settlement Patterns in the
New World,' New York.
Sanders, William T., 1957, 'Tierra y Agua', PhD dissertation,
Harvard University.
Sanders, William T., 1965, 'Cultural Ecology of the Teotihuacan
Valley', Department of Anthropology, Pennsylvania State
University (mimeo).
Sanders, William T., 1968, Hydraulic agriculture, economic
symbiosis and the evolution of states in Central Mexico, in
'Anthropological Archeology,' Washington.
Sanders, William T., and Price, Barbara J., 1968, 'Mesoamerica:
the Evolution of a Civilization,' New York.
Steward, Julian H., 1955, Some implications of the Symposium,
in Steward, J.H. (ed.), 'Irrigation Civilizations: a Compara-
tive Study,' Washington.
White, Leslie A., 1949, 'The Science of Culture,' New York.
White, Leslie A., 1959, 'The Evolution of Culture,' New York.
Willey, Gordon R., 1953, 'Prehistoric Settlement Patterns in
the Viru Valley, Peru,' Washington.

NOTE

* Expanded from 'Prehispanic Irrigation Agriculture in Nuclear
America', in M. Fried, ed., 'Explorations in Anthropology,'
New York, Crowell, 1973, pp. 225-33, 242-3.

SELECTED FURTHER READINGS: PART III

M. Harris, 1968, 'The Rise of Anthropological Theory', London,
Routledge & Kegan Paul, ch. 22.

M. Harris, 1971, 'Culture, Man and Nature,' New York, Thomas
Y. Crowell, ch. 17.

E. Mandel, 1971, 'The Formation of the Economic Thought of
Karl Marx', London, New Left Books, pp. 116-39.

W.P. Mitchell, 1973, The hydraulic hypothesis: a reappraisal,
'Current Anthropology,' 14, pp. 532-4.

B. Moore, 1962, 'Political Power and Social Theory', New York,
Harper Torchbooks.

F. Tökei, 1972, Contribution à la nouvelle discussion sur le
mode de production asiatique, 'Nouvelles Etudes Hongroises',
pp. 80-95.

G.L. Ulmen, 1975, Wittfogel's science of society, 'Telos', 24,
pp. 81-114.

R.L. Winzeler, 1976, Ecology, culture, social organization and
state formation in Southeast Asia, 'Current Anthropology',
17, 4, pp. 623-40.

K.A. Wittfogel, 1969, Results and problems of the study of
oriental despotism, 'Journal of Asian Studies', 28, pp. 257-65.

K.A. Wittfogel, 1972, The hydraulic approach to Pre-Spanish
Mesoamerica, in F. Johnson (ed.), 'Chronology and Irrigation',
vol. 4 of 'The Prehistory of the Tehuacan Valley', University
of Texas Press.

PART IV
THE CONTEMPORARY DEBATE
ON THE AMP

EDITORS' INTRODUCTION

The silence imposed upon the discussion of the AMP among
Marxists broke in the late 1950s. Today there exists an extensive literature on the subject by scholars from many parts of
the world. Broadly speaking, this literature falls into three
categories, although there may be considerable overlap among
these categories in any single work.

First of all, there has been a notable increase in marxological
studies and a wider circulation of Marx's writings on pre-
capitalist social formations, on colonialism, and of his correspondence with the Russian populists. We have witnessed a renewed
interest in Marx's characterizations of 'non-European' society,
the sources from which he worked, and in whether or not Marx
operated with a well defined concept of a specific mode of production, the Asiatic.

Second, the recent discussions of the AMP have coincided
and fed into a wider, re-examination of historical materialism.
Theoretical justifications for modifying, developing, or dis-
carding a concept of the AMP are grounded in more fundamental
explorations of the epistemological and theoretical status of the
concepts of 'mode of production', 'social formation', 'class',
'relations of production', 'exploitation', etc.

Finally, the renewed debate has spawned numerous individual
studies of particular historical societies. In the more fruitful
of these endeavours, archeological, historical, and/or ethnographic material are used to illustrate and develop the concept.
In others, a specific definition of the AMP is counterposed to
material on ancient China, Mesopotamia, pre-Columbian America,
etc. and conclusions are drawn as to the applicability of the
AMP to a particular society, or its more general validity.

Although questions concerning the specificity of the AMP
have largely become intellectual or academic concerns, the ten-
sion between theory and ideological or political practice evident
throughout the career of the concept, have by no means sub-
sided. However, today the terms of the controversy are far
more complex, given the divisions among the world's communist
parties and other groups claiming Marx's legacy.

Our selections are presented in the chronological order in
which they appeared, with the exception of the excerpt from the
final chapter of L. Krader's 'The Asiatic Mode of Production' -
undoubtedly the most thorough-going investigation of the sources
and writings of Marx and Engels on the subject. Krader's
systematic outline of the AMP serves as a conclusion or rather
as a starting point for further research into pre-capitalist
state formations and the theory of 'civil society'. Krader has
attempted a synthesis of what have become two opposing tendencies in the interpretation of Marx's AMP. We can designate
these divergent views as the 'primitivist' and 'modernist'
perspectives after the fashion of earlier debates on the economy
of ancient Greece and Rome. For the primitivist, the AMP grasps
a totality in transition from a previously classless form to a
class-organized formation. The basic defining characteristic
of the AMP is the contradictory coexistence of communal property
and labour, and the exploitation of the communities by a higher
unity, the state. The central Marxian text for the primitivist
tendency is the 'Formen' and the sorts of societies that fall
under its purview are the primarily archeologically known
societies of China, Egypt, Mesopotamia, the Creto-Mycenaean
and Etruscan civilizations of the Mediterranean, and the pre-
Columbian states of America.

F. Tökei, the Hungarian Sinologist, and M. Godelier, the
French anthropologist, were key figures in the formulation
of the primitivist perspective. Whereas E. Welskopf, one of the
first scholars in a socialist country to signal the importance of
the 'Formen', argues that Marx envisioned the AMP as a separate historical epoch on a par with the ancient, feudal, and
capitalist modes, Tökei interprets the AMP as a qualitatively
different transitional form, situated between primitive communism and the ancient mode of production. Whereas in the
West this transitional form was rapidly superseded, in the
East it became 'bogged down'. Godelier initially saw the AMP
as a contradictory transitional form. However, he interprets
the AMP as a structural model rather than a conclusion about
human social evolution; the AMP is a tool to explore the different processes by which inequality is introduced into classless
societies. He distinguishes two possible forms of the AMP, one
in which the higher community is directly involved in the conditions of production, and one in which it appropriates surplus
labour or product in the sphere of circulation. Furthermore,
stagnation is one of the possible outcomes, others include the
transition to slavery or directly to feudalism.

Le Thanh Khoi's examination of Vietnam history from the
tenth to the nineteenth century challenges the viability of the
primitivist interpretation of the AMP. However, a common epistemological problem clearly emerges from Le Thanh Khoi's and
J. Golte's critical appraisal of the usefulness of the concept for
understanding Vietnamese and Inca society respectively. This
is the question of an exclusive one-to-one correspondence

between a mode of production - in this case the AMP - and a
particular society at a specific point in time.
   The notion that knowledge of the economic, political, and
ideological structures of a given historical society can be produced within the concept of a single mode of production in
perhaps a carry-over from a vision of historical materialism is
a lineal stage theory to be applied to the history of nations.
In combating this idea, Althusser and Balibar (1970) introduced
the concept of social formation to analyse societies in terms of
an articulation of two or more modes of production (Godelier,
1971a, pp. 63-4). In practice, the temporal and spatial boundaries of a social formation have been politically defined, a
practice that has been recently criticized (Friedman, 1976,
pp. 5-7). This practice is illustrated in H. Islamoğlu's and
C. Keyder's examination of the Ottoman social formation, which
they define as an articulated whole incorporating petty commodity
production and merchant capital, but dominated by the AMP.
Here, the AMP is not considered a transitional form, but a
'distinct articulation of political, ideological, and economic
levels in which the unit of reproduction is defined by the extent
of political authority'. Like Banu's rechristening of Asiatic
relations as 'tributary', Keyder and Islamoğlu centre their
definition of the AMP in the mode of surplus appropriation.
Integrating their analysis within a conceptual framework derived
from I. Wallerstein's distinction between world empires and
world economies Islamoğlu and Keyder start from the premise
that the Ottoman Empire was once a unit of reproduction, later
to be transformed into a dependent peripheral area of the world
capitalist system. Their own description of the Ottoman social
formation, particularly their depiction of differential dependence
on long-distance trade of various centres of the empire and
segments of the population throw some doubt on their initial
premise and invites the exploration of world economies distinct
from and/or prior to the world capitalist system.
   Unlike the cultural materialist strategy which seeks the
specifying elements of a type of society in the natural or technological productive forces, both the modernist and primitivist
interpretations of the AMP focus on the relations of production.
However, for the modernist, exemplified by Islamoğlu and Keyder,
relations of production are very much equated with the mode
of surplus appropriation, whereas the primitivist centres on
property relations (e.g. Tökei) or a combination of property
relations and methods of surplus appropriation (e.g. Godelier).
   The development of these opposing tendencies within the
renewed discussions has led Chesneaux (1968) to warn against
deploying the concept as some sort of residual category, and
provided the rationale for Hindess's and Hirst's attempt to
demonstrate the logical inconsistency of the AMP as a mode of
production. Although they have more recently advocated dropping
altogether the concept of mode of production, in *Pre-capitalist
Modes of Production* (1975) they argued for a rigorous logical

construction of modes of production. For each mode of production, there must be a 'correspondence' between a set of relations
of production and a set of forces of production. They rejected
the concept of the AMP: first, because of its 'arbitrariness',
i.e. the idea that a single set of relations of production (the
'tax/rent couple') could correspond (empirically?) to two distinct sets of forces of production (village communes and peasant
farms). Secondly, they maintained that the tax/rent couple
could only be conceptualized when the state is a necessary
condition of production, an impossibility from the perspective
of Engels's analysis in which the state presupposes the existence
of classes distinct from it, and illogical if based upon a func-
tionalist theory of the necessities of irrigated agriculture.

As for the first reason for rejecting the AMP, there are several
problems with the concept of 'correspondence' which the authors
admit in their subsequent autocritique (1977). However, in their
second justification for eliminating the AMP, they appear to have
skirted the main problem, not just as it affects the AMP, but
for the development of a Marxist theory of pre-capitalist
societies. First of all there is the question of what is meant by
a 'condition of production'. From their analysis of feudalism,
it would appear that they are referring to the control of
particular means of production vital to the reproduction of the
totality (i.e. mills, storage facilities, etc. by the feudal lords).

But does the appropriation of surplus on the part of one
group necessarily imply their control of specific means of production? This question has directed recent anthropological
research. The solution to this question has involved the introduction of the concept of 'reproduction'. Appropriation of the
surplus-product in many pre-capitalist formations is not necessarily accomplished through the control of specific 'material'
means of production, but of means of 'reproduction' which may
include such 'things' as prestige goods, women, and the supernatural. In fact on this basis, Friedman would distinguish
'Asiatic' states as ones where the appropriation is effected by
those with a monopoly over the supernatural conditions of reproduction.

In their dismissal of the AMP, Hindess and Hirst refer to its
incompatibility with Engels's thesis on the origin of the state
in his 1884 work which drew heavily upon Morgan. The summon-
ing forth of Engels as some sort of theoretical backstop not only
ignores the contemporary debate on the conceptualization of
classes in non-capitalist and particularly tribal societies, but
smacks of reimposing the limits of orthodoxy.

The re-emergence of the concept of the AMP among Marxists
in the late 1950s and the works of some of its more polemical
advocates and opponents can no doubt be put down to external
factors such as the Twentieth Party Congress and the Sino-
Soviet split. The continuing debate on the status of the concept
has coincided with a period of dramatic political changes: the
defeat of the United States in Indo China; increasingly voiced

dissidence within the USSR; the development of Eurocommunism, the meteoric rise of the OPEC nations, etc. Whereas the beginnings of the debate centred on revising a recipe-like evolutionary theory, today it is part of a much more wide-ranging re-examination of Marxist concepts. The movement has changed from trying to explain single societies in terms of the AMP, to analysing a social formation as an 'articulation' of modes of production, and finally, to analysing larger systems of reproduction prior to capitalism which include Asiatic or tributary relations of appropriation.

PROBLEMS OF PERIODISATION IN ANCIENT HISTORY*

Elisabeth C. Welskopf

In his works on India (1853 and 1856) and in his preliminary
studies for 'Capital' (the so-called 'Grundrisse'), Marx more
thoroughly analysed the factors, common and distinctly specific,
to the conditions in the Ancient Orient and Classical Antiquity.
These works and the conclusions drawn by Marx in them, are
of prime interest for our problem. Marx wrote this extensive
manuscript (the 'Grundrisse') between October 1857 and
November 1858, during the great economic crisis of the capitalist world. It was a study for his own information in preparation
for 'Capital'. He proceeds . . . to put forward the 'patriarchal,
ancient, and feudal' stages of development. In the first chapter
of this work, the 'Chapter on Money', we find the comment
'regardless of the character of this super- and sub-ordination:
patriarchal, ancient or feudal' (1857-8a, p. 159). Marx sees all
these super- and sub-ordinate relationships as distinct from
the capitalist social formation in which the determinant relations
of production are brought about by private exchange. To
characterise the 'super- and sub-ordination' Marx uses the
concepts of 'spontaneous or political': the concept of 'spontaneous'
may be applied to the relations of kinship and inheritance,
originally so influential in human society; 'political' refers to
the influence of external economic power. In the later passages
of his work, however, Marx clearly analysed the secondary
influence of these factors in the realisation of economically based
relations. In the 'Chapter on Capital', he goes on to examine
the property forms and corresponding essential relations of
slavery in the Ancient Orient and Classical Antiquity. Here
we are above all interested in the conclusions Marx drew from
his own analyses for periodisation of history.
  In terms of the conditions of despotism, Marx speaks of 'Asiatic
land forms' (ibid., p. 472), of 'oriental despotism' as a system
(ibid., p. 473), of the 'specifically oriental form' of property
(ibid., p. 477) and of Asiatic history as a 'kind of indifferent
unity of town and countryside' (ibid., p. 479). He characterises
the specifically oriental property in the following way (ibid.,
p. 494):

In the Asiatic form (at least predominantly), the individual
has no property but only possession (ibid., p. 484). The
Asiatic form necessarily hangs on more tenaciously and for the
longest time. This is due to its presupposition that the individual does not become independent *vis-à-vis* the commune;
that there is a self-sustaining circle of production, unity of
agriculture and manufactures, etc. (ibid., p. 486). Slavery
and serfdom are thus only further developments of the form
of property resting on the clan system. They necessarily
modify all of the latter's forms. They can do this least of
all in the Asiatic form . . . slavery here neither suspends the
conditions of labour nor modifies the essential relation (ibid.,
p. 493). In the oriental form this *loss* [of landed property –
E. Welskopf] is hardly possible, except by means of altogether
external influences . . .
Marx characterises the conditions of Classical Antiquity as
distinct from these economic characteristics of despotism. In
the former, private property appears as the foundation; the
loss of landed property through the development of usury is
not only possible, but a widespread phenomenon. As the division
of labour between agriculture and handicrafts develops, the
unity of the village economy is broken down; the town controls
the countryside and slavery, in the form of private movable
property, becomes the dominant relation of production. It is
precisely this comparison between the ancient and the ancient
oriental conditions that enables Marx to identify their specificities.
How did Marx periodise world history after these studies?
In the 'Contribution to the Critique of Political Economy'
which Marx wrote in 1857-8, i.e. immediately afterwards, we
find a comment on this: 'In broad outline, the Asiatic, ancient,
feudal and modern bourgeois modes of production may be designated as epochs marking progress in the development of
society' (1859, p. 21).
The determinant outline for the Marxist-Leninist view of
history is clearly stated in this passage by Marx: that is, the
idea of 'epochs marking progress in the development of society'.
At this point, the conditions of the original form of society are
not spelt out, since they had not yet been researched. Marx
divides the universal history of 'antiquity' into an 'Asiatic'
and an 'Ancient' epoch. These designations are obviously postulated concepts, the first, a geographical, and the second, a
time-related, concept. The concept of the 'Asiatic mode of production' was a provisional term for Marx. It arose from the fact
that the majority of societies that had developed the epoch of
the 'patriarchal' mode of production were situated in Asia.
Corresponding to the afore-mentioned four-stage division of the
history of class divided society, Marx, in the same work, writes
of the 'bourgeois', 'feudal', 'ancient' and 'oriental' economies.
It is important for us to note that Marx regarded the economic
relations of despotism as a particular epoch in history, in which

the first peoples raised themselves above the primitive conditions of the original form of society and established great corporate groups. Marx returns to this division of 'antiquity' into two epochs of universal history in 'Capital'. There he speaks of 'Asiatic, ancient, etc. modes of production' (1867, I, p. 83). In Marx's writings, posthumously bound as the third volume of 'Capital', we find the same division in relation to commerce in a note. . . .

As we know, in 'Capital' I Marx also writes about 'society based on slave labour' (1867, I, p. 209) without going into the historical delineation of this period in detail, since in this passage he is only concerned with the comparison and differentiation of wage labour and slavery.

In 1876, a good decade after the appearance of 'Capital' I, Engels took up the polemic against Eugen Dühring. In this work he also touched upon the question of ancient oriental despotism. In his preliminary studies he noted (1878, pp. 418–19):

Hence this much is correct in the whole theory of force that hitherto all forms of society needed *force* to maintain themselves and to some extent or other were even established by force. This force, in its organised form, is called *state*. So we have here the banal idea that as soon as man rose above the wildest conditions states existed everywhere and the world did not wait for Dühring to learn this.

But state and force are precisely what all hitherto existing forms of society have had *in common*, and if I should try to explain, for instance, the oriental despotisms, the republics of antiquity, the Macedonian monarchies, the Roman Empire and the feudalism of the Middle Ages by stating that they were all based on *force*, I have explained nothing as yet. The various social and political forms must therefore be explained not as due to force, which after all is always the same, but as due to that *to which force is applied*, as due to *that which* is being robbed - the products and productive forces of the epoch in question and their distribution, resulting from themselves. It would then appear that oriental despotism was founded on common property, the antique republic on the cities engaged in agriculture, the Roman Empire on the latifundia, feudalism on the domination of the country over the town, which had its material causes, etc.

Engels made this note that the ancient despotism depended on communal property, not for publication, but in the course of his preliminary studies. It should be understood in the context of the passages from 'The German Ideology' and the 'Grundrisse' on the role of communal property in the Ancient Orient, where such property is represented by the higher unity, the despot. It is significant that Engels introduces ancient oriental conditions separately and emphasises the role of power under these conditions. He locates these relations of production in periods when power had already formed itself into the power of the state. As regards ancient societies, Engels points out

certain differences in the nature of the Hellenic *polis* and the
Roman Empire without going into their common characteristics.
Finally, upon closer analysis and appropriate comparison, Engels's
statements in the preliminary studies for 'Anti-Dühring' appear
to share a common denominator with the four-stage division of
'The German Ideology' and Marx's four-stage division of class
society.

Marx never had the time to revise further his four-stage
division of the history of class society or to fix the concepts in
their final form. In the writings he left, particularly in the
'Grundrisse', it certainly appears that he characterised the
'ancient Asiatic' mode of production as the mode of production
of patriarchal economic despotism (and general latent
slavery). . . .

The discussion over periodisation that has gone on for decades,
has nevertheless led towards a gradual clarification. Some
propositions have generally been renounced, since they obviously
do not agree with the reality put before us by modern specialist
research. No one now, as far as we know, supports the classification of ancient oriental societies as a form of 'primitive society',
i.e. a formation without antagonistic production relations or class
conflict. The attempt to class the Ancient Orient and Classical
Antiquity together as one formation, in which the dominant
relation of production, without exception, was slavery (as
personal property) has also apparently been abandoned under
the pressure of the available evidence.

The core of the continuing debate on the basis of scientific
socialism centres upon the following questions.

Advocates of scientific socialism designate each social formation
according to the reigning relations of production of the period,
from which the inherent historical law of its structure and its
development over time is derived. Slavery, as a relation of
private and personal property, is considered to have stimulated
development and to have ultimately been the determinant relation
of production in the states of Classic Antiquity (in the Hellenic
polis and in Rome). Slavery is defined as a private property
relationship between man and man by Engels, Lenin, and Stalin.
Marx also used the expressions slave and slavery only in this
sense. The chief characteristics of the slave relationship are
the expropriation of land and the appropriation of the person,
in so far as the concept is not modified for expediency by a
particular attribute. In the structure of the ancient oriental
countries, slavery of the type found in Classical Antiquity did
not prevail since the irrigation system led to centrally directed
communal labour by the resident population. This form of labour
became determinant for all relations. In the Ancient Orient,
the landed property of the king formed the basis of production.
Hereditary possession remained with village communities of
subjugated producers who were only occasionally mobilised (Marx,
1857-8a, pp. 472-3). Furthermore, the development of these
countries had a different tempo and a different result. Their

relations of production did not lead to a destructive dissolution
from within but ended in stagnation lasting thousands of years
if they were not brought to an end by external political or
economic intervention. From what perspectives does it seem
justified to include the Ancient Orient, which developed in the
context of specific property conditions as an antagonistic form
of communal labour on dike construction and road systems, into
the category of a 'society based on slavery'?

On this question, the following views are open to discussion:
1   The conditions of Classical Antiquity are taken as the model.
The conditions of the Ancient Orient are only a primitive stage
of those found in Classical Antiquity, without sufficient significance to warrant their treatment as an independent social
formation. Against this view, one can adduce proof from all the
arguments that Marx uses for the universal historical role and
the particular structure of Asiatic countries. Marx's arguments
have not suffered but have gained from modern research.
Nevertheless, it is a fact that under despotism, slave relations
as relations of private property already existed and were gaining ground, without ever finally coming to prevail. Struve in
his preface to 'Chrestomathie . . .' correctly points this out.
But we also find an essentially similar process in subsequent
periods: feudal relations developed within the pale of the slave-
owning society of Classical Antiquity, capitalist relations
developed within feudalism. Continuity in development on its
own does not justify our classifying a society according to the
nature of its future conditions.
2   The conditions of the Ancient Orient define a specific epoch.
In principle, the development of the Hellenic polis via the world
of the Hellenistic states through to Rome is no different from
that of the Ancient Orient, i.e. a particular chance development
in some countries. This viewpoint is contradicted by the course
of history over the following 2,000 years, where specific progress is associated with Classical Antiquity. It is also contradicted
by Marx's and Engels's analyses of the specific property relations
in the social formation of Classical Antiquity.
3   Finally, we could dispense with the more usual and narrower
definition of slavery as a relation of private property, and include the 'latent' slavery existing under despotic domination in
the concept of slavery. The latter would have to be greatly
expanded and more highly abstract if it were to include both
fixed or resident and mobilised producers in the relations of
servitude. Employing such a general concept enables one to
deal with the varied development of individual countries in a
more or less purely synchronistic fashion. But in this way,
we lose the profound insights into the inherent historical con-
formity of both the ancient oriental and the ancient classical
development that Marx and Engels revealed; we move between
abstractions and chance events. If we assume a comprehensive
view of the Ancient Orient and Classical Antiquity, only the
solution suggested by Struve or even more so that put forward

by Ranovitch seems to conform to what we know and make way
for the understanding of the essential historical connections
gained by Marx and Engels. To the best of my knowledge, a
scientific definition of slavery in the appropriate comprehensive
sense has never been tried and any attempt at it brings the
dangers of softening the boundaries of the concept and emptying
it of its specific characteristics simply because such qualitatively
different economic conditions are reduced to their common
denominator.

If we decide not to combine the Ancient Orient and Classical
Antiquity within a single category, we can no longer deal with
their separation in such a way that the conditions of the Ancient
Orient are classified with 'primitive society', i.e. a formation
without contradictory relations of production and without class
antagonism. On the basis of the results of modern research,
this distinction inevitably leads us to the four-stage process of
antagonistic social formations as Marx first conceived it.

Marx's and Engels's basic understanding of the progressive
character of economic - and with it, general social - development
is maintained in the four-stage process of antagonistic relations
of production. This more deeply defined differentiation of the
Marxist conception, in our opinion, appears to best contain
historical reality. This periodisation by Marx, in which the
ancient patriarchal despotism appears as a separate epoch,
has nothing whatsoever to do with the idea of 'feudalism' in
the Ancient Orient. On the contrary, by its precision, it
facilitates the discussion with new ideas. Marx never spoke of
feudal property, of 'vassals' or of 'serfdom' in the Ancient
Orient, but took pains to develop a terminology appropriate to
a stage of development preceding that of the slave-owning
society of Classical Antiquity. The term 'satraps' for the
dependent governors, princes or officials of the despot, for
example, may still be very inadequate, but it shows an attempt
to find a specific system of concepts for a section of world
history, the structure of which shows specific, necessary, and
unrepeatable cohesion in the state of the development of the
productive forces, the relations of production, political and
juridical institutions, and the world of ideas in religion, art
and science. To this extent, Marx's conception and attempt at
a specific terminology seems to us an indication of the direction
to be taken.

Discussions about the correct characteristics and correspond-
ingly correct classification of despotism in the line of historical
epochs is important not only for the history of antiquity, but
also for the history of the Middle Ages or feudalism. Marx
comments that not only slavery - as a relation of private
property - but also serfdom, i.e. the definitive form of servitude under feudalism, can modify the 'Asiatic forms' least of
all (1857-8a, p. 493). The concepts of 'feudalism', vassalage,
bondage, and serfdom arose in world history, not on the foundation of despotism, which stagnated and hindered the development

of oppositions, but on the basis of the destroyed slave-owning society of Classical Antiquity at the edges of its territory. In so far as parallelly designated 'feudal' relations became linked with despotic forms of servitude, they were modified, as Marx says. For the historian this is not an unknown fact and is one well worth noting. Elsewhere we have already referred to the fact that Marx sees the final dissolution of ancient oriental despotic forms based upon the village system in India, as only taking place with the infiltration of English weaving-products. Engels locates the breakdown of despotism under feudal conditions in the Byzantine Empire, after the conquest by the Turks. The particular differences and transitions were scarcely researched in the time of Marx and Engels, and even today not enough research on these questions has been carried out.

BIBLIOGRAPHY

Engels, F., 1878, 'Anti-Dühring', London, Lawrence & Wishart, 1969.

Marx, K., 1867, 'Capital, I', London, Lawrence & Wishart, 1970.

Marx, K., 1859, 'A Contribution to the Critique of Political Economy', London, Lawrence & Wishart, 1971.

Marx, K., 1857-8a, 'Grundrisse: Foundations of the Critique of Political Economy', Harmondsworth, Penguin, 1973.

NOTE

* From 'Die Produktionsverhältnisse im Alten Orient und in der griechisch-römischen Antike', Berlin, Akademie-Verlag, 1957, pp. 428-31. Translated by J. Gordon-Kerr.

THE ASIATIC MODE OF PRODUCTION*

Ferenc Tökei

Marx's 'Forms which Precede Capitalist Production' (a part of
the 'Grundrisse' (1857-8),) can provide a definitive answer to
the problem of how well grounded Marx and Engels's category
of the AMP was, and the extent to which this category constituted a theory that was organically linked to a whole series
of other developments in historical materialism. This text of
Marx analyses property forms and social forms from the perspective of the emergence of capitalist relations. . . .
    The first form of landed property is tribal communal property;
the basic presupposition of both (communal) property and
(personal) possession by the individual is membership in the
community. This form of property is clearly discernible among
nomadic pastoralist tribes for example. Upon settling down,
modifications in this form of property depend on a number of
different and fortuitous factors. At this point the 'Asiatic'
modification of landed property within the primitive community
appears. . . .
    Marx grasps that what is essential in the modification of tribal
communal property takes place in the stage of agriculture. Thus
disappears this apparent contradiction according to which the
Asiatic peasant could be considered as the owner of the basic
conditions of labour, while the despot appropriated the surplus
product because of his monopoly of these very conditions. The
appearance of exploitation was based on public property. We
shall subsequently return to the rather complicated question of
whether the Asiatic despot and his officials ought to be considered an exploiting class; however, for the moment we would
like to emphasize that Marx possesses the key to the understanding of class relations in Asiatic society. . . . With reference
to real village communities, on the one hand, and an imaginary
tribalism on the other, Marx also shows why the appropriation
of the surplus product is transformed into exploitation. As we
shall see, this process goes along with the fact that the tribal
aristocracy becomes a class as the tribal communities which it
embodies disintegrate. The true richness of this 'rough'

construction by Marx only becomes apparent when it is contrasted with the other forms of property. . . . Marx imagined
that the realization of tribal communal property could take very
different forms: communitarian forms, more or less democratic
or despotic forms, forms in which agriculture was carried out by
families or by the community as a whole, etc. However, despite
what seem to be very marked differences, Marx in his analysis
of property relations, centred only on whether membership in
the community was the presupposition of ownership or possession, and if it were, the property form was considered an example of landed property of the primitive tribal form. . . .

Processes of development beginning at different periods,
in different places and under different historical conditions ,
have produced three different forms of property: Asiatic,
Ancient, and Germanic. The starting point for each of these
three processes of development is tribal communal property.
Asiatic societies retained it in an unchanged form and furthermore stabilized and gave it legal expression. The Greeks and
Romans developed the Ancient form of property and the Germanic
peoples, the feudal form, from it. These three processes of
development – at first glance isolated from one another – have
at the same time created the links of universal development,
moreover they could not but succeed one another. Although
the primitive tribal community determined by nature is the precondition, the importance of this foundation varies in these three
forms. The co-ordination, or to be more precise, the succession
of these three forms, according to this perspective, delineates
the emancipation of the individual and his severing of the
'umbilical cord' that tied him to the natural community. In Asia
this severing never took place; in European antiquity it
occurred by separation from the Ancient form, while under
European feudalism, it became the basis of society. . . .

The development of one or another of the three forms of
property will depend upon a whole range of external and internal
factors and their specific combination. Only the development
of the 'Asiatic' form was a function of the abundance of available land, a situation not at all historically specific to antiquity.

If there is in fact an abundance of uncleared land (a situation
common in ancient Asia) it follows from the precondition of
tribal communal property – i.e. from tribal membership – that
production cannot dissolve the form of property, on the contrary
it preserves the latter. On the other hand, the realization of
the precondition of private landed property, i.e. the severance
of the individual from the 'community's umbilical cord' requires
a set of specific historical factors, among which land scarcity
plays no minor role. From Marx's analysis it clearly emerges that
landed property is freed (in the course of its development)
from being part of the community, understood as the precondition
of its existence, i.e. private feudal landed property cannot be
directly generated from tribal communal property. Such independence of the individual from the community can arise only

with the absolutely necessary contribution of the Ancient form
of property. (In European history, this happened in such a
way that the Germanic peoples conquered Ancient society, where
private landed property already existed.) Thus, in antiquity
the Ancient form of property is the only possible form of the
development of private landed property. At the same time – at
the level of universal development – it is an indispensable step
that can neither be neglected nor skipped: one is in the presence
of the transition towards private feudal landed property. This
becomes absolutely clear if we try to give a conceptual and clear-
cut form to the Marxist analyses. Given the fact that ownership
or merely possession of land in the three forms is mediated by the
community, the essence of the three forms of property can be
fixed in a formula representing the reciprocal relationship between
the individual, the community, and the land. In this way, the
essential relations of the three forms of property can be represented as in the diagram.

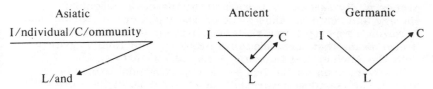

In representing these relations, it sufficed to draw out the
basic correlations of historical contingency, namely that the
origin of the three forms of property had been the tribal community. (Furthermore, from the perspective of the other relations, this was in no way accidental and ought to be represented
in other diagrams as well.) The diagram displaying the essential
relations of the three forms of property clearly shows that these
three forms are phases in the general development of property
relations. Tribal communal property is diametrically opposed to
feudal landed property (despite all attempts to show that Asiatic
societies and feudalism are akin). But the occurrence of this
total reversal of factors required, both in principle and in
historical practice, a separation or the dialectical negation of
communal property, i.e. the ancient form of property. The
diagram of the first form of property requires no further
explanation: the individual only relates to the land in his capacity
as member of the tribal community; there is no private landed
property, no direct relationship between the individual and the
land at this point. Paraphrasing Marx, the diagram of the
Ancient form can be interpreted in the following way: As a
member of the community-as-state, the individual possesses
his share of public land; furthermore, he is already the owner
of another piece of land and as such, he is a member of the
community-as-state, possessing his share of public land. Finally,
in the 'Germanic' form, the individual is already the owner of
land and it is only in this capacity that he becomes possessor of

the remnants of public land which play a subordinate part.
Clearly, the three diagrams represent three phases in the process of general development through to feudalism in the course
of which there is a tendency towards private landed property
completely detached from the precondition of the community.
(With feudalism, this developmental tendency changes;
capitalism initiates the dissolution of small peasant property
and attempts to nationalize large landed property; the latter
attempt will only be fulfilled by socialism.)

An objection to Marx's analysis and our diagrams could be
raised: they are too unilateral in so far as no mention is made
of the class conditions in Asiatic despotism and that neither the
slavery of antiquity nor medieval serfdom is taken into account.
Marx's outline, however, in no way overlooks these elements;
nevertheless, the property forms set out are considered as the
starting point of internal correlations and historical motion.
Slavery and serfdom are considered secondary to the forms of
property. By secondary is meant that the development of
slavery and serfdom takes place on the basis of the Ancient and
Germanic property forms sketched. But on the other hand,
it also means that these elements break up and finally destroy
both the Ancient and Germanic forms of property. . . .

However, even on the basis of tribal communal property,
slavery and serfdom necessarily develop; it is all the more
inevitable that it is slavery which develops on the basis of
Ancient property and serfdom which develops on the basis of
Germanic property. Now we have come to the problem of discovering how the conditions of the different property forms
can be reproduced. In other words, the problem is to understand the functioning of the law implied by the diagrams in
the course of reproduction (and gradual disintegration). On the
one hand, this means considering the development of the
'secondary' elements. On the other hand, given the development
as a whole, we have to examine the workings of the law in the
specific 'Asiatic' period, in the second phase of antiquity and
throughout the Middle Ages. . . .

Glancing at the diagram of the 'Asiatic' property form, we
immediately see that neither slavery nor serfdom could in themselves change this form. If one community subjugates another
or individuals are uprooted from their community, the conquered
individuals can only be incorporated as property of the community.
If a family head or official who comes into possession of slaves
or serfs is himself still attached to the 'umbilical cord' of the
community, if his power rests on communal property - as was the
case throughout Asia - then the conquered individuals enter
into the same relation with him as does every member of the
community with its representatives. This explains why in the
Orient we find no basic social differentiation between slavery
and serfdom or between slave-serfs and 'free' commoners. They
are all identical members of the community and equally under
obligation to the community's representatives. This 'general

slavery' (Marx), however, can neither be called ancient slavery nor serfdom, but merely patriarchal dependence, which is based upon tribal communal property and is known as patriarchal slavery. . . .

According to Marx - due to the scarcity of land - the reproduction of the ancient form of property required expansionist wars which were followed by enslavement. Now, this slavery developed under conditions very different from those of Asia, since it was communities consisting of individual owners which took prisoners. A struggle developed over whether the slaves should be incorporated as communal property or as private property. In this struggle the private ownership of slaves inevitably won out and was soon followed by the enslavement of former owners. . . .

If we now look at the diagram of the Ancient form of property from the perspective of slavery, trade, etc. - i.e. the elements that necessarily emerged from and opposed the original structure - we find that the development and ultimate disintegration are but the embodiment or historical unfolding of the relations determined in the basic diagram. In fact, this double form of property already implies a contradiction between the two forms - communal property and private property. It naturally follows from the process of production and reproduction that this contradiction historically unfolds, deepens, expands, and effects further evolution. This is why the above diagram of the Ancient property form - applied here to the historical development of Ancient society - can also be expressed in the following manner: Initially (with the founding of cities) the individual was a possessor of ager publicus and a proprietor of his own plot, only in so far as he was a member of the (patrician) community. But soon proprietors emerged who, as owners of private property, were excluded from the community and therefore from possession of public lands. These were the plebeians who struggled to be admitted to the community by virtue of their holding private property. They won this struggle, although with their admission, the original community was transformed and a new one appeared (the populus Romanus). Now, clearly the emergence of new owners, representing the most developed form of private property and their economic and political victory, primarily came about as a consequence of the original form of property. These factors were accelerated through the development of trade and a money economy and finally consolidated through the institution of the Ancient form of private slave-holding.

In fact, the difference between the forms of Ancient property, set out in the basic diagram, can be distinguished in the following way. In the former, possession and ownership rest on the precondition of membership in the community (of the founders of the city) while in the latter, the precondition for membership in the community is ownership. From this perspective, the earlier form is close to the 'Asiatic' form of property while the new form approaches the Germanic. All the essential relations of Ancient society and Graeco-Roman history rest upon the unity and conflict between these contradictory relations of property. Given the fact that the city founders 'distinguished themselves separately as possessors of urban lands on the one hand, and private proprietors on the other', they lay down the basis for the appearance of more developed forms of private property. Clearly, trade and the money economy are the principal factors that favoured the 'newcomer' and 'alien' owners. Finally, in the hands of rich new private owners, slavery develops to the point that community membership is no longer the precondition for slave-holding. This explains why slavery in the Ancient sense of the term, i.e. the private ownership of slaves, could only develop on the grounds of the Ancient form of property, i.e. in the Greek and Roman cities, and why it later became the basis of production. Among all the forms of property, only the Ancient enabled the development of slavery. However, as we have shown, the Ancient form at a later stage inevitably developed slavery as the crucial means of its own reproduction and, simultaneously, of its destruction. It is all the more important to emphasize the close relationship between the Ancient form and slavery, as it has recently become fashionable to challenge the slave-based nature of Ancient Greek and Roman society.

Clearly following from what has been stated, serfdom could only develop on the basis of 'Germanic' property, in the same way that feudal property is the necessary precondition for capitalist development. . . .

Marx's way of analysing the process of the simultaneous reproduction and breaking up of the different property forms clearly shows that he conceptualized the forms of property as expression of the conditions of production, the economic base, upon which a particular mode of production is founded. His outline provides a straightforward formulation (1857–8a, p. 495):

> The original unity between a particular form of community (clan) and the corresponding property in nature, or relation to the objective conditions of production as a natural being, as an objective being of the individual mediated by the commune – this unity, which appears in one respect as the particular form of property – has its living reality in a specific *mode of production* itself, a mode which appears both as a relation between the individuals, and as their specific active relation to inorganic nature, a specific mode of working (which is always family labour, often communal labour). The community itself appears as the first great force of production; particular kinds of production conditions (e.g. stock-breeding, agriculture), develop particular modes of production and particular forces of production, subjective, appearing as qualities of individuals, as well as objective (ones).

In the same text, Marx provides the following summary (ibid., p. 496):

> *Property*, then, originally means – in its Asiatic, Slavonic, ancient classical, Germanic form – the relation of the working

(producing or self-reproducing) subject to the conditions of
his production or reproduction as his own. It will therefore
have different forms depending on the conditions of this production. Production itself aims at the reproduction of the
producer within and together with these, his objective conditions of existence. This relation as proprietor – not as a
result but as a presupposition of labour, i.e. of production –
presupposes the individual defined as a member of a clan or
community (whose property the individual himself is, up to
a certain point). Slavery, bondage, etc., where the worker
himself appears among the natural conditions of production
for a third individual or community (this is not the case,
e.g. with the general slavery of the Orient, *only* from the
European point of view) – i.e. property no longer the relation
of the working individual to the objective conditions of labour –
is always secondary derived, never original, although [it is]
a necessary and logical result of property founded on the
community and labour in the community. It is of course very
simple to imagine that some powerful, physically dominant
individual, after first having caught the animal, then catches
humans in order to have them catch animals; in a word, uses
human beings as another naturally occurring condition for his
reproduction (whereby his own labour reduces itself to ruling)
like any other natural creature. But such a notion is stupid –
correct as it may be from the standpoint of some particular
given clan or commune – because it proceeds from the development of *isolated individuals*. But human beings become individuals only through the process of history.
We would like to point out two features of this summary. The
first is the original communal precondition of the three forms of
property, i.e. the historical fact that the Asiatic form, as well
as the Graeco-Roman and feudal forms, developed out of tribal
property (the Asiatic being the least 'developed' form). Basically
this is an expression of the laws of universal development. From
the perspective of universal development, this occurrence is
not entirely the result of chance, since it expresses in a concrete
way that all forms of private property have necessarily been
preceded by the phase of communal property. (This applies both
to particular cases as well as to universal development. We have
not depicted this law in our diagrams since, on the one hand,
it is only completely lacking in the diagram of the Germanic form,
and because it is contained in the succession of the three forms
from the point of view of universal evolution.) Therefore,
Ancient slavery and Medieval serfdom can only emerge through
the Ancient or 'Germanic' negation of the original communities
and tribal property. The second feature must be accorded special
emphasis: Ancient slavery and Medieval serfdom 'where the
worker appears as part of the natural conditions of production
for a third person or a community' can only develop from the
conditions of 'Graeco-Roman' or 'Germanic' property. Only the
state patriarchal form, and therefore Asiatic empires, which

rely upon the communal ownership of slaves – or serfdom if you
prefer – can arise from the conditions of tribal communal
property.

However, this form is adapted and integrated within a situation
of general subjugation or the subjection imposed by 'general
slavery'. Consequently, the AMP can in no way be considered
a slave or feudal society, or even an Asiatic subcategory or
variety of these social forms. For, from the perspective of the
mode of production, it exhibits fundamental and radical differences in the form of property and its 'living reality'. Thus, in
agreement with Marx, we considered absolutely justified the use
of the category of the AMP and its distinctness from both the
Ancient and the feudal modes of production.

We should now consider the place of the AMP within the
universal development, the three phases of which are determined
by the three forms of property preceding capitalism. Clearly,
it would be more appropriate not to call the first form of property
'Asiatic', but tribal or primitive tribal property. (Marx most often
refers to it as tribal and when speaking of its concrete historical
manifestations, uses the terms 'Indian', 'Slavonic', etc. as well
as 'Asiatic'.) As early as 'The German Ideology', Marx and
Engels wrote that the first form of tribal property provided the
natural foundation of societies of fishermen, hunters, and
pastoralists. 'At the very most' (höchstens) this form of property
survives into the stage of agriculture. Now in another passage
of this outline of Marx's, we read that a specific mode of production develops for each specific type of relations of production
(for instance, for pastoralism, or for agriculture). The meaning
of Marx's remark is clear: with the emergence of agriculture,
i.e. 'this specific type of relations of production', the former
mode of production, i.e. the primitive community, must be
modified and a new mode of production must emerge superseding
the primitive community. We might ask, what for Marx is this
new mode of production which necessarily accompanies the
emergence of agriculture. In our opinion, Marx and Engels felt
that the Ancient mode of production corresponded to the stage
of agriculture. As early as 'The German Ideology', Marx and
Engels accepted that tribal communal property, the natural basis
of the primitive community, only exceptionally survived in the
stage of agriculture. According to Marx, the evolution of the
division of labour provides the rationale for distinguishing two
degrees of development even at the beginning of the stage of
agriculture: The degree manifested by the oriental unity of
manufacture and agriculture, and the degree of their separation.
The oriental unity of manufacture and agriculture hardly
accelerates the development of the pre-existing mode of production;
on the contrary, it is likely to conserve it by preventing its
complete disintegration through the very slow – although progressive – development of the means of production and other
factors. However, the separation between manufacture and agriculture not only leads to the total disintegration of the mode of

production of the primitive community, but simultaneously generates and develops the Ancient mode of production. In this way,
Engels's theory of 'the great social divisions of labour' is an
organic part of Marx's and Engels's intepretation of history.
It is valid both for general development and for the specific
developments entailed by it.

But - as with laws of the history of general evolution - this
law must likewise be understood as one of typical evolution involving real progress. However, besides progress, there are
cases of stagnation, even regression, which occur for a specific
reason - or more exactly development does not follow the *typical*
path due to the absence of specific historical preconditions.
Thus, Engels essentially used the insufficient division of labour
to characterize the 'high civilizations' of America as societies
which in the course of evolution had deviated from the typical
path, and had taken a special path which landed them in the
Middle Status of Barbarism.

To determine the place of the AMP in history, our starting
point must be that in its property relations, this mode of production is inseparable from the primitive community; the AMP
is constructed upon the foundations of the conditions of property
of the primitive community, and consequently, it uses and
modifies the institutions of the latter. It remains to be seen from
what form and manner this modification and negation of the
primitive communal order takes and what results are produced.
The passages cited from Marx and Engels show that they considered the appearance of exploitation as the essence of the
'Asiatic' modification of the primitive community; the historical
situation of the AMP depends precisely on determining the nature
of exploitation. . . . It clearly emerges that Marx and Engels
considered that exploitation rested upon tribal communal property,
and not upon private landed property. The appropriation of
ground rent based upon communal landed property is patriarchal
exploitation, a rudimentary, ambiguous, and contradictory form
which, in the course of typical development, could not survive
for very long (only during the heroic period of the Greeks and
Romans and only at the very beginning of feudalism among the
Germanic peoples). It is a transitional form of exploitation which,
in the line of typical development, is relatively quickly displaced by other forms of exploitation anchored in certain
fundamental forms of private property relations: in the forms of
slavery in antiquity and in forms of feudal exploitation in the
European Middle Ages. (Of course, at the same time, these forms
are interspersed among the typical forms of exploitation, but
they are nevertheless displaced as they are completely subjected
to the new forms for which they provide support.) Transitional
patriarchal exploitation alone corresponds to the transitional
family division of labour. Consequently, the structure of society
is necessarily of a transitional type, limited to 'the extension of
the family'. This brings us to the most controversial of all these
questions: How can the society of the AMP be conceptualized
as being divided into classes?

It is clear that the democratic or despotic forms of the tribal communities mentioned by Marx, depend on the degree to which migrations, intermarriage, conquests, growth, etc. modify the natural character of the original tribal community. It is equally unquestionable that such modifications of the original tribal community vary in direct proportion with the growth of the community through the subjugation of other increasingly distinct communities, a process that inevitably results in the development of increasingly larger and 'superior' communities, while at the same time the previous communities are generally preserved. Where they survive, the small communities are naturally closer to the originally democratic form than the increasingly larger communities, since the latter have to employ more and more despotic methods, if only to have their supremacy acknowledged and to maintain it. Thus, we can understand the universally known fact that states resting upon Asiatic village systems - whether founded by an indigenous dynasty or by other nomadic conquerors - almost always employ the most despotic methods of government. In fact, in these empires normally only the village communities' Gentile organisation survived intact; the 'higher' communities (originally tribal, then federalist, later regional, provincial, state, etc. organizations) were transformed into tax units organized from above, and, from the perspective of the village communities below, were none other than communities of 'general slavery' or general exploitation. Essentially this growth proceeded in the natural fashion, since uncultivated land was in abundance so that only external obstacles (for example, the border of another country) could hamper this continuous and even spontaneous expansion of the network of villages. There is no need to show that through this method of expansion, the contradictions growing within the economy and the society are 'resolved' in such a way that they increase and consolidate encapsulation in transitional forms. Now, in despotically organized and expanding empires, a division of men into two large groups takes place: non-producers and direct producers. The non-producers do nothing but receive taxes in the name of the growing communities and appropriate the surplus product from the direct producers, isolating themselves (becoming increasingly aristocratic) from the peasants of the village communities. The relations of these two groups of men to the conditions of production, primarily to the land, are different; the peasantry cultivates the land and the aristocracy of officials (through the appropriation of surplus product) behave as if they were land-owners and had the monopoly of the factors of production. Of course, this is not private landed property: it is only as a whole - as the state - that the aristocracy behaves as land-owners, since membership in the 'community' is the indispensable precondition in the relationship to the land. Nevertheless, these two forms of patriarchal possession are equally determined by two different modalities of community membership. The peasantry can hold

and cultivate the land because of its attachment to the village
community, while the aristocracy of officials can do so by the
quality of being distinguished members of the community: the
representatives and embodiment of the community. These two
different relationships to the land, according to the two types
of membership to the community, can be illustrated as in the
diagram.

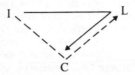

This diagram no longer represents the community of the tribal
communal property, but the modified case which 'at the very
most' serves as the foundation of the AMP. In this way the
relation of ground rent, made possible by membership in the
community, can be unequivocally applied only to the peasants of
a village community. The dotted line indicates the possession
of land or, more exactly, the possession of the ground rent of
the aristocracy of state officials. Although this is not a relation
of private landed property, ownership is made possible by
being a member of the community, a fact that enables it,
through the patrimoniality of functions, to take on concrete forms
of realization which are close to private landed property. But
is this dual quality of communal membership, and consequently
of the relationship to the conditions of production, sufficient to
characterize these two separated groups of men as two differ
ent, even opposed, social classes? This double relationship is
already found in the primitive community: the chiefs and other
(public) officials of the most primitive tribes and clans are
similarly distinguished members of the community and thus
masters over the factors of production, in contrast to the
ordinary members who find themselves reduced to servitude.
Nevertheless, the relationship represented by the dotted line
can only be drawn in the diagram of the actual 'Asiatic' form
of tribal communal property. For in the most democratic form
of communal property, this relationship that undoubtedly exists
at the base does not appreciably affect the whole mode of production. This is because in the conditions of the primitive
community, family heads and other public officials serve the
truly common interests of actual communities. They fulfil true
functions as officials and the surplus product set aside for
them is indirectly returned to the community; therefore one
cannot speak of exploitation. On the other hand, the 'great
communities' of Asiatic empires are too distant from the naturally
determined communities; they are always secondary, fictively
organized communities. In fact, these communities are systems of
taxation, which only take on a concrete form when looking at
the top of the system, since they are only the communities of the

officials set up upon the basis of usufruct; when looking at
the base, the communities are abstract and at the very most,
communities of 'general slavery'. The exact meaning of the
diagram on page 259 is: peasants in the village communities hold
land on the basis of tribal communal property, directly through
their attachment to their actual village community, whereas the
tribal organization is in fact disintegrated and only preserved
in the form of the system of taxation. Thus the state aristocracy
operates through representing the 'communities' organized from
above. Therefore, the direct producers and the appropriators
of ground rent do not belong to the same community, although
belonging to the community is the precondition of their possession. In this sense, the former are ordinary members of actual
village communities, whereas the latter represent communities
that have been organized for and by themselves. Thus in the
presence of the AMP, it is justified to speak of exploitation
and to consider state officials as an exploiting class. In this
way, on the basis of tribal communal property, a class society
is formed. This is why the AMP is fundamentally different from
primitive society and its forms of typical disintegration. At the
same time, the AMP is the basic contradiction of oriental society,
through which all the special features of the Orient can be understood. Why, if not for the need to demonstrate the unity and
reality of 'imaginary tribalism' (Marx), could the 'real despot'
and his officials be led to build palaces and pyramids, temples,
and tombs? The absence of private property once again becomes
'the key to the oriental heavens' (Marx).

   Although the first class structure in history, that specific
to the AMP cannot be considered as the typical first class structure because it is rather doubtful that this structure represented
social progress. Indeed, rather than dissolving tribal communal
property, this class structure preserves it. As with the division upon which it rests, this type of class structure should
also be considered transitional; it emerges for a shorter or
longer period when every primitive society is dissolving. But
in the course of typical evolution, it never becomes stable or
even lasts any length of time, since its rapid disappearance is
that which ensures true evolution. Its transitory state can
be seen in the fact that this type of class structure does not
merely mitigate slavery through forms of patriarchal dependence,
but in fact dissolves it into the patriarchal conditions of 'general
slavery'. Its transitory state is expressed in the fact that
neither the patriarchally exploited class nor the patriarchally
exploiting class have even approximate boundaries: the ordinary
peasant of a village community is also a family head and the
patriarch of his own small family; every low-ranking official
is the patriarchal slave of the representative of various 'superior
communities'; finally, the despot at the top of the pyramid rules
over all his subjects with the power of ownership. This situation
greatly complicates the class structure specific to the AMP; this
should not surprise us, if we take into account that so simple and

extremely polarized a class structure as we find in capitalism
has never existed before in history. The more ancient the
social formation examined, the more complexities are encountered.
This multidirectional stratification offers a natural starting
point for the most diverse analyses of class conditions in Asiatic
societies. However, clearly on the basis of tribal communal
property, any relationship of patriarchal dependence is a result
of the fact that the men of one group are no more than ordinary
members of the community, while the others behave as representatives of partly real, partly fictitious, communities. We are
inevitably led to say that class antagonisms in Asiatic societies –
a transitory fundamental asocial formation however – take
place between ordinary members of the communities and 'public'
state officials.

At this point we would like to stress that the scientific analysis
of the class position of the Asiatic bureaucracy (this term refers
not to a managerial function, but to the deeper sense of the
state official's usufruct resting upon communal property) has
been greatly hampered through the attempts to see 'a new class'
in the stratum of socialist officials which have been made by both
present-day anarchists and revisionists. The question of the
AMP is connected with these naive ideas in the sense that these
theoreticians maintain, that if a class society could have arisen
on the basis of tribal communal property, then a class society
could equally arise on the basis of socialist communal property.
This occurs in oriental and socialist society in the same way,
i.e. through the alienation of public functions and their transformation into a state bureaucracy. In this way, the newly liberal
Mr Wittfogel has used the term 'hydraulic society' to refer to
socialist societies in his recent book. Many Marxists in their
struggle against such opinions find it necessary to challenge the
principle that a class society could have possibly ever arisen on
the basis of communal property, i.e. that officials can ever act
as a class. However, this argument is challenged by the facts
and historical evidence of Asiatic state despotism and the explanation of these facts, maturely reflected upon from Marx's and
Engels's philosophical perspective. If one followed such a line
of reasoning, oriental societies can only be conceptualized as
slave-based or feudal societies. Now, as we have shown, such
an attempt is not only contrary to Marx's and Engels's findings,
but deprives the Marxist categories designating social formations
of all scientific value. In this way, these categories can be
applied to all societies with traits superficially resembling slave-
based or feudal formation. This form of 'Marxist' argument will
result in the same anti-historical error as the speculations of a
Djilas. It is not a case of declaring that tribal communal property
or the state founded upon it cannot provide the basis of a class
society. The extent to which the ambivalent rather than typical
dissolution of primitive society inevitably leads to a similarly
ambivalent and transitory contradiction has to be examined. At
the same time one must look at the extent to which socialist

state property's fostering the creation of 'a new class' is impossible.

In the foregoing analysis, we have tried to show, from various angles, that the oriental path of development – or to be more precise, oriental stagnation – was a natural process. One should not be surprised that this development did not follow the typical path of universal evolution, since no people has followed this historically typical path of development to its ultimate end – for this could only result from universal evolution. Hence oriental evolution cannot be effectively compared to the evolution of any European society, since in *every* case of European development sudden halts, distortions, ambiguities and regressions can easily be brought to light. The evolution of the Orient must not be compared to the whole of European evolution, since the latter has essentially coincided with universal evolution up to the period of capitalism. . . .

Returning to the problem of the historical place of the AMP, the question should be raised of how the 'Asiatic' form of property – besides the fact that it was necessarily constituted naturally – provided the indispensable precondition for further development, i.e. the development of the Ancient form of property. Marx's and Engels's explanation demonstrated that the Asiatic form of property, as tribal communal property, is not only the natural, but the indispensable precondition for all evolution. Nevertheless, in Asia the preservation of this tribal communal property, the foundation of the AMP, is in no way the precondition of the Ancient and Germanic forms of property – in the same way that the evolution of the Germanic form is unimaginable without the Ancient form of co-operation, or that capitalism can only develop out of feudal society, or that socialism could only have emerged from capitalism. The ambivalent dissolution of primitive society and the stabilization of its transitional forms in the Ancient Orient before the Hellenic era, i.e. the whole of evolution based upon the AMP prior to the Greeks, can not be termed the precondition of Greek civilization in so far as a whole set of factors contributed to its development and lay down the foundations for the emergence of private landed property. The developmental tendency of the AMP, however, is not progressive if it leads to non-development and stagnation, as is the case when any transitional type is stabilized and not permitted to wither away. For example, clearly the patriarchal form of slavery based upon communal property is far from pointing the way to the private ownership of slaves. On the contrary, it severely hampers the development of Ancient slavery. Absolutely all the essential traits of the AMP that stabilize transitional forms can be viewed in the same light. Thus, any more or less Asiatic aspect of ancient property or society is exactly what hampers development; it only encourages development in as much as any regressive force is apt to do. According to the entirely coherent view of Marx and Engels, becoming bogged down in transitional forms results not in truly typical development, but in the setting in of stagnation and regression. Thus, being bogged down in this way arrests typical development at a transitional stage, and at the same time represents a deviation from the typical path, and to some extent a new path. It logically follows that only the development of capitalism and thereupon socialist development can integrate these ambivalent and semi-barbarian civilizations within the universal evolution of mankind.

On the basis of the preceding analysis, we can draw some final conclusions from the forms of property defined by Marx considered here. First of all, as a type, the AMP is a transitional formation, notably found between the primitive communal mode of production and the Ancient mode of production; with respect to the property relations, the AMP is most closely linked to the primitive communist formation; even from the perspective of social divisions, it does not belong to the categories of primitive community, slave society, or feudalism (unless we want to deprive these categories of their real meaning). At the same time, glancing at our diagram, it is clear that the (clearly expressed or naturally implicit) tendency by those using the category of the AMP in the 1920s, which interpreted it as a fourth basic formation preceding capitalism, was totally wrong. The diagram clearly shows that a fourth basic formation preceding capitalism is in principle unimaginable, since the 'Asiatic' and the 'Germanic' forms are the opposite of each other, and the Ancient form is essentially the unity, struggle, and reciprocal effect of the characteristic relations of the 'Asiatic' and the 'Germanic'. If a fourth form of property prior to capitalism is unimaginable, then the AMP cannot be considered a basic formation. We must once again affirm the transitional state of formation and underline that the diagram based upon Marx's terms in no way contradicts Engels's theory of the basic formations. On the contrary, one can only understand the category of the AMP and accord it its place in history on the basis of the theory of the basic formations. There is another side to the coin. If the category of the AMP is eliminated, the development of the Orient will remain incomprehensible. The theory of the basic formations will likewise equally be deprived of its true meaning. The category of the AMP is an organic and fundamentally important part of Marx's discoveries. Its repudiation would be harmful to political economy and to the Marxist philosophy of history. Marx and Engels solved the basic problems related to this category; we can appreciate the correctness of their solution daily in our own time and even in our political life.

NOTE

* From 'Sur le Mode de production asiatique', Budapest, Akadémiai Kiadó, 1966, pp. 36–68 passim.

THE ASIATIC MODE OF PRODUCTION*

Maurice Godelier

HYPOTHESIS OF THE NATURE AND LAWS OF DEVELOPMENT OF
THE ASIATIC MODE OF PRODUCTION AND THE CONCEPT OF
TYPICAL LINE OF DEVELOPMENT OF HUMANITY

The nature of the Asiatic mode of production
With the concept of the Asiatic mode of production, Marx has
shown us societies within which particular village communities
are subject to the power of a minority of individuals who represent a higher community, the expression of the real or imaginary
unity of the particular communities. This power at first takes
root in functions of common interest (religious, political, economic) and, without ceasing to be a functional power, gradually
transforms itself into an exploitative one. The special advantages
accruing to this minority, nominally as a result of services
rendered to the communities, become obligations with no counterpart, i.e. exploitation. The land of these communities is often
expropriated to become the ultimate property of the king, who
personifies the higher community. We therefore have exploitation
of man by man, and the appearance of an exploiting class without the existence of private ownership of land.
    In our view, this picture presents a form of social organisation
characterised by a contradictory structure. This form of organisation combines community structures and the embryo of an
exploiting class. The unity of these two contradictory elements
rests precisely on the fact that it is in the name of the higher
community that the individual communities are exploited by this
minority. A society characterised by this contradiction is thus
presented simultaneously as a final form of classless society
(village communities) and an initial form of class society (a
minority exercising state power, a higher community).
    Thus we are putting forward the hypothesis that Marx, with-
out having been completely aware of it, described a form of
social organisation specific to the transition from classless to
class society, a form which contains the contradiction of that
very transition.

In our view, this hypothesis allows us to understand why the concept of Asiatic mode of production is referred to more and more to illuminate the periods and societies of ancient Europe (Creto-Mycenean or Etruscan monarchies), of Black Africa (kingdoms and empires of Mali, Ghana, Bamoum kingship, etc.) and pre-Columbian America (great Mesoamerican or Andean agricultural civilisations). A common element appears throughout these many specific realities - a common structure which combines community relations and the embryo of class and is related to the same situation of transition to class society. Because of this relation between the situation and structure it is possible to explain the geographical and historical universality of the form of social organisation which emerges when the conditions for the transition to class society develop; maybe at the end of the fourth millennium BC in the case of Egypt with the transition of the tribal Nilotic societies first to monarchies and then to a unified empire,(1) or in the nineteenth century with the birth of the Bamoum kingdom in the Cameroons. By providing many examples of societies in the process of transition to class organisation the archeological and ethnological knowledge accumulated since the nineteenth century provides the concept with a field of application which Marx or Engels could not have envisaged. In becoming more and more widely applicable both in time and space, the concept no longer applies exclusively to Asia. It may therefore be necessary to abandon the use of the adjective 'Asiatic'.

The emergence and different forms of the Asiatic mode of production
Within the terms of this general theoretical hypothesis, the second task is the systematic study of the nature of transition to class society, of the emergence of the conditions of transition.
   For Marx, the Asiatic mode of production is linked to the need to organise major economic projects beyond the means of particular communities or isolated individuals and constitutes the precondition for productive activity for these communities. In this context, appear forms of centralised power which he called 'oriental despotism',(2) a term familiar since the seventeenth century. The state and the dominant class intervene directly in the conditions of production; in the organisation of major projects, the correspondence between productive forces and relations of production is made direct.
   Even if this hypothesis provides the key to the typical most developed forms of the transition to the 'Asiatic' mode of production, it does not exhaust all the possible circumstances. We propose the addition of a second hypothesis to Marx's. Assume the possibility of an alternative route and another form of Asiatic mode of production in which a minority dominates and exploits the communities intervening in their conditions of production not directly, but indirectly, by appropriating surplus labour or produce as profit. In West Africa the emergence of the

kingdoms of Ghana, Mali, Songhay, etc. was not the result of
the organisation of major projects but seems to be linked to the
control exercised by the tribal aristocracies over intertribal or
interregional trade in rare commodities, gold, ivory, hides, etc.,
between black and north Africa.(3) In Madagascar, alongside
the Imerina kingdom which was based on irrigated rice agriculture and had put the marshes in the Tananarive plain to good
use, there appeared the kingdom of Sakalava based on nomadic
herding and the trading of cattle and slaves.(4)

Our hypothesis allows us to explain the emergence of a dominant
class in agricultural societies based neither on great agricultural
works nor on herding without incurring the difficulties or contradictions aroused by the expressions 'nomadic feudalism' (mongolian feudalism, etc.).(5)

Comparing these two forms of Asiatic mode of production –
with or without major projects – proves that there is a common
element: the appearance of an aristocracy exercising state power
and consolidating the bases of its class exploitation by appropriating a part of the produce of the communities (either in labour or
in kind). What the presence or absence of major projects determines is the appearance of a bureaucracy and an absolute centralised authority known by the vague and antiquated term
'despotism'. Thus, we do not consider it necessary to search
mechanically with Wittfogel for gigantic, mainly hydraulic,
projects, a bureaucracy and a strongly centralised authority
in order to rediscover the 'Asiatic' mode of production.(6)
Rather, the theoretical task is to draw up a typology of the
various forms of this mode of production, with or without major
projects, with or without agriculture, and at the same time to
draw up a typology of the forms of communities within which
this mode of production grows up. Thus it might be possible to
reconstitute many models of the way in which inequality is
introduced in classless societies and leads to the appearance of
antagonistic contradictions and the formation of a dominant
class. The collaboration of historians and anthropologists would
be essential to this task.

We have tried to define the structure and certain forms and
conditions of the appearance of the Asiatic mode of production;
now we must approach the problem of the laws of evolution of
this social formation.

*The dynamic and the laws of evolution of the Asiatic mode of
production*
If the appearance of the Asiatic mode of production indicates
the emergence of an initial class structure whose outlines are
still flexible, it also indicates the regular appropriation of part
of the communities' labour by this class, i.e. the existence of
a regular surplus. From the point of view of the dynamic of
the productive forces, a society's transition to the Asiatic mode
of production would not mean the beginning of stagnation but,
on the contrary, would be a progressive sign for the productive

forces. If Pharaonic Egypt, Mesopotamia, the Mycenean kingdoms, and the pre-Columbian empires belong to the Asiatic mode
of production, then here we have evidence that it corresponds
to the most brilliant civilisations of the metal age, to the period
when man definitively wrenches himself away from the economy
of land occupation and once and for all passes to the domination
of nature, invents new forms of agriculture, architecture,
mathematics, writing, trade, currency, law, new religions, etc.
In our view, the Asiatic mode of production originally meant not
stagnation, but the greatest possible progress of the productive forces accomplished on the basis of the previous communal
forms of production. This is amply confirmed in the works of
the great archeologists Childe and Clark.(7)
     If the Asiatic mode of production originally indicated a pro-
gressive step for the productive forces, what is the law of its
evolution? We believe that it is the law of the development of its
internal contradictions, just as in any other social formation.
The internal contradiction of the Asiatic mode of production is
that of the coincidence of community structures and of class
structures. The Asiatic mode of production would evolve by
the development of its contradiction towards forms of class
societies in which community-based (communal) relations have
less and less reality because of the development of private
property.
     The Asiatic mode of production, like any other social forma-
tion, would stand for stagnation only if it could not be left
behind, when its contradictions fail to develop and its structure
is petrified, leaving the society floundering in a state of relative
stagnation. Although in each case the nature and the timing of
a step forward would depend on specific circumstances, it
would always mean the obliteration of the former modes of communal social organisation; failure to make this move would on
the contrary mean maintaining them.
     This permanence and the stagnation that accompanies it are all
the more threatening to an 'Asiatic' society, based as it is on
self-sustaining communities without any radical separation of
agriculture and industry and, if the land is available, likely
to allow for their demographic growth by splitting into daughter-
communities which will perpetuate the same traditional forms of
production and social life. It is this development which is described in Marx's famous text on Indian communities which has
become dogma for the proponents of the secular stagnation of
Asia (1859, I, pp. 338-9):

> The simplicity of the organisation for production in these
> self-sufficing communities that constantly reproduce them-
> selves in the same form, and when accidentally destroyed,
> spring up again on the same spot and with the same name -
> this simplicity supplies the key to the secret of the unchange-
> ableness of Asiatic societies, an unchangeableness in such
> striking contrast with the constant dissolution and refounding
> of Asiatic states, and the never-ceasing changes of dynasty.

The structure of the economic elements of society remains untouched by the storm-clouds of the political sky.
Moreover, in so far as the state's exploitation of the communities
takes the form of a massive levy of revenue in kind, the structures
of production can stabilise since there is no incentive to create
a market. Because it is possible for the state to use peasant
labour, the possibilities for the development of a market are
limited and the transformation of the productive forces is slowed
down. Besides, these forms of exploitation can be so intense
that they hold back the development of production for a long
time.
Aside from this evolution of the Asiatic mode of production
towards sinking and stagnation what forms can this development
take when its internal contradiction is developed? - forms which
lead to its fragmentation because of the emergence of private
property. We think there are at least two possible forms of this
disintegration.
One would lead to the slave mode of production via the ancient
mode of production. That would be the path taken by the Graeco-
Romans. It would lead to societies based on the combination of
private property with commodity production. In this combination
lies both the secret of the 'Greek miracle' and of the expansion
of the Roman Empire; it also explains the uniqueness of this
line of development, the typical character of its class struggles
between free men, and of their exploitation of slave labour.
Aside from this well known path, we think that another one
exists, one which would lead slowly, with the development of
individual ownership, from certain forms of the Asiatic mode of
production to certain forms of feudalism, without going through
a slave stage. The appearance of individual ownership within
the communities or of personal estates belonging to the aristocracy would transform the communities and, at the same time,
the forms of their exploitation by this aristocracy. There is a
slow transition from the collective exploitation of the communities
to an individual exploitation of the peasants. This evolutionary
process seems to be the most frequent and to correspond to the
transition to a class society in China, Vietnam, Japan, India,
Tibet. . . .(8)
While we cannot substantiate these hypotheses here, we can
draw attention to the light they may throw on the development
of the last centuries of Inca society. These hypotheses are in
tune with Métraux's interpretation of the late development of
private estates belonging to the emperor and his caste. The
estates were settled by the Yana, people attached to the nobles
and the notables of the kingdom by links of personal, and no
longer collective, dependence:(9)

> The increasingly important part played by the *yanas* under
> the empire is only to be explained on the grounds that their
> efficiency was superior to that obtained by the traditional
> system of forced labour. By depriving them of some of their
> members, the Inca weakened the communities and initiated

a revolution which, had it continued, might well have changed
the structure of the empire. It would have tended to become,
instead of a grouping of largely autonomous rural collectives,
a kind of 'pre-feudal empire' in which nobles and officials
would have owned great estates worked by serfs or even slaves.
This evolutionary path towards a certain form of feudalism is
not only the most frequent but the simplest; unaccompanied by
a great development of commodity production and currency, it
does not break with the 'natural' forms of the economy and for
a long time preserves the combination of agriculture and industry.
Besides, in so far as it remains necessary to control and organise
major projects during this transition to private property, central
authority plays an important part and the domination by the
state and the monarch of the 'feudals' and peasants gives these
'feudalisms' a 'particular' shape where features of the Asiatic
mode of production persist.(10) Because of this and other
characteristics, extreme caution must be exercised in comparing
these 'feudalisms' produced gradually by the Asiatic mode of
production with the Western form of feudalism born of the de-
composition of the slavery mode of production. The main difference between them and the West is that they slowed down the
development of commodity production and prevented the appearance and the triumph of industrial capitalism. The case of the
Meiji revolution in Japan needs to be studied separately. However, it is undeniable that the industrial base, modern productive
forces, and methods of organisation were imported from the
Western capitalist countries and were not developed in the
Japanese feudalism within which had appeared a certain form of
commodity capitalism.(11)

Of the two forms of development of the Asiatic mode of production, one towards a slave system, the other towards certain
forms of feudalism, it is the former which seems to us more and
more unique and exceptional, and here we differ from the dog-
matic conceptions of many writers. The Western line of development, far from being universal because it is found everywhere,
appears to be universal because it is found nowhere. The mistake of Marxists has usually been the desire to discover a slave
mode of production everywhere and failing this, to create it
in order to rediscover it. If this is so, why did Marx and Engels
consider the Western line of development to be 'typical' of the
development of humanity? How are we to understand the universality of what now appears to be unique? Is it a remnant of the
capitalist world's superiority over the rest of the world, a disguised form of racism, a pseudo-science? This final question
brings us to our last hypothesis concerning what is understood
by a 'dominant or typical' line of development of humanity.

*The forms of disintegration of the Asiatic mode of production
and the 'typical' line of development of humanity*
To recognise a 'typical' form of development implies a previous
awareness of the 'general line' of this development and of the

nature of its movement as a whole. Is it possible to grasp the
general nature of the movement of history in retrospect?

Marx and Engels attempted to do this. We think that since their
time no new knowledge has come to light which invalidates the
essential point of their conclusions. The general movement of
history has been the transition of the majority of peoples from
a classless social organisation to class societies. That is the main
point. This emergence requires the development of the inequality
inherent in the appropriation of the means of production, and this
inequality itself presupposes the disintegration of ancient communal solidarities based on co-operation through labour and
living kinship relationships.

Thus, in retrospect, the movement of history appears as the
indissoluble unity of the development of two contradictory
elements of social reality: (a) the general development of means
of dominating nature and ensuring the survival of an ever-
increasing species; (b) the progressive breakup of communal
solidarities and the general development of inequalities between
individuals and groups.

Engels made this contradiction central to understanding the
nature of 'civilisation' (1884, p. 174): 'Since the exploitation
of one class by another is the basis of civilisation, *its whole
development* moves in a *continuous contradiction.*'(12)

Even though it is necessary to abandon the old categories
of the nineteenth-century Anglo-Saxon anthropology (the
succession of the three stages: savagery, barbarism, civilisation) because of its vague and ambiguous nature and because
of all its ideological implications, and to replace it with a division
between classless and class societies, Engels correctly described
the whole movement of transition from one to the other as the
fundamental fact of history.

If this is the over-all movement of history, then the 'typical'
line of development of humanity is that through which the contradiction of the maximum development of the productive forces
and of class inequalities and struggles is realised.

In recognising the typical line among the various lines of evolution of societies, the criteria consist in looking for the time and
place in which the greatest progress of the productive forces
took place. The answer is obvious and without mystery; it is
the line of evolution which gave birth to industrial capitalism,
the origin and basis of the most modern and efficient forms of
production, of the transformation of nature. Industrial capitalism
has appeared only in the line of evolution set in motion by the
Greeks. The decisive character of this line of evolution is that it
has ensured the maximum development of the productive forces,
thus providing immense possibilities for the exploitation of man
by man. This development cannot be explained by the appearance
of private property alone. It existed in China, Vietnam, etc. In
addition, it is necessary for private property and commodity
production to be combined. (13) Only this combination created
the most favourable conditions for technical progress, while

revealing itself to be incompatible with the functioning of the
former solidarities of communal life; it substituted the search for
private profit for the submission to common interest, by breaking
off the often sacred collective link of the individual with the land
of his ancestors.

It seems that this combination appeared for the first time in
a pure form among the Greeks: 'Here lies the root of the entire
revolution that followed' (Engels, 1884, p. 111).

The Romans took it up and generalised it, giving it its universal juridical expression with the theory of the 'jus utendi
et abutendi' which became the legal model in commodity societies
based on private property.

The uniqueness of the line of development of the Graeco-
Roman societies is becoming clearer. It consists not in having
overcome certain forms of Asiatic mode of production, and having
done it maybe earlier than other peoples, but in having overcome
them in moving towards a mode of production based on the combination of private property with commodity production.

Similarly, what makes Western feudalism unique, beyond its
similarities of form with what are called the feudalisms of Turkish,
Chinese, African, Japanese, etc., what prevents its confusion
with them and the basis for their essential difference, is that
it alone created the conditions for the appearance of industrial
production and world trade. It alone has allowed the forms of
natural economy to be surpassed.

Finally, by allowing and imposing the creation of a world
market, industrial capitalism has made a universal history
possible by subsuming all the less developed societies under its
development, which is that of the Western capitalist societies.

Moreover, industrial capitalism alone has opened up the possibility of socialism, first in theory, then in practice.

Thus, the Western line of development is typical because it
is unique in developing the greatest progress of the productive
forces and the purest forms of class struggles and also because
it alone has created the preconditions for Western and all other
societies to pass beyond the class organisation of society.

So it is typical because in its particular development it has
obtained a universal result. It has provided the practical base
(industrial economy) and the theoretical conception (socialism)
to extricate itself and all societies from the older and newer
forms of exploitation of man by man. It provides the whole of
humanity with the conditions for the solving of a universal
problem posed since the appearance of classes: how to ensure
the maximum development of the productive forces without the
exploitation of man by man? It is typical because it has value as
a 'model' or 'norm' because it provides possibilities which no
other single history has offered and gives other societies the
possibility of saving themselves the intermediary stages. (14)

This perspective provides Engels's remarks in 'Anti-Dühring'
with their full import (1878, p. 334):

But if . . . division into classes has a certain historical justification, it has this only for a given period, only under given social conditions. It was based upon the insufficiency of production. It will be swept away by the complete development of modern productive forces. And, in fact, the abolition of classes in society presupposes a degree of historical evolution at which the existence, not simply of this or that particular ruling class, but of any ruling class at all, and, therefore, the existence of class distinction itself has become an obsolete anachronism. It presupposes, therefore, the development of production carried out to a degree at which appropriation of the means of production and of the products,·and, with this, of political domination, of the monopoly of culture, and of intellectual leadership by a particular class of society, has become not only superfluous but economically, politically, intellectually a hindrance to development. This point is now reached.

The true universality of the Western line of development lies in its specificity and not outside it, in its difference from not in its resemblance to other lines of development. The unity of universality and of singularity is a contradiction but this contradiction is in life not in thought. When the unity of this contradiction is not recognised, we can take one of two equally futile paths:

1 a multitude of societies survive and evolve side by side, each locked in its historical singularity by the intransigent scholar. Nothing is comparable with anything else and history remains a mosaic of bits deprived of global coherence;

2 on the other hand, one can try to see the same process everywhere with the singularities erased and then history becomes the more or less successful application of universal forms to which it is necessarily subject. In fact, these forms which are sought everywhere are nothing more than those of the Western line of evolution which must be sought everywhere since the possibility of many lines of evolution has already been denied.

The typical character of this line of development then finds its root, not in itself, in its own specificity, but in a necessity external to history. But, as we know, the opposite of an external necessity is an internal finality. Within such a perspective, history was a future without surprises, a reality created in advance through which humanity, from its very first entry into primitive communism was bound to emerge one day into true communism. This second path was chosen by many Marxists, especially after Stalin's treatise on the laws of historical development in 'Dialectical and Historical Materialism', in which primitive communism, slavery, feudalism, capitalism, and socialism necessarily follow each other.

Marx, however, had already warned against this error in the 'Contribution to the Critique of Political Economy',(15) where he specified that (1859, p. 211)

What is called historical evolution depends in general on the fact that the latest form regards earlier ones as stages in the

development of itself and conceives them always in a one-
sided manner, since only rarely and under quite special conditions is a society able to adopt a critical attitude towards
itself.

From this point of view, socialism appears to be a modern mode
of production, just as incompatible with the former pre-capitalist
modes of production as capitalism itself and perhaps even more
incompatible since capitalism could make use of the old relations
of exploitation within the countries it dominated, whereas
socialism cannot do this.

Having begun by looking for a lost and even rejected Marxist
concept, we have tried to reconstruct it through Marx's and
Engels's texts without prejudging its scientific value. Once reconstructed, it was still necessary to find out why it had been
lost. Our search led us to discover unmysterious reasons: the
relationship between Morgan and Engels, and the state of the
most advanced archeological, linguistic, and anthropological
knowledge of the second half of the nineteenth century. Pushed
into the background by Engels's brilliant analysis, the concept
faded away, reappeared to some extent towards 1927-30 after
the failure of the Chinese revolution, then was totally rejected
until the night when the renegade Wittfogel dug it up to turn it
into a war machine against socialism. At the same time, imputated
of the Asiatic mode of production, deprived of the hypothesis
of plural forms of transition and evolution towards class society,
Marx's model of the evolution of societies ceased to be an open
system of hypotheses to be verified and became a closed system
of dogmas to be blindly accepted.

Historical materialism cleansed itself of its scientific substance
from within and stood like a new philosophy of history, an ideal
world in which the philosopher contemplated the historical
necessity which moved humanity from primitive communism to
true communism. On a practical level, the divorce between
anthropology and history, between Western and non-Western
history, seemed increasingly definitive. By a strange paradox,
innumerable facts have emerged to press scholars to revive a
dead concept. If this concept indicates a social formation corresponding to the contradiction of certain forms of transition from
classless to class society then perhaps we have rediscovered a
historical reality which demands and forms the basis for the
collaboration of the anthropologist and the historian (or the
archeologist). In order to understand the specific contradiction
of the Asiatic mode of production, it is necessary to be simultaneously an anthropologist to analyse the community structures
and a historian to account for the embryo of exploiting classes.
The disjointed bits of historical and anthropological knowledge
could be reconstructed around this contradictory reality to form
a unified whole of anthropological knowledge.

We think that with its revival the Asiatic mode of production
has mortally wounded old superseded statements, theoretical
corpses which crumbled at the slightest shock, since they were

always only pretending to be alive: like the existence of a universal stage of slavery, the impossibility of bypassing stages. But this resurrection is and must be more than a return to Marx since it would then be a return to a superseded stage of historical science. We have attempted, therefore, to put the concept back into working order so that it would become capable of dealing with the problems posed by the comparative archeology, anthropology, and history of today. We have suggested a structural definition of the Asiatic mode of production, assumed a relationship between the structure and certain situations of transition to class society and, at the abstract level, grasped the theoretical possibility of a wider field of application than Marx had been able to predict for the concept. But to move forward it may be necessary to abandon the geographical adjective 'Asiatic', to define rigorously the old word 'despotism', and to search cautiously for 'major projects' and 'bureaucracies'.

It may be necessary to consider stagnation as a possible, but not the only form of evolution of the Asiatic mode of production and to imagine many possible forms of its disintegration whose motive forces should be looked for. We have suggested the hypothesis of an evolution of the Asiatic mode of production towards certain forms of feudalism and considered this path to be a more frequent form of transition to a true class society than the Western one. The latter increasingly appears to be simultaneously unique and universal, to have developed to their highest point the characteristic features of a class society, the domination of man over nature and the domination of man by man. Thus, in the final analysis, we think that it is not only the concept of Asiatic mode of production which must be put back into working condition, but the very notion of historical necessity, of historical law. Unless this is done the works of historians will set off blindly, threatened tomorrow with yesterday's fate and, at another level, social practice will develop without knowing where it came from, where it can go, or how to get there.

Of course, the analyses and hypotheses we have suggested will be disputed or confirmed by a wider discussion. To accept them without proof would be to abandon the name while retaining the spirit of dogmatism. On the contrary, to look for an Asiatic mode of production in such and such a history, without having previously posed the problems of the theoretical status of this concept, is to be nothing but a well-meaning positivist. We therefore suggest research in the following directions:

1   Is it possible to reconstitute the different processes through which inequality is introduced in classless societies and leads to the formation of a ruling class (this question is asked of historians and anthropologists)?

2   Is it possible to draw up a typology of the different forms of Asiatic mode of production with or without major projects or agriculture, etc. and deal with the problem of a typology of the forms of communities by analysing the forms of appropriation of land, the origin and nature of aristocratic and royal powers, etc.?

3 Is it possible to describe different forms of development of the Asiatic mode of production towards class societies?

4 What is the process which introduced a commodity economy among the Greeks and the Romans? How is it possible simultaneously to take the 'Greek miracle' seriously and to deidealise it?

5 What is the relationship between the concepts of Asiatic mode of production and of military democracy?

These studies will demand the creation of a rigorous language and perhaps the rapid adoption of less loaded, more precise, terminology in place of the expression 'Asiatic mode of production'. (16)

NOTES

* From 'The Concept of the "Asiatic Mode of Production" . . .', in D. Seddon, ed., 'Relations of Production', London, Cass, 1978, pp. 240-52.

1 W.B. Emery, 'Archaic Egypt', London, Pelican, 1962, pp. 28-104.

2 F. Venturi, Oriental Despotism, 'Journal of the History of Ideas', vol. 1, 1963, pp. 133-42.

3 See J. Suret-Canale, 'Afrique noire', Paris, Editions Sociales, vol. 1, 1961, p. 112.

4 See G. Condominas, 'Fokon'olona et collectivités rurales en Imerina', ed. Berger-Levrault, Paris, 1960, p. 29; P. Boiteau, 'Contribution à l'histoire de la nation malgache', Paris, Editions Sociales, 1958.

5 See Vladimirtsov, 'Le Féodalité mongole', 1948. Comments by A. Belenitsky, Les mongoles et l'Asie centrale, 'Cahiers d'histoire mondiale', 3, 1960; J. Harmatta, Hun society in the age of Attila, 'Acta Archeol. Acad. Sci. Hung.', 1952.

6 See J. Maquet's objections to Wittfogel: Une hypothèse pour l'étude des féodalités africaines, 'Cahiers d'études africaines', no. 6, 1961.

7 G. Childe, 'Social Evolution', London, Fontana, 1961, where Childe was trying to complete Morgan's model by integrating the great oriental civilisations of the Bronze Age. For Grahame Clark, 'World Prehistory', see my account in 'La Pensée', 1963, vol. 7.

8 A. Stein, 'La Civilisation tibetaine', Paris, Dunod, 1962, pp. 97-103.

9 A. Métraux, 'The Incas', London, Studio Vista, 1965, p. 100.

10 See. L. Simonovskaia, Deux tendences dans la société féodale de la Chine de la Basses Epoque, 'International Congress of Orientalists', Moscow, 1960.

11 Among a great wealth of documentation see the works of the Japanese Marxist, Takahashi, La place de la révolution du Meiji dans l'histoire de Japon, 'Revue historique', December 1953; The transition from feudalism to capitalism, 'Science and Society', 1952, vol. 4.

12   Emphasis M. Godelier. There is therefore no possible misunderstanding on the use of the term 'civilisation' by Engels.
It does not indicate a hidden racism or the badly disguised
confession of a 'moral' or intellectual superiority. And this
attitude is shared by many anthropologists who have lived
with so-called 'savages' or 'barbarians'.

13   It is the fact of commodity production which provides the
key to the scientific study of capitalism, the final stage in
the development of Western societies. This was emphasised
by Marx in the opening words of 'Capital' (1867), repeating
those of the 'Contribution' (1859): 'The wealth of those
societies in which the capitalist mode of production prevails,
presents itself as "an immense accumulation of commodities",
its unit being a single commodity. Our investigation must
therefore begin with the analysis of a commodity' ('Capital',
I).

14   It is in this light that Marx's famous letter to Vera Zasulich,
8 March 1881, is to be understood.
But does this mean that the historic career of the agricultural community must inevitably lead to this result?
Certainly not. The dualism within it permits of an alternative: either the property element in it will overcome the
collective element, or the other way round. Everything
depends on the historical environment in which it is
found. . . . These two solutions are *a priori* possible,
but of course, they both require very different historical
environments.
In the second version of this letter, Marx is more specific
concerning historical environment:
Its historical environment, the existence of capitalist
production, provides it with the ready-made material
conditions for co-operative labour organised on a vast
scale. It can thus incorporate the positive points elaborated
by the capitalist system without going through it. . . .
It can gradually supplant plot agriculture by industrial
agriculture thanks to machines. After having first achieved
normality in its present form, it can become the immediate
starting point for the economic system towards which
modern society is directed, and change thoroughly without
beginning by committing suicide (translator's note: most
of this is my translation from the French). See Marx's
and Engels's preface to the second Russian translation of
the 'Manifesto'.

15   See M. Godelier, Economie politique et philosophie, 'La
Pensée', October 1963.

16   We would like to mention A. Caso's important article, Land
tenure among the Ancient Mexicans, 'American Anthropologist', vol. 65, no. 4, August 1963, pp. 862-78, on land-
ownership among the Aztecs. We read this text after having
written our analysis of the Morgan-Engels relationship and
it appears strikingly to confirm our analysis. According to

the author, Aztec society combined the features of a communal society, tribal with communal land-ownership and those of a class society dominated by an aristocracy holding religious, political, and military power and controlling the state (p. 875). The king owned lands 'not as an individual, but as a functionary' (p. 868). The clergy and the military were kept by tributes and forced labour, levied on communities of free men. Apart from this state ownership, the nobility and the king owned private demesnes exploited by 'serfs attached to the domain' (p. 870). The existence of such social inequalities and of an aristocratic private property may, according to the author, be understood 'as long as we do not try to do so in terms of Iroquois organisation or Roman property' (p. 874). He concludes: 'It astonishes us that such false conclusions could have been arrived at as those of Morgan (1878) and Bandelier (1880) which were in vogue during the first quarter of this century' (p. 863).

Let us mention Gibson's two articles on La transformation des communautés indiennes en Nouvelle Espagne de 1500 a 1820, 'Cahiers d'histoire mondiale', no. 3, 1955, and particularly The Aztec aristocracy in colonial Mexico, 'Comparative Studies in Society and History', vol. 2, no. 2, January 1960, pp. 169-97 in which the author criticises (p. 171, para. 5) Bandelier's conclusions concerning the Aztecs in 'On the Social Organisation and Mode of Government of the Ancient Mexicans', Cambridge, March 1880; Bandelier is a follower of Morgan's; and F. Katz's 'Die sozialökonomische Verhältnissebeiden Azteken im 15 und 16 Jahrhundert', Chs 3 and 10, Berlin, 1956.

THE 'TRIBUTARY' SOCIAL FORMATION*

Ioan Banu

There are two questions which must not be confused: the first
is the extent to which the solution to the problem of the 'Asiatic'
('tributary') social formation is resolved in the Marxist classics;
the second is the viability of the concept today in the light of
our knowledge of new historical facts.

As to the first question, in contrast to many participants in
the debate, I personally feel that Marx and Engels deliberately
left the problem open. They never felt it possible to set out a
definitive view of the 'Asiatic' mode of production, its distinctive
characteristics, or its sphere of application as a concept. Nor
did they state whether or not it could be legitimately considered
a separate mode of production. In my opinion, this is why they
did not employ the concept in certain cases, for instance in
Engels's 'Origin of the Family' or in their enumeration of historical types of antagonism and class struggle in the 'Communist
Manifesto'.

In my work I have tried to define the characteristics of the
'tributary' social formation by using those classical works which
mention the AMP. However, my intention was not to forget the
other works, but to separate out and highlight the concept
for the reader.

I disagree equally with both those who reject the notion that
this concept was constantly present in the thought of the founders
of Marxism and as well with those who think that Marx and
Engels exhausted this subject.

With respect to this problem, let us not confuse the question
of the standpoint actually taken by the Marxist classics with the
position they would have taken, had they known of the historical
material discovered after their death, material which would have
enabled them to fill in the gaps which they themselves acknowledged, and have a clearer view of certain of the more controversial aspects.

The aim of contemporary Marxists to explain new scientific
data in the spirit of Marx, Engels, and Lenin is praiseworthy.
However, they should not attribute more to the masters than they

actually said, even if they are firmly convinced that today, in the light of new discoveries, it is possible to take a definitive stand where the Marxist classics found such a clear-cut position impossible in their day.

As for the second question, the new historical data seem to firmly justify the conclusions which Marx and Engels hesitated to make. For example, our knowledge of the history of ancient oriental philosophy gained over the past century, notably in Europe, has provided very valuable evidence supporting the concept of the specific 'tributary' mode of production elaborated by Marx and Engels.

Returning to the debate, we see that those in favour of the concept sometimes put forward definitions so rigid that they can hardly stand up to the facts, and therefore complicate the discussion. For example, it is maintained that according to Marx, stagnation is an absolute law for all societies of a 'tributary' (Asiatic) type. How then can it be explained that the founders of Marxism referred to the transition from the AMP to other modes of production in their consideration of certain societies? Chinese historians, confronting the millennial historical evolution of their society, would have difficulty in applying such a concept to their country's ancient history, if the latter was defined by reference to a non-evolving stage, among other criteria.

Some sort of 'rigidity' is inevitable in any definition, but one should not forget that a category expresses what is general in a 'pure' form, while in the reality of social life, the confrontation of the general with the particular requires an extremely dialectical flexibility, taking into account the great number of objective variables and the different degrees of evolution. With respect to the AMP, Marx emphasises that the same economic base exhibits 'innumerable gradations and variations' in different places.

On the other hand, we know that the appearance of a new social formation does not entail the immediate disappearance of all the elements of the previous formation. Indeed, certain elements can remain for a long time. The fact that each economic formation is defined by its dominant, although not exclusive, characteristics, does not mean that we forget the other features.

Another important point: I think, as did Marx and Engels, that all societies where village communities are found should not be treated in the same way. There are instances where these communities appear as the highest stage of the primitive commune, within the framework of that community, without therefore becoming the foundation of another socio-economic formation.

The appearance of private property - as Marx describes it in the drafts of his letter to Vera Zasulich - is tied to an evolution of the agrarian community towards the stage of slavery and not towards a 'tributary' type of society.

In other societies, we find qualitatively new features connected with the agrarian communities (for instance, social conflict and the state); these features require the recognition of a 'tributary' to the slave formation. However, in other cases, there is no stage of slavery. In some instances, even though the primitive communist formation has been transformed into slavery, 'tributary' types of features that appeared in the final stage of primitive communism remain as survivals in the new society based on slavery and even carry on into feudal society which follows the latter. The outcome depends on concrete historical factors which vary from society to society. Now, for many participants in the discussion, tributary-type features are ALWAYS *either* a variant of the primitive social formation or separate social formation, or a variant of the slave social formation, etc. This is a mistake.

NOTE

*   Translated from Les Premières Sociétés de Classes, 'Recherches Internationales' (Paris), 57-8, 1967, pp. 251-3 (originally published in 'Revista de Filosofie' (Bucharest), nos 2 and 3, 1966.

A CONTRIBUTION TO THE STUDY OF THE AMP: THE CASE
OF ANCIENT VIETNAM*

Le Thanh Khoi

There have been many discussions of the notion of the 'Asiatic
mode of production'. This notion has given rise to too many
generalizations based upon insufficiently accurate knowledge of
the societies to which it has been thought applicable. No doubt
these generalizations have been useful in drawing up a working
outline; however, in my opinion, it is now time to concentrate
on monographic analyses so that our reasoning may be more
solidly based.

   This brief article, based upon the case of ancient Vietnam
(tenth to nineteenth centuries), is intended as a contribution
towards deepening the terms of the debate. The example of
ancient Vietnam is all the more interesting since, far from
confirming the use of the concept of the 'Asiatic mode of production', it contradicts certain characteristics generally attributed to the notion, particularly the absence of private property
and even (for some authors) the absence of social classes. On
the other hand, Vietnam does not reveal the classic features of
'feudalism' either; over the period from the tenth to the nineteenth century there is no observable evolutionary tendency
toward 'feudalism' in this sense. On the contrary, the inverse
is true, due largely to the strengthening of the literati-official
class.

In the tenth century, Vietnam already exhibited the main features
of a nation-state. For over 1,000 years, this state had been
established in an area approximately extending from what is now
the Chinese border to Hoanh-son beyond which lay the Champa.
Although Vietnam's long incorporation as a part of the Chinese
empire entailed not only the diffusion of Chinese institutions,
language, and writing, but also continuous immigration and inter-
breeding, it did not result in the society's loss of its national
language. While the *nôm* system of writing only appeared in the
thirteenth century, in the eighth century we find the oldest
evidence of common names: the famous name *Bô cai* (father and

mother) with which the people rewarded one of the heros of the
Phung Hung struggle for independence.
      Likewise, despite the strong cultural influences from the
North, Vietnam retained a number of traits from the Hung-vuong
civilization (seventh to third century BC): tattooing, the chewing
of betel, the laquering of teeth, the ancestor and spirit cult,
not to mention rice cultivation which the Viet, themselves,
diffused to China.
      Of course political unification did not take place overnight.
After the fall of the Tang empire, Vietnamese kings constantly
struggled against the centrifugal tendencies of local chiefs in
both the delta and the Middle and High Region. Only in the mid-
eleventh century, more than a hundred years after independence,
did the Ly dynasty succeed in almost entirely eliminating regional
bases of power.
      Nevertheless, disintegrative tendencies continued to exist, but
in another form. Initially the royal family only trusted its closest
confidants. It divided up the country into military districts entrusted to sons, brothers, and allies. Domains were handed out
to reward generals and to bind them to the royal family. Although
the extension of these domains progressively diminished, the
practice lasted until the fall of the monarchy, that is, until
1883, when Tu Duc substituted rents for the last of aristocratic
landed property which was handed over to the villages to be
divided among the inhabitants.
      Several historians characterize ancient Vietnam as 'feudal',
extending this term to cover the period since the Han Chinese
conquest, since China of this period was considered 'feudal'.
The preceding period was considered to be 'based upon slavery'.
However, the notion of a 'feudal system' runs up against a
number of problems. In order to resolve these problems, various
Marxist scholars have revived the long forgotten concept of the
'Asiatic mode of production' and have recently proposed to apply
it to Vietnam and China (see C.E.R.M., 1969). . . .
      We know that for Marx the 'Asiatic mode of production' corres-
ponded to one of three types of property born of the breakdown
of primitive society, constituting transitional forms in the evolution towards class societies. (The two other types were the
'Germanic' mode where communal property was the functional
extension of individual holdings, the individual being a member
of the community by virtue of individually holding land; and the
'Ancient' mode, where the right to ager publicus and the private
ownership of land and slaves coexisted.)
      The Asiatic mode is characterized by the existence of village
communities, still partially organized on the basis of kinship,
in which the individual has usufruct of communal landed property
by virtue of his membership in the community. It is distinguished
from the primitive community by the development of forms of
exploitation benefiting a ruling class, and from a slave-based
system by the absence of private property and individual freedom. The individual is free despite the fact that he is obliged

to pay tax and participate in corvée labour. It is precisely the
state's ability to avail itself of abundant peasant labour that
restricts the development of slavery. The form of corvée is
unlike that found in a feudal regime where the peasant is individually dependent upon the lord. . . . We don't find lords,
but officials in this case: tax and corvée are levied in the name
of the state, exploitation is collective. Marx spoke of the
'general slavery' of individuals subjugated in this way.

However, exploitation does not exist in so far as services are
rendered in exchange and consequently a part of the appropriated
surplus returns to the community.

In the Asiatic mode of production, the state (the ruling class)
fulfils economic functions. It developed with the necessity for
great public works (the realization of which required the combined and co-ordinated effort of the village communities which,
given their isolation, were doomed to fail in such a task). The
state, according to Engels in 'Anti-Dühring', is 'first of all,
the entreprenur responsible for the collective conservation of
irrigation works along the river valleys; without such works any
form of agriculture is impossible'.

This 'economic High Command' oversaw other functions, as
J. Chesneaux has remarked (1969, pp. 26–7), for instance:
the control over fallowing lands, the upkeep and guarding of
roads (the villages were never entirely self-sufficient); the
military protection of villages from raiding nomads or invading
foreign armies; the direct state takeover of some sectors of
industrial production beyond the capacity of village communities, e.g. mining or metallurgy (the state foundry).

At first sight, this outline would appear to suit Vietnamese
reality better than the notion of a 'feudal system'.

Two factors influenced the appearance and consolidation of
the Vietnamese state: the struggle against foreign rule, and
the necessity for large-scale hydraulic works. . . .

During the 1,000 years of Chinese occupation, national resistance movements occurred one after another, beginning with
the rebellion of the year 40 when the people of Giao-chi, Cuuchân,
Nhât-nam, and Hop-phô answered the call to arms by the Trung
sisters and rose up and seized sixty-five fortresses. After
Vietnam achieved its independence, China tried to re-establish
its rule several times; Vietnamese leaders were only able to
triumph over the aggressor by joining together in national unity.
This was the case in the struggles of Ngô Quyên against the
Nam Han, of Ly Thuong Kiêt against the Song, of Trân Hung-dao
against the Mongols, of Lê Loi against the Ming, of Nguyên Huê
against the Tsing. Hô Quy Ly, however, lost the war of 1406
because he was unable to bring about national unity. In addition
to the threat from China, the need to struggle against Cham,
Khmer, and Lao invasions also resulted in the people drawing
closer to the monarchy.

Generally, the nobles loyally fulfilled their duty during dire
times, coming to the assistance of the royal family with their

private armies (when they had them, i.e. under the Ly and the
Trân). Of course, their behaviour was not without a material
motivation. . . .
When the monarchy was weak, the nobles did not hesitate to
set themselves up as independent lords by subjugating village
communities. But they never attained the degree of authority
held by European feudal lords . . . [for the following reasons]:
on the one hand, their domains were quite small; on the other
hand, the principle of non-hereditary grants enabled the
royal family, once again in strong hands, to take back such
property. Finally, the number of serfs and slaves who could
serve in private armies was limited. Furthermore, Vietnamese,
unlike European nobles, did not have a monopoly over arms.
The Far Eastern Empire was open to all capable of taking it (we
will return to this point later). The royal family, for its part,
had begun to build up a national army very early on. Before
the mid-eleventh century, the Ly instituted a 'conscription
register' in every village, from which soldiers were called up.
   The national character of Vietnam has often been misunder-
stood by European authors who only see the borrowing of its
institutions and official language from China and acknowledge
the latter's sovereignty.
   In fact, the Chinese state apparatus was not copied, but
adapted to the Vietnamese situation. Vietnam was purely a
nominal vassal of China. Vietnamese kings, aware of their weak
position, always sought to save face, not because they were in
an actual state of political dependence, but out of respect to
Chinese culture and the adoption of a diplomatic stance towards
the greatest power of the Far East. . . .
   Indeed, neither the ceremony of investiture nor the payment
of tribute, usual features of vassalage, injured their status.
The investiture of the sovereign was merely a ritual act like
legal recognition under modern law. It did not entail any form
of subordination in either internal or external affairs (in which
the Chinese suzerain only intervened when asked to do so by
his vassal). . . . In a similar fashion, periodic tribute, another
indication of vassalage, bears no relation to the normal interpretation of this phenomenon, because 'of the strong common
interests of the feudal leaders of Vietnam and of China' (Chesneaux,
1955, p. 31). Like investiture, tribute was a socio-political
ritual, a 'game' based upon obligations to give, receive, and
return gifts, like the Indian potlatch or the Trobriand kula. In
return for the gold, ivory, incense, cinnamon, pearls, and other
prestige goods, the suzerain was required to reciprocate with
even more sumptuous gifts.
   The Vietnamese king firmly asserted his independence because
his jurisdiction derived not from the suzerain, but from 'the
mandate of heaven', i.e. ultimately from popular consent. His
people similarly asserted their independence as a state, as
the no less 'civilized', however distinct, 'South' as opposed to
China, the 'North'.

The second factor determining the development of a unified
and centralized state in Vietnam earlier than in Europe was the
necessity for large-scale hydraulic works.

The northern delta, the cradle of Vietnamese society, has been
created by the fertile alluvial soils of the Red River; if flooding
maintained the land by elminating erosion, it also threatened
harvests, villages, and human life. On the other hand, rice,
an aquatic plant, requires a humid environment; drought is as
destructive as flooding for it. Thus, there is the ever-present,
vital problem of combating both an excess and insufficient
supply of water. The damming of rivers and the irrigation of
rice-paddies were tasks which all dynasties considered their main
duties; the neglect of these duties always signalled the decline
of a dynasty. The annals always record dike construction and
canal excavation, which in addition to their importance for
irrigation, also constituted a superb local, regional, and national
communication network. . . .

Thus, the state appears as a 'unit of assembly', co-ordinating
the efforts of the local communities to maximize the benefits from
large-scale hydraulic works. In Marx's and Engels's view, on
the basis of this power of function, the state was transformed
into an instrument of exploitation, through the appropriation of
all the lands; the local communities only retained possession of
the land to which an individual had access through the double
mediation of his membership in the local community and through
the 'higher community', the supreme landlord; no form of
private property could arise.

Now, this picture is fundamentally vitiated by a misunderstanding of the reality and ideology of Asia. Marx's and Engels's
mistake, and of their European and oriental followers who
mechanically repeat their ideas, has been to transpose the
Roman notion of property to the Far East and to deduce from it
monarchical 'despotism' and the 'general slavery' of individuals,
who are considered merely as tenant farmers of the sovereign.

In actual fact, the notion of 'property' in the Far East, as
elsewhere, is inseparable from general philosophical conceptions.
Originally, we find the metaphysical idea that 'Heaven produces
beings and objects and the Earth nourishes them' (Maspéro,
1950, p. 203). Land is given by Heaven for all men to possess.
The land therefore belongs to the people and not to the sovereign
who is only present to preside over the general welfare. He
'speaks, not as the landlord which he is not, but as the sovereign,
enacting for each level of the hierarchy, the right of appropriation of the land which belongs to all' (Maspéro, 1950, p. 204).
Regardless of his power to dispose of the land, neither he nor
his subjects should allow its disuse or deprive his people of it.

This is the very goal of the heavenly mandate. The sovereign
must make the cosmic and terrestrial orders correspond. In
other words, he only reigns for the good of his subjects. When
he fails in his purpose, provoking through his injustice or
tyranny, misery and disorder, he loses the heavenly mandate

and the people exercise the right to rebel. 'Majestic Heaven has
no favorites, it favors only virtue. The people's favor is not
unchanging, they only love beneficent rulers' (Confucius).

In the agricultural countries of the Far East, good harvests
are precisely the main indication of the public welfare and they
are insured through the protection and irrigation of the rice-
paddies. Any disaster is interpreted as the fault of government. . . . Responsible for the ills that might befall the people,
the monarch publicly confesses his unworthiness for which he
must attempt to make amends. If he does not correct himself and
the ills continue, the fall of his house is nigh. Thus was the law
of dynastic change in Vietnam. We can see that Vietnam was not
truly an 'Asiatic despotism' given the very nature of the heavenly
mandate and the people's right to rebellion.

This conceptualization of the people's 'sovereignty' and the
state's role in the economy are incompatible with the Roman notion
of property implying the right of exclusive and unrestricted use,
possession, and disposal. The sovereign cannot do what he
pleases with the land; he is limited to playing an economic role
(whereas neither the slave nor the feudal state, where there
are numerous private owners, intervene in the production process). Nobles and representatives of the Buddhist religion
receive grants of land because they assume the role of guardian
of the public material and spiritual welfare, not because they
have sworn allegiance to the king. They return the land once
they have fulfilled their function. Since they represent a small
proportion of the population, they only control a limited amount
of land. One of the reasons why there is not a feudal system in
Vietnam is because benefits are small since most of the land must
be left to the people.

The village commune held land which it periodically redistributed
among its members since it had to ensure them the minimum for
survival. Commune members had to cultivate the land or else
the commune could take it back and give it to others. In the same
way, if a commune had more land than people to work it, and
another commune, too little, the sovereign could transfer the
excess land to the second commune. Because the state fulfils
economic and military functions, the tax paid by the direct producers is not rent.

The oriental conceptualization alone explains why private
property emerged very early on and continued to develop, whereas the Marxist conceptualization denies the existence of private
property. We have evidence of the existence of private property
from the twelfth century.(1) In 1135, to protect private property,
Ly Thân-tông decreed that 'those who had sold rice-paddies and
ponds must not, on pain of punishment, seek to buy them back
at a higher price' (Pham Thi Tam and Ha Va Tan, 1967, p. 26).
His successor, Anh-Tông, took other measures to protect private
property in 1143 (Ha Va Tan, 1967, IV, p. 26). Large private
land-owners who make temple offerings appear in the thirteenth
century. In 1248, one of them gave more the 100 mâu to Bao-ân;

his generosity is immortalized on a stele of this temple (Pham Thi Tam and Ha Va Tan, 1963, p. 21). When 'Quai vac' dike was built in 1248, king Trân Thai-tông ordered that owners of land affected by the construction of the dike were to be compensated. Had the king been the supreme landlord, he would not have had to pay such compensation for this was a case of expropriation for public use.

In actual fact, the development of private property stems from the principle of social utility. Uncultivated land belongs to everyone. The person who develops the land becomes its owner. On the other hand, if he abandons the land, he loses the ownership of it to the person who cultivates it. Lê Loi's ancestor became a large land-owner by clearing an entire area at the foot of Lam-son. By encouraging the development of the Mekong Delta, the Nguyên created a class of large land-owners who were to be their most loyal supporters against the Tây-son. Besides this 'original' manner of acquiring property, other ways emerged as the economy developed: sales by indebted or hungry peasants, mandarins 'and notables' fraudulent or violent usurpation of publicly and privately held land especially during troubled times.

Where does a purely deductive statement divorced from reality, on the AMP, like the following by Ferenc Tökei (1966, p. 72), lead us?

> How could private landed property arise when no class or strata of 'Asiatic' society could have desired an end to the previous form of property since only tribal communal property could insure tax revenue for the aristocracy of officials? As for the peasants in the village communities, the community also helped them, besides which they could hardly imagine another manner of existence . . .

Now, private property was created precisely because it opened more possibilities for exploitation than public property for the mandarins and notables. Public property gave them temporary title to the land, to a salary composed of taxes or rent extracted from the peasantry. Private property, on the other hand, was hereditary and large land-owners drew greater benefits both in terms of political power (due to the availability of private armies of serfs and clients) and of economic exploitation (since they could raise the rents at will, while taxes were set by the state). Finally, the combination of public holdings and private domains was a possibility; as the money economy developed, the state could pay an increasingly larger proportion of its officials' salaries in money rather than in kind. As for the peasant, he also had an interest in private plots which either were not taxed or subject to less tax than on public lands. In any case, the peasant remained a member of the village community to which he could turn in times of need.

To sum up, many of the features (notably 'general slavery', 'oriental despotism', 'the absence of private property' and 'stagnation for millennia') that Marx and Engels associated with the Asiatic mode of production do not fit the facts. There were

several real classes: the aristocracy (up to the Nguyên), the
peasants, and slaves - not an 'embryo of classes'. The category
of literati is less clear. In fact, a distinction should be made
between two sorts of literati. There were those who, having
passed their examinations, became officials enjoying political
authority and were thus able to acquire some of the means of
production (contrary to the bourgeois system, here it is the
office which appropriates the revenue). But there were those
literati who, having failed their examinations, returned to live
among the people from whom they differed very little in terms of
material welfare. Although both sorts of literati were steeped in
the same culture, they had different interests. The simple
literati who shared in the peasants way of life, often took to
their defence against political authority, either verbally or in
writing. As for the commoner class of land-owners, we do not
know their statistical significance given our present state of
knowledge; however, their strength in the South in the eighteenth century is known. For some authors (Godelier, 1970,
pp. 133-4),

> the essence of the Asiatic mode of production is the combined
> existence of *primitive communities*, still organized on the
> basis of kinship and where communal property prevails, and
> *state power* which expresses the real or imaginary *unity* of
> these communities, *controls* the use of essential economic re-
> sources and *directly appropriates* a proportion of the labour
> and product of the communities over which it rules.

It is 'one of the forms of transition from classless to class society'
(ibid.). But this definition overlooks the basic fact that social
classes do exist, not just the state and the village communities,
but within the communities which are hardly 'primitive'. This
is a class society and not a 'transitional form'. Class exploitation is realized both through communal forms of ownership and
possession of land,(2) and directly through aristocratic and commoner private property. Nor can we term a situation lasting
one or two millennia, depending on the country, a 'transitional
form'. The non-emergence of social classes over such a period
would be incomprehensible.

   In conclusion, in our opinion, the concept of the Asiatic mode
of production requires so many essential corrections that it
would be better to abandon it and to construct another as far
as our analysis of non-European societies permits.

NOTES

* Translated from 'La Pensée', no. 171, 1973, pp. 128-9, 133-4,
  135, 136-40.
1 This does not mean that private property had not existed at
  an earlier period.
2 The periodical division of communal land is done by notables
  who keep the best parts for themselves. On the other hand,

the latter are privileged when contrasted with ordinary
peasants: they are partly or entirely exempt from tax,
corvée labour, and military service.

BIBLIOGRAPHY

C.E.R.M., 1969, 'Sur le mode de production asiatique', Paris,
    Edition Sociales.
Chesneaux, J., 1955, 'Contribution à l'histoire de la nation
    vietnamienne', Paris.
Chesneaux, J., 1969, 'Sur le mode de production asiatique',
    C.E.R.M., 1969.
Godelier, M., 1970, Preface to 'Sur les Sociétés précapitalistes',
    Paris, Editions Sociales.
Maspéro, H., 1950, 'Etudes historiques', vol. III, Paris.
Pham Thi Tam and Ha Va Tan, 1963, Vai nhân xet vê ruông dât
    tu huu o Viêt Nam thoi Ly Trân, 'Nghiên cuu lich su', no. 52,
    July.
Pham Thi Tam and Ha Va Tan, 1967, 'Toan thu', IV, Hanoi.
Tökei, F., 1966, 'Sur le mode de production asiatique', Budapest,
    Académiai Kiadó. (Parts of Tökei's study are included in this
    book.)

THE ECONOMY OF THE INCA STATE AND THE NOTION OF THE AMP*

Jürgen Golte

In a society, the productivity of labour expresses the level of
the development of the productive forces. The productivity of
labor itself depends upon the development of the means of
labour, knowledge of nature, the producers' skills, the organisation of labour, and natural conditions.
Although all these factors are present in every process of
social development, their combined effects are specific to each
society. The primary factor explaining the increase in the social
productivity of Inca society was the organisation of the labour
force which enabled a more effective use of the natural conditions; the development of the means of labour no doubt played
a secondary role in this increase.
The Andean environment can be characterised by its extreme
diversity. Between the coast and the tropical forest, we find
a series of natural landscapes with different climatic, edaphic,
botanical, and zoological conditions as we rapidly climb and
descend from sea-level to 6,000 metres.
Even before the emergence of the Incas, Andean societies
made use of nearly all the ecological zones. They had developed
the necessary know-how and means of labour to use the distinct
natural environments for their reproduction. In Inca society,
the most effective use of natural conditions was achieved through
the distribution of the population among the different ecological
levels according to the possibilities of exploiting these environments and the social needs of consumption. On the one hand,
this rationalisation in the use of the labour force consisted in
the permanent shifting of populations from overpopulated regions
to underpopulated and, consequently, underexploited areas. On
the other hand, there was a temporary displacement of people
to ecological levels different from their original habitat, with
a view to producing goods for the original social groups' consumption (see Murra, 1972; Cobo, 1956, pp. 109-11).
Another type of intensification of the social productivity of
labour was achieved through distributing work according to the

labour power available at different times of the year in relation
to the seasonally varying intensity of agricultural work. In Inca
society, during the months between planting and harvesting,
labour power was recruited for public works, e.g. the construction of footpaths, irrigation canals, roads, storage facilities, etc.
which themselves contributed to labour productivity (Murra,
1956, pp, 167-203). On the other hand, semi-finished goods,
particularly wool, produced at one ecological level, were distributed for completion to the more numerous and seasonally
unoccupied population of other levels.(1)

Finally, co-operative labour was organised according to its
social effectiveness. For example, the labour of several groups
was brought together to build irrigation systems which would
benefit only one of these groups (Cieza de Léon, 1967, pp. 76-7).

Since the allocation of labour within this economic system was
aimed at the reproduction of the society as a whole, it had to
look after the reproduction of the parts making up this totality.
Thus, systems of distributing consumer goods to different sections of the society which corresponded to systems of assigning
tasks in reproductive activities were developed.(2) On the other
hand, an infrastructure permitting the movement of people and
goods was created.(3)

Of course, private ownership of the means of production in
a society so organised could be counterproductive since it
would reduce the availability and possibilities of taking advantage
of the means of production which had to be available to the
society as a whole.(4)

In the following pages, we shall see how the relations of production and corresponding system of distribution in Inca society
enabled this particular advancement of the productive forces.
Inca society was organised in different superimposed levels which
permitted an overriding centralised administration and facilitated
the allocation of resources, the organisation of labour and the
distribution of goods, despite the relatively low level of technology. In our examination of the relations of production and
system of distribution we shall ascend from lower to higher
levels of organisation which enable us to see the functioning of
the system as a whole as well as the development of state institutions and the emergence of social classes within this development.

At the local level, we find simple reciprocity of labour in the
relations between domestic units organised within a system of
kinship. These reciprocal relations emerge from the requirements of the productive process, e.g. groups which break the
ground with the taclla. In this reciprocal relationship (ayni),
members of two domestic units do the same sort of work on the
land assigned to each of them on alternate days. The fruits of
labour belong to the possessors of the land, who are responsible
for feeding those who work on it. More complex systems of
reciprocity in which several domestic units take part (Diagram 2)
are derived from this system of simple reciprocity (Diagram 1).
Such more complex systems also occur within a kinship system
(Golte, 1973, pp. 29-31; Golte, 1974, pp. 492-3).

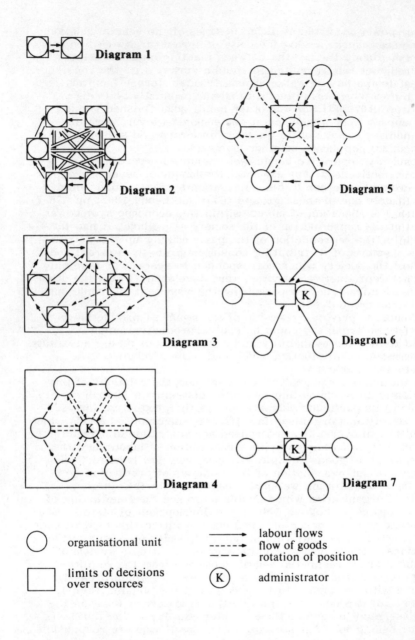

Diagram 1

Diagram 2

Diagram 3

Diagram 4

Diagram 5

Diagram 6

Diagram 7

○ organisational unit

□ limits of decisions over resources

——→ labour flows
- - -→ flow of goods
– – –→ rotation of position

Ⓚ administrator

Generally, the local-level economy includes several types of cultigens and can include the raising of animals. For some of these activities only a small number of persons are required. Such tasks requiring few people, such as stock-raising for example, are rotated between the different family groups of the local unit, while the remainder of the population carry out agricultural work. Responsibility for organising the rotation of tasks belongs to an individual designated because of his position within the kinship system. This person divides his working time between agricultural and administrative labour. The more complex the administrative work becomes, the more the person is released from the immediate process of production. In this case the remaining domestic units carry out the agricultural work of the person organising the system of reciprocal duties (Diagram 3).(5)

The local social units are organised into ethnically defined units which usually control different ecological zones on the slopes of the Andes. Since all domestic units consume products of different ecological levels, there must be access to these levels. This is resolved through a system derived from complex reciprocity. Domestic units from different local groups take turns working lands at different ecological levels assigned to the ethnic unit and administered by its leader. The fruits of labour are gathered and redistributed by the latter and thus reach all the domestic units (Diagram 4). In this system there is a rotational division of labour among agricultural activities and a fixed division of labour between those who carry out agricultural work and those who administrate the allocation of tasks and distribution of the product. As the number of administrated units grow, so the administrator is all the more detached from the others. The administration of work becomes the domination of those who work and the redistribution of goods appears as the distribution of favours and gifts (Diagram 5) (Murra, 1972; Golte, 1970; 1973, pp. 31-5; 1974, pp. 494-502).

The administrator, i.e. the leader of the ethnic unit, is responsible for the organisation of labour in the construction of public works and for their control. As the embodiment of the public, he is the person who decides the public interest. No doubt he can interpret his private interest as the public interest, for example to obtain luxury consumer goods.(6)

At the highest level of the development of the relations of production, the administrator occupies the position of a public work. Service performed for him and his consumption appears as a collective duty. He is assigned a personal retinue (yana) who must work directly for him and his relatives and who are exempt from other productive activities (Diagram 7) (Murra, 1966a).

The totality of these relations of production and distribution are simultaneously found within the ethnic units at different levels of organisation. Social differentiation stems primarily from the division of labour between the immediate producers, on the

one hand, and the administrators of work and the distribution
of goods, on the other hand.

In its relations with the ethnic units, the Inca state bureaucracy makes use of their division of labour and developed class
structure. The Inca bureaucracy, at the state level, fulfils
functions similar to those performed by the leaders of the ethnic
unit at the regional and local levels. It perfects the systems of
land allocation, labour organisation, shifting of population, the
distribution of goods and accounting; its power is based upon
such activities. The Inca state bureaucracy takes over ethnic
groups as levels of organisation. However, it regroups them
into numerically defined levels of organisation to facilitate administration. In the same way that ethnic units provide their
leaders with a retinue of servants, the Incas remove large groups
of persons for their service (yana, aclla, cañari, lucanas).
Furthermore, they undertake large-scale public works with labour
power recruited from the ethnic units during the periods between
planting and harvesting. In similar fashion, they raise armies
to expand the area under their domination. Public interest is
defined as that which perpetuates and reproduces the society.
Thus at the state level there is a certain ambivalence in these
relations of reproduction. For example, the redistribution of
products, the construction of roads, irrigation systems, footpaths and depositories for redistributing grain, etc. are works
in the interest of the direct producers as are many aspects of the
worship of deities believed to be indispensable to ensure production. These same works sometimes also benefit the dominant
group in so far as they assure that the level of surplus production is maintained. The dominant group also benefits from
other activities such as the construction of palaces, the manufacture of luxury goods: the products of silversmiths, weavers,
and potters, and particularly from the work performed by the
yana and other servile groups.

Thus, the development of the productive forces in Andean
society through advances in the organisation of labour power
led to a specific class division. Those who administered labour
power and the distribution of goods became a privileged class,
since in performing their functions they managed to have themselves reinterpreted as the embodiment of the collective interest;
as administrators they were able to legitimate this interpretation.

According to Marx, Asiatic society is basically defined by
three elements: the absence of private landed property, self-
sufficient village communities founded on a combination of agriculture and household craft industry, and the fundamental
importance of the state due to its performance of functions delegated by the self-sufficient village communities, by which a
class division between those controlling the higher unity and
the village-organised population is established. The interrelationship between these classes consists in the payment of
tribute or the performance of labour by the villagers and the
fulfilment of higher order functions by the group embodying the
higher unity (1857-8a, pp. 472-3).

In order to evaluate the usefulness of the notion of the AMP
for the analysis of Inca society, two concepts require further
explanation: the concept of village self-sufficiency and that
of the nature of the functions delegated to the higher unity.
Marx himself states that the Asiatic village community 'contains
within it all the conditions of reproduction and surplus production' (ibid.). Given the 'combination of handicrafts and agriculture within the village', i.e. the lack of a division of labour
beyond the village level, there is no exchange between villages
or development of an opposition between town and country
(Hobsbawm, 1964, pp. 27-34). The result of village self-
sufficiency and the absence of a division of labour beyond the
village level is the stagnation of Asiatic societies. Although
the higher unity may nominally own the land, it in fact belongs
to the village communities which exploit it directly. Villagers
may work either in family groups or collectively.

The relation to the higher unity and ultimately the despot is
derived from the fact that he is the higher proprietor, i.e. for
the individual villager, the higher unity mediated by the village
community, appears as the natural presupposition of his reproduction (Marx, 1857-8a, p. 473):

> A part [of the communes'] surplus labour belong to the higher
> community . . . and this surplus labour takes the form of
> tribute, etc., as well as of common labour for the exaltation
> of the unity, partly of the real despot, partly of the imagined
> clan-being, the god.

A second aspect of the relationship of the village community
to the higher unity is the existence of 'collective conditions of
real appropriation through labour: aqueducts, roads, etc.' which
'appear as the work of the higher unity' (ibid.). Here we have a
contradiction with the notion of village self-sufficiency, a contradiction not very well resolved in Marx's writings. If such collective
conditions exist and if they are the means of production which
not only 'appear' as the work of the higher unity, but in fact
are solely realisable by that unity, then the village community
does not 'contain within it all the conditions of reproduction and
surplus production'. Unfortunately, most of the recent interpretations of the AMP have not seen this point as problematic.(7)
However, it would seem a crucial problem in applying the notion
of the AMP to Inca society.

The Incas' position within the state is based upon the organisation of labour, the rational allocation of land, and the distribution
of semi-finished products and consumer goods. These functions
performed by the state bureaucracy already existed on a lesser
scale among the ethnic units. Inca administration uses the
officials of the latter, who had emerged through the performance
of these functions, and organises a similar system on a larger
scale, encompassing the society as a whole.

In this system neither the village community nor the ethnic
unit are self-sufficient - in contrast to the Indian village
community which Marx used as a model in developing the notion

of the AMP (Sofri, 1971, pp. 15-37). Rather, the characteristic feature of the Inca economy is that the productive forces are no longer organised at the local or regional level, but on a state-wide scale. In the state economy, the village community and the ethnic unit are parts of a hierarchy of units rationalised into a hierarchy, numerically organised to facilitate both the organisation of labour and the distribution of goods.(8)

The ownership of land in a system of this type necessarily tends to be state ownership. In the Inca state, state property is expressed through its allocation of lands to different levels of the organisational and distributive hierarchy which is organised numerically. Each organisational level (state, suyu, province, hunu, etc.) is allotted a range of the resources in the different ecological levels. Privileged administrators organise work on the lands assigned to the organisation units at different levels of the hierarchy: they are also in charge of distributing the products of labour. Work itself is organised according to the manpower requirements, i.e. it may be carried out by the organisational unit collectively or by delegated groups (mitmaq).

The organisers of the different levels derive their power and privileges from their function as administrators of labour and the distribution of products. The larger the organisational unit, the grander the privileges which includes freedom from having to do immediate productive labour, access to more consumer goods, polygyny, and the assignment of servants (yana) - the number of whom increases as one moves up the administrative hierarchy.

Marx himself mentions Peru in his sketch of the AMP. Recently Métraux and Godelier have concurred in the relevance of the concept to the Incas (Marx, 1857-8a, p. 473, Godelier, 1973a, b; Métraux, 1961, 1972).

Métraux's image of the Inca village communities appears to be very much influenced by the idealisation of late-nineteenth-century and contemporary peasant communities of the Peruvian indigenist movement which has also transferred this idealisation to their conceptualisation of the Inca state. However, such a picture of the village community does not correspond to the position and economic organisation of the people under the Inca state. Research over the past fifteen years has thrown light upon the interdependence of the different levels of organisation within the system of vertical control over ecological levels, and on the emergence from this system of the social formation encountered by the Spaniards upon Conquest.(9) Thus, Métraux's interpretation which starts from an opposition between the ayllu - the village community united through kinship ties and the communal ownership of land - and the state, cannot be sustained. For Métraux, 'It is clear from all this that the communities of the Empire were self-sufficient and produced a surplus permitting the noble caste and officials to live in comfort and luxury' (1961, p. 102). From this viewpoint it is not difficult to conclude that Inca society corresponds to the notion of the AMP.

In confirming this correspondence, Godelier draws heavily upon Métraux and Murra's early works, i.e. he starts out from an opposition between the ayllu and the Inca state. In his second work on the subject, 'The non-correspondence between the form and content of social relations' (1973b) Godelier outlines more thoroughly his theory of the development of the Inca mode of production and its emergence on the basis of the incorporation of ethnic units in the state through military means. However, the mode of production of these ethnic units is left unclear. He appears to associate it with primitive community; thus the Inca mode of production must appear to him as something which takes the form of a classless tribal society only to erect a state and class-organised society, taking tribal forms and transforming their content through military expansion. For Godelier, the necessity to produce a surplus capable of assuring the conditions of domination emerges from the Incas' conquest of other ethnic units. In fact, expansion and the development of the productive forces are interdependent. However, we have tried to trace how the characteristics present in the Inca mode of production already exist in the ethnic units. Only in this way can we understand the origins of Inca expansion which endeavored to enlarge control over ecological levels and manpower, i.e. the origins of Inca class society are to be found in a specific development of the productive forces among Andean societies and not in military expansion.

In conclusion, the application of the notion of the AMP to Inca society results in certain problems. In many traits, Inca society resembles societies which have been described as Asiatic; it differs fundamentally, however, from the model outlined by Marx in that the village communities were not self-sufficient. One could argue that the village communities and ethnic units were in large measure self-sufficient since they produced most of the goods necessary for their reproduction. However, this view would overlook precisely those factors that are the expression of the development of the productive forces characteristic of Inca society.

NOTES

* Translated from unpublished paper given to XLI International Congress of Americanists, Paris, 1976.

1 For example, see the declaration of Juan Xulca, kuraq of Auqumarca (Huánuco): 'the Inca gave them wool for cumbi cloth' Ortiz de Zúñiga, 1967, p. 48; or Polo de Ondegardo, 1961-17, p. 128 (eds translation).

2 In particular, livestock products (Polo de Ondegardo, 1940; p. 135), coca, and other products of both extremes in the ecological zones were distributed (Golte, 1970).

3 The road system and the great central depositories for every sort of product distributed throughout the kingdom were the

main mechanisms enabling these transfers; however, continuous censuses and kipu calculated statistics were a precondition for the administration's movement of goods.
4　The right to dispose of land did not prevent assigning rights
of usufruct to organisational units at different levels. The
important point is that in the last instance, the state could
decide upon land use when it felt that the needs of the
society as a whole were at stake. In fact it took advantage
of this right to dispose of land.
5　The rotation of tasks among different domestic units -
mitachanakuy - did not include administrative work. Yupanako,
i.e. the carrying out of the work of another who is working
for the group in another area or the type of work applied
in both the case of rotational and fixed division of labour.
6　The division between private and public interest in such cases
is entirely inadequate and arbitrary.
7　For example, see Hobsbawm, 1964, pp. 33-4; Ahlers et al.,
1973, pp. 21-46; Godelier, 1971; Sofri, 1971; Bartra, 1969;
Melotti, 1971, pp. 53-63. Chesneaux (1969, pp. 41-3) refers
the problem to a number of studies but offers no solution.
8　The decimally organised hierarchy was introduced late,
but the previous dualistic system maintained in the south of
the Inca state very much lent itself to the allocation of tasks
and the distribution of products. Matienzo characterised the
ethnic leadership in the dualistic system in the following way:
'The job of the caciques and headmen is to rest, drink, count,
and distribute. They are very capable in this, far more so
than any Spaniard. They were certainly a sight, slowly counting out their many-coloured stones' (Matienzo, 1967, p. 21).
Juan Xulca, kuraq of Auquimerca (Huánuco), tells of how the
system of decimal organisation worked: 'no one who was capable of work was left without it. The main headmen did no
more than tell the Indians what they had to do. The chief
headman distributed tribute to the other headmen of each
quaranga, who in turn distributed to those of each pachaca'
(Ortiz de Zúñiga, 1967, p. 47).
9　See primarily the works of Murra (1960, 1962, 1964, 1966a,
1966b, 1967, 1968, 1970a, 1970b, 1972, especially the bibliography); Espinoza (1963, 1969-70, 1971); Morris (1972);
Golte (1970, 1973, 1973a, 1974).

BIBLIOGRAPHY

Ahlers, Ponner, Kreuzer, Orbon, and Westhoff, 1973, 'Die vorkapital
istischen Produktionsweisen', Erlangen.
Bartra, Roger, 1969, El modo de producción asiático en el marco
de las sociedades precapitalistas, in Chesneaux et al., 1969.
Chesneaux, J., et al., 1969, ,El Modo de producción asiático',
Mexico.
Cieza de Léon, Pedro de, 1967, 'El Señorio de los Incas', 2a parte
de la Crónica del Peru, Lima.

Cobo, B., 1956, 'Historia del Nuevo Mundo', vol. 2, Madrid.
Espinoza S.W., 1963, La guaranga y la reducción de Huancaya, 'Revista del Museo Nacional', XXXII, Lima, pp. 8-80.
Espinoza, S.W., 1969-70, Los mitmas yungas de Collique en Cajamarca, siglos XV, XVI y XVIII, 'Revista del Museo Nacional', XXVI, Lima, pp. 9-57.
Espinoza, S.W., 1971, Los Huancas, aliados de la Conquista, 'Anales Cientificos de la Universidad del Centro del Peru', Huancaya, no. 1, pp. 3-407.
Godelier, M., 1971, 'Teoria marxista de las sociedades precapitalistas', Barcelona.
Godelier, M., 1973, 'Ökonomische Anthropologie. Untersuchungen zum Begriff der sozialen Struktur primitiver Gesellschaften', Reinbek.
Godelier, M., 1973a, Der Begriff der 'ökonomischen Gesellschaftsformation': Das Beispiel der Inka, in Godelier, 1973, pp. 92-100.
Godelier, M., 1973b, Die Nichtentsprechung zwischen Formen und Inhalten sozialer Beziehungen. Erneute Reflexion über das Beispiel der Inka, in Godelier, 1973, pp. 281-92.
Golte, J., 1970, Algunas consideraciones acerca de la producción y distribución de la coca en el estado inca, 'Verhandlungen des XXXIII Internationalen Amerikanistenkongresses', Band II, Stuttgart, pp. 471-8.
Golte, J., 1971, Bauern in Peru, 'Indiana', supplement 1, Berlin.
Golte, J., 1973a, El concepto de sonqo en el runa simi del siglo XVI, 'Indiana', 1, Berlin, pp. 213-18.
Golte, J., 1974, El trabajo y la distribución de bienes en el runa simi del siglo XVI, 'Atti del XL Congresso Internazionale degli Americanisti', vol. II, Genoa, pp. 489-505.
Hobsbawm, E.J., 1964, Introduction to 'Karl Marx: Precapitalist Economic Formations', London.
Marx, K., 1939, 'Grundrisse der Kritik der politischen Oekonomie', Moscow.
Matienzo, J. De, 1967, 'Gobierno del Perú (1567)', Paris-Lima.
Melotti, U., 1971, Marx e il Terzo Mondo, 'Terso Mondo', anno IV, nos 13-14, Milan.
Métraux, A., 1961, El Imperio de los Incas: despotismo o socialismo, 'Diogenes', no. 35, Buenos Aires, pp. 87-109.
Métraux, A., 1972, 'Los Incas', Buenos Aires.
Morris, C., 1972, El almacenaje en dos aldeas de los chupaychu, in Ortiz de Zúñiga, 1972, pp. 383-404.
Murra, J.V., 1956, The Economic Organization of the Inca State, doctoral thesis, University of Chicago.
Murra, J.V. 1960, Rite and Crop in the Inca State, in 'Culture in History', Stanley Diamond (ed.), New York, pp. 393-407.
Murra, J.V., 1962, La función del tejido en varios contextos sociales del estado inca, 'Actas y Trabajos', 20 Congreso de Historia Nacional del Peru, Lima, pp. 215-40.
Murra, J.V., 1964, Una apreciación etnológica de la visita, 'Diez de San Miguel: Visita hecha a la provincia de Chucuito . . . 1567', Lima, pp. 421-44.

Murra, J.V., 1966a, New data on retainer and servile populations in Tahuantinsuyu, 'Actas y Trabajos', XXXVI Congreso Internacional de Americanistas, vol. 2, Seville, pp. 35-45.

Murra, J.V., 1966b, El instituto de investigaciones andinas y sus estudios en Huánuco, 1963-6, 'Cuadernos de Investigacion', I, Huánuco.

Murra, J.V., 1967, La visita de los chupachu como fuente etnológica, in Ortiz de Zúñiga, 1967, pp. 418-44.

Murra, J.V., 1968, La papa, el maiz y los ritos agricolas del Tawantinsuyu, 'Amaru', VIII, Lima, pp. 58-62.

Murra, J.V., 1970a, Current Research and Prospects in Andean Ethnohistory, 'Latin American Research Review', V, 1, pp. 3-36.

Murra, J.V., 1970b, Información etnológica e histórica adicional sobre el reino lupaqa, 'Historia y cultura', vol. 4, Lima, pp. 49-62.

Murra, J.V., 1972, El 'control vertical' de un máximo de pisos ecológicos en la economia de las sociedades andinas, in Ortiz de Zúñiga, 1972, pp. 427-76.

Ortiz de Zúñiga, I., 1967, 'Visita de la Provincia de León de Huánuco en 1562', vol. 1, Huánuco.

Ortiz de Zúñiga, 1972, 'Visita de la Provincia de León de Huánuco en 1562', vol. 2, Huánuco.

Polo de Ondegardo, J., 1916, 'Relación de los fundamentos acerca del notable daño que resulta de no guardar a los indios sus fueros . . . Colección de libros y documentos referentes a la historia del Peru', series I, vol. 3, Lima, pp. 45-188.

Polo de Ondegardo, J., 1940, Informe sobre la perpetuidad de las encomiendas del Perú, 'Revista Histórica', XIII, Lima, pp. 125-96.

Sofri, G., 1971, 'El modo de producción asiático: Historia de una controversia marxista', Barcelona.

THE OTTOMAN SOCIAL FORMATION*

Huri Islamoğlu and Cağlar Keyder

In Ottoman history a new definition of the object of study is in
order, since the present paradigm is characterized by an inability
to integrate - both in structure and in time - its various areas
of concentration. Its problematic confines it to a dichotomous
temporalization of Ottoman history, consonant with the implicit
theorizations underpinning historical research. A hybrid institu-
tionalist functionalism on the one hand, and a crude modernization
perspective on the other, provide the framework for most present
research in Ottoman history. Our attempt in the following section
will be to advance a totalizing framework, seeking to integrate
both the diverse elements of the structure into an intelligible
whole, and to bring together the two disjointed temporalities
of the sixteenth and the nineteenth centuries by means of a
periodization centered on the concept of peripheralization.

This totality is the theoretical construct of 'social formation'
through which the Ottoman Empire will be studied. The Ottoman
social formation was characterized by a dominant Asiatic mode
of production,(1) in which the control of the central authority
over the production and appropriation of surplus constituted
the crucial mechanism of reproduction. The articulated whole
was reproduced according to the requirements of this mode,
but it also incorporated forms of petty commodity production,
and, of course, merchant capital. In its later stages, the social
formation contained 'feudalized' areas as well, which, however,
remained subordinate to the division of labor imposed by the
ruling class, concretized in the state. We shall now discuss how
these different economic practices were integrated through the
political control of the state.

The Asiatic mode of production is characterized by independent
peasant production in which the peasants do not form autonomous
units but constitute components of a larger unit, the limits of
which are defined by the extent of the authority of the state.
The peasant producer is integrated into the larger unit through
the delivery of his surplus in the form of taxes to the state,
and through the ideological-juridical apparatus that provides the

matrix for the state's extraction of agricultural surplus. Thus, this integration ensures a political determination of the division of labor within the system.

The ideology of the ruling class focused on the state was consonant with this political control over the economy. Thus the peasants came to believe (the belief being reinforced by juridical practice) that all land belonged to the Sultan; and that the state, in order to preserve the eternal order, did not permit accumulation of land. In the cities, a similar ideological function was served by the hisba regulating guild practices. Even mercantile activity was covered in ideological prescriptions: riba, speculative profiteering, was prohibited, although merchants as a class received protection and privileges due to their contribution in 'increasing the wealth of the land'.

The state is not confined to the political instance; it is the dominant vertical element that integrates the system in the Asiatic mode of production, cutting through the economic, political, and ideological levels. By means of its administrative apparatus, it controls tax revenues and therefore the production of agricultural surplus.(2) In the original conception of the Ottoman Empire, this function of the state was performed through the institution of timar.(3) Timar was the generic term accorded to a system of land grants distributed for the purpose of raising and supporting a provincial army. The beneficiaries of the grants were state officials empowered to collect the traditional product-tax of öşr,(4) in the area designated as their timar. These grantees (sipahis), in exchange for the privilege of collecting taxes, had to support and deliver during time of war a specified number of mounted soldiers. In addition, they functioned as the administrative cadre representing the central authority in the smallest villages. Together with the kadı, the judiciary representative who was empowered to apply the edicts of the seriat and the Sultanic law, the sipahis formed a bureaucratic layer, locally reproducing the political and ideological functions of the state.

The sipahi collected the traditional tax in kind, which necessitated the prestation of corvée-labor from the peasants when the product which was due to the sipahi was carried to the nearest market. The sipahi, in order to fulfill his obligations of provisioning the mounted soldiers with horses and arms, was in need of cash, which he obtained through selling the surplus product. Sipahis were accorded tax-collection rights for a specified period. Although the office was often extended from father to son, the kadı, who was answerable only to Istanbul, acted as a check on the possible abuses of the sipahi.(5) The timar system, however, ceased to be operative after the sixteenth century. We shall later discuss the reasons for its demise and the system of tax-farming which replaced it.

The integration of urban crafts and trade into the social formation was also effected through state control. This control translated into a determination of the economic division of labor which ensured the delivery of surplus to the seat of its utilization in the required product-form.

In the Ottoman social formation, urban craft production was
undertaken by guilds under strict state regulations (hisba).(6)
The state controlled the production process, that is, allocation
of raw materials; quantity and quality of the goods produced;
and it also fixed prices. This supervision of urban production
was in part achieved through the mechanism of the state's control over internal trade. The state directed the movement of raw
materials and enacted prohibitions on their export in order to
assure that the guilds were supplied with the raw materials they
required. Manufactures, on the other hand, were sold at author-
ized markets. The state's control over input and output prices,
as well as over the scale of production, meant that capital accumulation within the guild structure was effectively prevented. It
also meant that the state both through the taxation of manufactures sold in the markets, and through its position as the chief
purchaser of artisanal production, laid claim on the surplus
during the process of circulation.

Although trade in general was not subject to hisba regulations
that governed the guilds, the state imposed strict controls over
internal and external trade. Internal trade which ensured the
functioning of a highly monetized economy with a developed
division of labor was organized through a system of regional and
interregional markets. These markets were established with the
state's official sanction and their location was administratively
determined.(7) The regional markets were weekly markets
formed in villages and small towns. They served multiple functions. State officials who received their salaries in the form of
taxes in kind converted a portion of their revenues into cash;
peasants sold part of their produce in order to obtain money required for the payment of certain taxes; urban guilds as well
as rural craftsmen purchased their raw materials and foodstuffs,
and sold their manufactures. There were also urban markets
where guilds could purchase their raw materials. Other markets
were formed on trade and army routes to meet the requirements
of long-distance caravans and troops. Inter-regional markets
(fairs) held annually or bi-annually, especially in the Balkan
provinces of the empire, provided a site for the exchange of
goods from the more distant regions of the empire. The state
promoted all these markets, ensured the safety of merchants and
their wares along the trade routes and in the market places,
and it closely supervised the business transactions.(8) Markets
were also a major source of revenue in the form of market taxes
(bac pazar) determined in proportion to business transacted.
These revenues either accrued to the state officials who were
the recipients of land-grants or were farmed out to officials
who competed strenuously for these privileges. In addition to
market taxes, internal trade provided the state with revenues
in the form of customs dues, which were farmed out to state
officials and to great merchants,(9) The Ottomans, consistent
with their policy of promoting internal trade, abolished most
of the customs (actroi) in Anatolia and the Balkans predating

Ottoman rule. However, they established some major customs
such as those in Istanbul and Divarbakir both as sources of
revenue and as means of controlling the movement of goods.
    The concern for the direction of the movement of goods was
a prime consideration in the state's control over internal trade.
In addition to supervision of markets, this control was effected
through the granting of concessions to merchants. The sale of
these concessions, of course, provided the state with revenue.
More significantly, however, the state, by limiting the right to
purchase goods from local markets to officially sanctioned merchants, ensured the movement of these goods towards the major
cities, especially to Istanbul. Thus the satisfactory provisioning
of Istanbul was assured through the state's control over internal
trade.(10) However, in the late sixteenth century, with the rise
in the prices of grains and raw materials in Western Europe,
the state found it more and more difficult to exercise a control over
merchants who preferred to sell these commodities in more profitable markets. Contraband trade carried the day. This crisis
whereby the state was unable to control the movement of goods
is well attested to by the increasing numbers of government edicts
prohibiting contraband and by the attempts to enforce the
prohibition on exports.(11)
    We have seen that the state's control over internal trade
achieved a number of systemic functions at once. First, the extent of mercantile activity and therefore mercantile accumulation
was thus controlled. Second, through restrictions, sanctions,
and prohibitions, the flow of commodities inside the Ottoman world-
empire was regulated. This assured a certain division of labor
designed to provide the surplus-receiving class with the commodities they required for their consumption. In the written law of
the empire, this last designation was reflected in the formulation of 'the necessity of provisioning Istanbul'. Third, and this
is the most important function from the point of view of the
articulation of the social formation, internal trade provided the
link between the Asiatic mode of production (with its peasant
producers and tax-collecting state officials) and petty commodity
production in the urban guilds. Thus merchant capital supplied
the concrete form of articulation which was ultimately effected
at the political level by the state.
    Given this crucial location of internal trade, it follows that
the weakening of political control over its functioning would
threaten a disarticulation of the system. In other words, the
reproduction of the system would no longer be guaranteed. In
fact, as we shall later discuss, this tendency to disarticulation
asserts itself when internal trade escapes administrative control
(a process which started through the incentives offered to contraband trade during the price inflation), and gradually articulates
with external trade. This development is also the story of the
peripheralization of the Ottoman world-empire.
    External trade in the Ottoman social formation was by no means
limited to luxury items but included the raw materials essential

for the guilds as well as foodstuffs. Here, too, the state's
control achieved systemic functions, similar to the ones we have
discussed for internal trade. This control was effected through
the sale of concessions which gave merchants the right to trade
in essential commodities. This mechanism, in turn, served to
provide the state with the money it always required. In fact,
merchant capital was integrated into the system as money capital
through the role of commercial concessions as well as tax-farming
and usury operations. Merchants as bankers were the state's
chief creditors, and as tax-farmers they advanced considerable
sums to the state.(12)

The Ottoman state, from its inception in the fourteenth century,
sought to establish control over international trade routes. This
consideration, in fact, largely determined the pattern of Ottoman
territorial expansion.(13) During the sixteenth century the
Ottoman monopolized the silk and spice trades and Bursa became
an entrepôt for East-West trade, where silk from Iran and spices
and dyes from India were exchanged for European woolens.
Akerman, Kilia, Kaffa on the Black Sea were entrepôts for
Northern trade with Poland, Russia, and Crimea. Here the products of the East and of Anatolia via Bursa and Istanbul, as
well as Mediterranean foodstuffs, were exchanged for the wheat,
furs, and iron goods of the North.(14) Italian cities - Venice,
Genoa, Florence - were intermediaries in the trade with Western
Europe. The control of this trade shifted amongst these states
depending on their political standings with the Ottoman government. In the sixteenth century, Ragusa gradually acquired a
monopoly position. Venice, however, despite the loss of its
supremacy in the Levant, continued to be active in spice and
silk trades until the seventeenth century.(15) In fact, the
Portuguese discovery of the sea route around the Cape, and the
consequent attempt to divert the spice trade to this new route,
did not change the volume of trade conducted along the traditional routes significantly.(16) Although Ottoman-Persian wars
and social upheavals in Anatolia in the late sixteenth century
might have had adverse effects on the traffic along the land route,
Alexandria and Aleppo continued to supply Europe with spices.
Only after the entry of the English and the Dutch into the Indian
Ocean in the later part of the seventeenth century, was the
spice trade diverted from the Ottoman territories.(17)

Transit trade in silk experienced a similar fate. With the disruption of the silk route through eastern Anatolia at the end
of the sixteenth century, Bursa ceased to be the major entrepôt
for this trade. This coincided with the entry of the English,
French, and the Dutch into Levant trade and the displacement
of Venice as Europe's chief supplier of silk via Aleppo. For a
brief interval until the mid-seventeenth century, European
merchants in Aleppo and İzmir turned to Syrian silk. But, when
they established their hegemony in the Indian Ocean, the Dutch
and the English came to supply Europe with silks of Persia,
Bengal, and China.(18)

This development, in turn, was to have multiple repercussions
for the Ottoman system. First, it signalled the loss of a significant source of revenue for the state in the form of customs
dues from the transit trade. Secondly, it resulted in the decline
of such cities as Bursa and Aleppo as centers for trade and
industry. Thirdly, the shift in trade routes signified the rise
of the Atlantic economy and a new mode of commercialization
which, as we shall discuss later, was to undermine the system.

The active participation of the English, French, and the Dutch
in Levant trade, beginning in the latter half of the sixteenth
century, can be characterized as a transitional stage in the
articulation of the Ottoman system with the world-economy. This
trade included such items as silk and spices as well as staples
for rising European industries such as cotton and wool. In the
1580s, the Ottoman government granted trading privileges to
the English who imported mohair-yarn, cotton, and silk, and
sold woolen broadcloth, tin, and steel. In the last quarter of
the seventeenth century, England was the largest trader in the
Levant, until it was overtaken by France who reigned supreme
until the end of the eighteenth century.(19) On the Ottoman
side, this is attested to by the increase in the trade with France
which consisted of exports of cotton and cotton thread and
imports of cloth. During this time İzmir and Selanik emerged as
major ports. The same period also witnessed a significant increase
in cotton exports to Central Europe over the land route.(20)

The increase in external trade, however, was not limited to
cotton exports. Beginning in the late sixteenth century, with the
rising demand for raw materials and grains and the consequent
price inflation in Western Europe, these commodities found their
way into European markets not through the regular channels of
external trade controlled by the state, but by means of an increasing volume of contraband operations. This period also
witnessed a significant growth in population. The Ottoman state,
faced with an increased demand for raw materials and for grains
in cities, imposed strict prohibitions on exports. In so doing, it
provided additional incentive for contraband. The result was a
weakening of the state's control over external trade, a development with multiple repercussions for the system. It meant short-
ages in grains and raw materials essential for the guilds, which
resulted in high prices for these commodities, an inflationary
trend further aggravated by the influx of European silver.
Guild production was threatened with competition from cheap
European manufactures. The palace as the major purchaser of
foodstuffs found itself faced with ever-increasing expenditures.

The increase in illegal exports was achieved through the penetration of merchant capital inside the agricultural sector. Even
state officials who held land grants engaged in the smuggling
of their produce. The volume of such trade and the consequent
development of commodity production in the empire depended, to
a large measure, on the dynamics of the growing capitalist economy
outside its borders. But the crucial point was that the change in

the structure of trade resulted in the eventual integration of
the Ottoman system into the world-economy as a supplier of raw
materials and importer of manufactures, that is, its peripheralization.
  As can be seen in this historical digression, external trade may
not be analysed as a factor inside the system dominated by the
political rationality of the state. The volatility of external trade
due to shifts outside the domain of control of the Ottoman state
was the principal variable undermining this rationality. Given
its geographical situation, the Ottoman Empire could not remain
impervious to developments in the privileged domain of 'hot'
history: Western Europe.

CONTRADICTIONS WITHIN THE SOCIAL FORMATION

The Ottoman social formation, as we have described it, contained certain contradictions which became manifest conjunctur-
ally. These contradictions, some of which we have already
mentioned, were inherent in the normal functioning of the social
formation. In ideological readings of Ottoman history, external
factors became the disrupters of the harmonious equilibrium that
had emanated from the Palace to the farthest reaches of the
Empire. In the approach proposed here, external factors are
comprehensible only through their mobilization of contradictions
existing latently within the social formation. Thus, for example,
shifts in the structural position of the Ottoman empire vis-à-vis
the world-economy became disruptive due to the prior existence
of merchant capital within the system.
  Since we presented the social formation as an articulation with
the Asiatic mode of production in dominance, we can immediately
identify two separate loci of contradictions: those embedded in
the Asiatic mode of production, and those resulting from its
articulation into the social formation.
  It was mentioned above that the production process in the Asiatic
mode of production does not create conflict or a confrontation
between the producers and the appropriators of surplus. The
individual peasant comes into contact only with the tax-collector,
who acts in the name of the central authority, only after the
production process, which he carried out as a free peasant.(21)
In this way the Asiatic mode of production is distinguished from
the slave, feudal, or capitalist modes of production; class conflict is not graspable in the production experience. For this
reason its principal dynamic is to be found in intra-class and not
inter-class conflict. In this analysis, class will refer to those
groups with a claim on the output through a determinate position
in the social structure.
  In the Ottoman social formation, only state officials had a claim
on the surplus. This surplus-receiving class consisted of various
factions corresponding to diverse functions in the state. These
factions were those of timar-holder, market inspectors, and tax-

farmers, all of which perform the function of revenue collection; and of ulema, judges, and bureaucrats in Istanbul, who performed the ideological, legal, and administrative functions of the state. Various stratifications inside the surplus-receiving class were possible depending on the specific conjuncture: for example, between the ulema and the bureaucracy, or between the military and the palace.(22). In fact, from the seventeenth century until the abolition of the Janissary Corps in 1826, the military was effectively independent of the control of the central authority, which permitted the janissaries to gain control over certain sources of revenue.(23) There is, however, another conflict over revenue which is more permanent and structural. It results from the institutionalization of tax-collection. The central authority requires local representation both for tax-collecting and purposes of ideological-political legitimation. Yet the authority vested in local representatives threatened to grow at the expense of the Palace. Stronger local representation created a potential threat. Thus, this intra-class conflict worked as a centrifugal force, tending to create local potentates who declared relative independence from the central authority.

Whenever the process of separation from the central authority was complete, a new unit was formed duplicating the larger unit. In the eighteenth century, for example, the ayans, gaining relative independence from the Porte, substituted their local authorities for those of the Sultan, and otherwise replicated all the functions vested in the central authority.(24) Not all ayans, of course, attained such autonomy. Most smaller ayans could do no more than cut into the surplus destined for Istanbul; they formed tributary arrangements with the Palace, merely increasing their shares out of the revenue. In such situations, the political and ideological institutions remained dependent on the central authority; the kadıs and municipal administrators were appointed by the palace. The ayans merely increased their share out of the surplus without altering the reproduction of the system. This development of the smaller ayans is perhaps closer to what has been termed 'feudalization'. Here, there is a parcellization at the economic level without a break-off from the larger political-ideological nexus of the state. It should be remembered, however, that this 'feudalization', or rather the localization of power, did not indicate a transition to feudalism. The local potentates remained politically subordinate to the central authority, and the division of labor inside the empire was not significantly disrupted, unless 'feudalization' coincided with commercialization.

The immediate translation of these developments inside the relations characterizing the Asiatic mode of production was a relative decline in the revenue collected by the central authority. The state depended on an efficient extraction of the surplus, not only because of its budgetary needs, but also because this prevented the rise of rival centers of power which required surplus for the process of institutionalizing their authority. Therefore, the relative decline in revenue doubly threatened the

nodal position of the state in the reproduction of the social
formation. A revenue crisis meant that the political-ideological
(and, of course, disciplinary) functions of the state could not
be formed effectively; and it also signalled the rise of competing
power structures.
We are not claiming that, historically, the main cause of revenue
crises was the local usurpation of surplus. In fact, from the late
sixteenth century on, revenue crises in the Ottoman state became
endemic, and measures designed to fight the deficitary tendency
seemed to dominate the policies of the Porte. The immediate
causes of the late sixteenth-century development were the growing
military expenses, the population increase in Anatolia, and the
price inflation. We will analyze the impact of these 'mobilizing'
factors at the end of this section.
It was mentioned above that the reproduction of the Ottoman
social formation involved a political control over the division of
labor which could only be implemented by limiting the degree of
functioning of merchant capital. The existence of merchant capital
naturally entailed the production of commodities for the market.
In fact it was through the close state supervision of the market
that merchant capital could be kept within politically desirable
bounds. However, the success of this political control, and
consequently the successful reproduction of the social formation,
required that the state as the locus of institutions enabling
political control remain powerful vis-à-vis merchant capital. The
tendency inherent in merchant capital is for its extended reproduction through incorporating a greater proportion of economic
activity inside the sphere of commodity production. It was this
tendency which threatened the reproduction of the Ottoman social
formation. In effect, this potential contradiction between the
existence of commodity production and the political rationality
of the Ottoman social formation took its toll when merchant capital
inside the Empire was integrated into the circuit of capital's
valorization in the world capitalist economy. It was then that
merchant capital expanded into areas formerly closed to commodity
production. As we shall show later, the peripheral transformation
that this expansion entailed could only effect certain evolutions
in the forces of production. Proto-industrialization, for example,
because of its potential competition with manufactured imports,
would not be one of the evolving organizational forms. As Tilly
and Tilly remarked in a survey of European historiography, the
absence of such proto-industrial activity is one of the differentia-
ting factors characterizing today's periphery.(25) The causes
of this absence are, of course, to be looked for in the process
of the destruction of crafts during the phase of their competition
with manufactured imports. On the other hand, petty commodity
production in individual farms and feudalized cash-crop production were forms that readily articulated with merchant capital.
We have dealt theoretically with two of the possible contradictions inside the social formation as we have defined it. Although
other contradictions may be specified, it can be asserted that their

outcomes as well will collapse into the categorization adopted above: tendencies either toward the parcellization of economic and political control, or toward the expansion of the realm of the market and consequently the weakening of political control over the economy.

It is, however, impossible to observe these tendencies in isolation from counteracting forces and from external factors which create the conditions for the tendencies to assert themselves in the historical-concrete space. We shall, therefore, analyze two such external factors which have been utilized by Ottomanists to explain either adaptations or decline of institutions in the Empire, and attempt to show their workings through the structure of contradictions inherent in the social formation. These factors which are of varying degrees of exogeneity all fall inside the same time period. The price inflation showed its effects at the end of the sixteenth century; population increase and the shift in the trade routes occurred roughly within the same time period. Hence it would be difficult to separate the effects of these factors in order to deal with each individually. For the purposes of exposition this less-than-ideal method will be adopted. We shall concentrate on population and prices; shifts in trade routes have been discussed above and will be touched upon in the context of price inflation.

The population growth that the Ottoman Empire witnessed during the later sixteenth century was characterized by a market increase in the numbers of nonproducers relative to producers and an increase in urban population relative to rural population.(26) The dynamics of this process is to be sought in the pattern of responses to demographic change which characterizes pre-capitalist formations. Given the inelasticity of agricultural production, despite an initial phase of extension of arable lands, these economies were unable to support an increasing population. The subsistence crisis first manifested itself in the rural sector and forced the flight of population into urban areas where the army and other administrative institutions, as well as towns themselves, were privileged recipients of the limited food supplies.(27)

In the Ottoman Empire a large number of the rural uprooted population found their way to Istanbul and other cities, or joined religious schools (medrese). The vagrancy of the new urbanites created problems, often solved through employment by the state. Medreses also created problems because the endowments of these pious foundations, already reduced under the impact of price inflation, fell behind the requirements of an expanded student body. Sometimes, the apprentice ulema took to extorting food from the peasants, and sometimes the medrese administration attempted to expand their tax base at the expense of the peasants.

Population growth was thus translated into an intensified struggle for tax revenues among the factions of the surplus-receiving class. This struggle became most pronounced at the

juncture where the system's inherent conflictual tendency was
located, that is, in the struggle between the central government
and provincial administrations. The increasing population swelled
the ranks of the central army, the janissaries, and of the Palace
employees, thus exerting a pressure on the existing state revenues.
The central government, in order to meet its increased expenditures, was compelled to seek additional sources of revenue in
the form of extraordinary taxes (avariz)(28) and through the
sale of tax farms. The provincial administrators, especially those
in Anatolia, on the other hand, found themselves surrounded with
inflated entourages when the rural unemployed gravitated towards
the occupation of irregular soldiers (levend).(29) Thus, unable
to meet their requirements with the revenues allotted to them by
the state, provincial administrators resorted to banditry in the
countryside. No longer supplying the central government with
mounted soldiers, they deprived the state of its share of tax
revenues. Moreover, their extortions in the countryside forced
peasant flights, resulting in upheavals and a decline in agricultural production.

The loss of control by the central administration over the
sources of revenue, and its appropriation by provincial administrators and the ulema in religious schools, were concretized in
the Celali(30) uprisings - a prelude to the system's 'feudalization'
in Anatolia.

A rise in the number of urban-dwellers accompanied population
growth. Existing towns became larger, and some villages developed into towns. This urbanization, however, was not entirely
caused by demographic pressure. At least in its initial phases,
population increase had resulted in an increase in the volume of
output and of the marketed surplus. Local markets and regional
trade had expanded both as a result of increased town population
and increased commodity production in agriculture. Together with
the weakening control of the central authority on provincial
administration, this development could have led to a complete
escape of merchant capital from under the political control of the
state. However, the devastation in the countryside following the
Celali uprisings, peasant flights, and re-nomadization checked
the tendency towards the full monetization of agricultural production.

Another development of the sixteenth century, partly caused
by population growth, and partly reflecting a world-wide
tendency, was the price inflation. The change in the structure
of trade which resulted in the increasing activity of merchant
capital, concretized in contraband trade, meant the flooding of
the Ottoman market with Spanish 'real'.(31) The widespread
usage of 'real' without being converted into Ottoman currency
(akçe) signalled the abandonment of the Ottoman system to
the movements of world prices. Thus the state lost control over
its currency in that it no longer had a monopoly over issuing
coins, and could not debase a currency which it had not minted.
Thus it was deprived of one of the traditional measures utilized

to counteract declining revenue. With the weakening of its
controls over external trade and an increase in contraband in
response to European demand, the state could not fulfill its
systemic function of fixing the prices within the world-empire.
The forming of prices outside of the world empire and the consequent divergence between decreed prices and world prices
was another factor contributing to the liberation of merchant
capital.

At the same time, the increase in prices, to which contraband
contributed, meant that state expenditures in money terms had
to increase as well, while certain taxes, which were customarily
assessed at 30-40-year intervals,(32) declined in real terms.
The revenue crisis beginning in the sixteenth century was
exacerbated through increases in expenditures. The most important increase was in the requirements of the military. Changes
in the technology of war shifted the balance from mounted to foot
soldiers, necessitating the maintenance of a permanently stationed
army. Thus the Palace had to house and feed (and pay a salary
to) a growing number of janissaries. These janissaries were
mostly stationed in Istanbul and it was therefore of crucial importance that they received their salaries promptly.(33) The
development away from mounted soldiers also confirmed the
outmodedness of the timar system as a mode of revenue extraction.
Tax revenue was now required in money form, and not as use-
specific cavalry.

The new mode of revenue collection which came to dominate the
fiscal system was tax-farming. The practice of tax-farming,
i.e. farming out of specified revenues to the highest bidder,
had been an integral part of the Ottoman tax system since its
inception. This system (iltizam) had always been the main source
of liquid funds for the central authority. It had, however, been
confined to certain sources of revenue such as customs dues,
the poll-tax levied on non-Muslims, and the sheep-tax. Beginning
in the seventeenth century the iltizam system was applied to
the traditional agricultural tax of öşr as well. Revenues on imperial domains, on vakıf estates, on former timar land, were all
farmed out.(34)

In the sixteenth century, when only certain revenues were
farmed out, great Jewish merchants of Istanbul served as
creditors to the state and thereby found a domain of valorization
where merchant capital and money capital were interchangeable.
They could lend the Palace money accumulated in trade and thus
obtain the privilege of further trade, while at the same time
earning interest on their loans.

The tax-farming system, so advantageous to the possessors
of liquid funds, proved to be well-suited to an environment
characterized by money fortunes accumulated during the inflationary period.(35) Thus the money revenue requirements of the
state coincided with the demands of the newly-rich, some of whom
were themselves highly placed bureaucrats, janissaries, and ulema.
Thus merchants were no longer the sole purchasers of tax farms.

By the end of the seventeenth century, tax-farming had rapidly spread, and had begun to cause worries in that the tax-farmers pursuing short-term maximization abused their granted sources of revenue. As a measure to counteract this possibility, a new system was devised in 1695, whereby tax-farms were granted on a lifetime basis (malikane), and the grantee acquired the right to manage his farm independently of state supervision, as well as the right to sell his expected revenue.(36)

The tax-farmer's main function was to provide the state with money. Therefore his contractual relationship with the state was defined in strictly pecuniary terms. In this relationship, the state was the recipient of money capital which was advanced prior to the collection of taxes. Hence during the process of realization of interest, the tax-farmer, unlike the sipahi, had no obligation to perpetuate the ideological-political relationship between the direct producer and the state. His concern was the maximization of returns on capital advanced. Although legally bound by the traditional rates of taxation, he sought constantly to trangress these bounds, and frequently did so, especially if he held an administrative position as well.

Tax-farmers provided the state with loans and obtained the right to collect the taxes, but they also introduced a new relation in agriculture: usury.(37) Usury never developed to an extent sufficient to destroy small peasant property. Yet it was a means to increase exploitation, and to bind peasants to feudalized units of production, for instance as share-croppers. Usury served to accelerate both capital accumulation and the destruction of the free state of the peasantry.

Both of these developments undermined the reproduction of the system which depended on the maintenance of small property in the countryside.(38) Money capital, like merchant capital, may articulate with all forms of pre-capitalist production.(39) Again like merchant capital, it also destroys the basis of the political reproduction of the system. It does this in two ways: by mediating the rise of ayans, and by accelerating commercialization.

Tax-farmers were legally (and literally) accountable to the central authority. While they were guaranteed military aid by local administrators to deal with reluctant tax-payers, they also had to present their financial accounts to be controlled by the bureaucrats. Of course, this checking mechanism could not operate when the tax-farmer and his auditor were the same person, or when the local administrator was also the tax-farmer. Through such identities, tax-farming was closely linked to the development of local potentates.

By the early eighteenth century, certain tax-farmers of local origin had transformed themselves into local potentates, or ayans. They had earlier been merchants, moneychangers, or provincial ulema, and had received tax-farms as sub-contractors to Istanbul merchants or high officials. These enriched sub-contractors were usually the local representatives of governors as well (mütessel-lim). Thus they could add political authority to economic accumulation and as such became known as ayans.(40)

On the other hand, the more important ayans, who created troubles during the eighteenth century, were provincial officials originally appointed by the Porte.(41) Taking advantage of a weak central authority, they sought to establish themselves permanently as governors. The mechanism whereby they could purchase tax-farms for the entire area under their administrative authority also allowed them to build up a client group of locally powerful sub-contracting tax-farmers. This tax-farming was important both as a device for economic enrichment and as a mechanism for building up political authority.

Another development which was accelerated by tax-farming was the emergence of commercial estates (çiftlik) in the eighteenth century. The onset of tax-farming had already broken the ideological reciprocity between the producer and the state, allowing for an intensified exploitation of the peasant. This potential, together with a growing European demand for wheat, maize, cotton, and tobacco, incited tax-farmers and especially malikane holders to produce commercially for the market at an expanded scale.(42) Large çiftliks were established in the Balkans, in Thrace (supplying the Istanbul market), and in Western Anatolia, where İzmir developed from a small coastal town to a major port of export trade within a century.(43)

Çiftliks were examples of commercial farming where enserfed peasantry or share-croppers were employed. Aside from the novelty of the labor organization, çiftliks also altered the traditional crop pattern. Cash-crops replaced subsistence grains. Owners of çiftliks accumulated commercial profits obtained through selling these crops to European merchants directly. These sales were carried out illegally, since administered prices inside the empire were always lower than prices obtainable in export markets. The eighteenth-century fairs in the Balkans, and similar developments in Anatolia during the latter half of the eighteenth and early nineteenth centuries, must be interpreted in this context.

From the point of view of the peasants, most of whom were under debt bondage, the çiftlik system meant a serious deterioriation in their status. There was an obvious increase in the rate of exploitation in economic terms.(44) Most of the taxes payable to the state did not cease and the dues demanded by the çiftlik-holders demanded half of the produce after the payment of öşr, in addition to labor services required of the peasant. Moreover, peasants under the çiftlik system lost security of tenure. Many Balkan çiftliks were minor fortifications guarded by armed men (kırcalıs) who were instrumental in forcing peasant flights, thus making available more land for commercial production.

From the point of view of the political system, the rise of the çiftlik was the most disruptive development. Commercialization of production and, more importantly, change in the status of the peasantry, both of which the çiftlik entailed, are necessary components in a process of peripheralization. With the çiftlik

organization, integration of the Ottoman system into the capitalist
world-economy attained an irreversible momentum. In the early
nineteenth century, the process of development of ayans was
reversed through a wave of centralizations, but the growth of
commercial agriculture continued without such reversals.

PERIPHERALIZATION AND THE COLONIAL STATE

Thus far, we have discussed isolated strands in the development
of the Ottoman social formation, leading to its dissolution as a
world-empire. As production became commodity production and as
these commodities began to enter the circuit of industrial capital
in its valorization process, peripheralization asserted itself.
Becoming a periphery is identical with entering the world division
of labor and thus ceasing to be a self-contained unit of reproduction.

The isolated strands of tax-farming, contraband trade, price,
and population movements gain significance when they are analyzed
as contributing to such an integration into the circuit of world
capital. It must first be mentioned that the timing of this integration varied from one specific region to another. The Balkans
became integrated into the European economy beginning in the
eighteenth century. For Egypt and the Levant, the process
gained momentum during the first quarter of the nineteenth
century. In Anatolia, the volume of trade increased significantly
beginning in the 1830s. Of course, exact dates are impossible
to determine for this process, the commencement of which, as we
showed above, can be traced back to the sixteenth century.

Nor is it possible to describe a single mode of peripheralization.
Again depending on the region, various forms emerged: commercial çiftliks in the Balkans, large cotton estates in Egypt,(45)
petty commodity production in peasant farms in Western Anatolia.
Labor organization ranged from share-cropping to family units
working on their own lands, via wage labor in capitalist farms.
While foodstuffs and raw materials were the major export commodities, hand-woven carpets in Western Anatolia also became a
peripheral product.

Trade was organized by foreign trading companies and merchants stationed in a few cities (Selanik, Istanbul, İzmir, Beirut,
Alexandria). Merchant capital of foreign origin entered into a
division of labor with native capital, in this case with ethnic
minorities. It was the Jews, the Greeks, and the Armenians, who
acted as intermediaries between British, French, Italian, and
German merchants and the actual producers.(46)

External trade of the Ottoman Empire increased in the first
three decades of the nineteenth century, especially after the
advent of steam-freighters in the Eastern Mediterranean.(47)
Growing imports of European manufactures meant a decline in
guild production and rural crafts, while raw materials were now
channeled to export markets.(48) After the 1838 trade treaties

with core countries and with the beginnings of railroad construction, penetration of merchant capital into even the most resistant region of the Empire, Anatolia, was accelerated.(49) Foreign and local merchant capital was articulated inside the legal framework through judiciary reforms. There were also attempts by foreign capital to set up capitalist farms in the Izmir region, although these remained insignificant.(50)

The history of foreign capital, public debt, and trade in the second half of the nineteenth century is typical of a peripheralization process, and is too well known to repeat here.(51) We will, however, briefly discuss the attendant changes at the political level.

These developments in economic integration were not automatically reflected at the political level. There was, in the initial stages, considerable dislocation between the new economic orientation of the empire and the old form of state appropriate to the Asiatic mode of production. This created contradictions which became manifest as the legal and political requirements of merchant capital clashed with the existing institutions. If we define a 'colonial state' as that form of state in the periphery which primarily serves the needs of merchant capital, the transformation in the politics of the nineteenth-century Ottoman Empire may be interpreted as the transition from a state mechanism of the Asiatic type to a colonial state. 'Modernization' and 'reform' of the Ottoman Empire should be evaluated within such a perspective.

It should not be forgotten, however, that this transformation did not occur simply at the instigation of merchant capital or more precisely at the instigation of the core states serving merchant capital. It also conformed to the economic requirements of the bureaucratic class and to the ideological inspirations of a ruling class closely articulated with 'Western-modern' ideas.(52)

With the transformation of the state into a colonial state, the Ottoman system lost its specificity. It was now characterized by the dominance of the capitalist mode of production both at the economic and the political levels. Therefore, it was no longer a proper unit of study. Its subsequent history could only be analyzed within the dynamics of the world capitalist system as an integral, albeit functionally-differentiated, component of this system.

NOTES

* From Agenda for Ottoman history, 'Review' (Journal of the Fernand Braudel Center for the Study of Economies), 1 (no. 1), 1977, pp. 37-55.
1 By the Asiatic mode of production, we mean a situation in which the agricultural producer is a free peasant and his surplus is appropriated in the form of taxes by the state. Only state officials belong to the class of surplus appropriators.

Thus the unit of reproduction is defined by the extent of the political authority. This situation reflects a distinct articulation of the political, ideological, and economic levels. Thus the Asiatic mode of production exists as a mode of production, provided that we accept the definition given in Althusser and Balibar (1970).
In the late 1960s and the early 1970s, there was a debate in Turkey on the nature of the Ottoman Empire. The debate started with Sencer Divitçioğlu, 'Asya Tipi Üretim Tarzı ve Osmanlı Toplumu' (Ottoman society and the Asiatic mode of production) (Istanbul Üniversitesi Yayınlarından, 1967), arguing that the Ottoman Empire was an example of the Asiatic mode of production. More orthodox participants in the debate rejected this view and coined terms as far-fetched as 'centralized feudalism'. This debate, however, was carried out within a narrow perspective without recourse to such essential concepts as 'social formation' or 'dominance'. Hence, even the most sophisticated contributions remained descriptive and ultimately empirical. This judgment can be extended to the entire literature (most of which appeared in 'La Pensée') generated by the concept of the Asiatic mode of production in the 1960s.

2   The argument in this section was originally formulated in Cağlar Keyder, The dissolution of the Asiatic mode of production, 'Economy and Society', V, 2, 1976, pp. 178-96.

3   Timar was not only for agricultural revenues. There were timars applying to revenues from mines, etc. See Halil İnalcık, 'Hicri 815 Tarihli Suret-i Defter-i Sancak-i Arvanid' (Copy of Official Notebook for the Province of Arvanid, year 815 Hicri)(Ankara, Türk Tarih Kurumu, 1954). For a description of the institution see Ö.L. Barkan, Les formes de l'organisation du travail agricole dans l'Empire Ottoman aux XVe et XVIe siècles. 'Revue de la faculté des sciences économiques de l'Université d'Istanbul', 1939-40. Halil İnalcık, 'The Ottoman Empire' (1973), and Gibb and Bowen, in 'Islamic Society and the West', I (1962), use the terminology proper to feudalism in describing the timar institution. For an account of the Seljuk origins see C. Cahen, 'Ikta', in H.A.R. Gibb et al. (eds), 'The Encyclopaedia of Islam', 2nd ed. (Leiden, Brill, 1960), pp. 1088-91. Timar systems did not apply to the collection of all revenue. On crown lands, the collection of taxes was entrusted either to tax-farmers or to salaried state officials.

4   Öşr was a traditional Islamic tax equal to one-tenth of the annual product. For a description of Ottoman taxes, see Halil İnalcık, 'Osmanhlarda Raiyyet Rüsumu' (Ottoman taxes), 'Türk Tarih Kurumu Belleten', XXIII, 92, 1959, 575-610, and Lüfti Güçer, 'XVI-XVII: Asirlarda Osmanli Imparator- luğunda Hububat Meselesi ve Hububattan Alman Vergiler' (The question of grains and taxes on grains in the Ottoman Empire during the 16th century) (Istanbul Universitesi Yayını

1075, İktisat Fakultesi 152). Gyula Kaldy-Nagy, The effects of the timar system on agricultural production in Hungary, in L. Ligeti (ed.), 'Studia turcica' (Budapest, Akademiai Kiado, 1971), pp. 241-8, describes the inception and effects of the timar system in Hungary.

5 Legal and administrative authorities were not so clearly distinguished. Actually, the sipahi was also empowered to enforce and collect penal dues. See Halil İnalcik, in T. Naff and R. Owen (eds), 'Studies in Eighteenth Century Islamic History: Decline or Change' (forthcoming). The sipahi-kadi relationship is discussed in Halil İnalcik, Adaletnâmeler (Legal statutes), 'Belgeler', II, 3-4, 1965, pp. 49-145.

6 For Ottoman guilds, see G. Baer, The administrative, economic and social functions of Turkish guilds, 'International Journal of Middle East Studies', I, January 1970, pp. 28-50. For hisba see Halil İnalcik, Capital formation in the Ottoman Empire, 'Journal of Economic History', 39, 1, 1969, pp. 97-140.

7 Suraiya Faroqhi, Sixteenth century periodic markets in various Anatolian sancaks: İçel, Hamid, Karahisar-i Sahib, Kütahya, Aydin and Menteşe, 'Journal of Economic and Social History of the Orient' (forthcoming), insists on this adminis-trative determination. Also see Lüfti Güçer, 'XVI-XVIII: inci Asirlarda Osmanh Imparatorluğunun Ticaret Politikası' (The commercial policy of the Ottoman Empire in the XVI-XVIIIth centuries) (manuscript, no date) for location and administration of markets. Markets were also found outside the settlement areas where they were usually larger. In the sixteenth century, the number of village markets in-creased, thus showing a response to economic environment. After the sixteenth century, periodic fairs increased in numbers both in Anatolia and the Balkans, indicating a greater commercial involvement of the peasantry. For a list of eighteenth-century fairs in the Balkans, see N. Svoronos, 'Le Commerce de Salonique au XVIIIe siècle' (Paris, Presses Universitaires de France, 1956). For the markets in the Balkans, I. Asdrachas, Aux Balkans du XVe siècle: pro-ducteurs directs et marchés, 'Etudes balkaniques', VI, 3, 1970, pp. 36-59.

8 The derbent organization was set up to protect the passes and the roads. Peasants were granted tax exemptions in return for their work maintaining roads, bridges, and passes. In time of the Celali uprisings and peasant flights, the maintenance of these routes was disrupted, threatening the security of merchants. For the derbent organization, see Cengiz Orhonlu, 'Osmanlı Imparatorlugunda Derbent Teskilatı' (Derbent organization in the Ottoman Empire) (Istanbul Üniversitesi Edebiyat Fakültesi, 1967).

9 See Mehmet Genç, A comparative study of the life term tax farming data and the volume of commercial and industrial activities in the Ottoman Empire during the second half of

the 18th century, A.O.E.S.I.E., 22-7 March 1976, Hamburg,
for tax-farms on internal customs revenues (mukataa).
10 For the palace's concern over the provisioning of Istanbul,
see Lüfti Güçer, XVIII. Yüzyul Ortalarinda Istanbul'un
İaşesi İcin Lüzumulu Hububatin Temini Meselesi (The provisioning of grains for Istanbul in the mid-18th century),
'Iktisat Fakültesi Mecmuasi', XI, 1-4, 1949-50, pp. 397-416.
Also see R. Mantran, 'Istanbul dans la second moitié du
XVIIe siècle' (Paris, Maisonneuve, 1962) and M.M.
Alexandrescu-Dersca, Contribution à l'étude de l'approvision-
nement en blé de Constantinople au XVIIIe siècle, 'Acta
Orientalia et Studia' (Société des Sciences Historiques et
Philologiques de la République Populaire Roumaine, Section
d'Etudes, 1957), pp. 13-37.
11 Contraband was carried out along the coastal areas, usually
by Greek ships which were stationed on the Aegean islands.
Its importance in commercialization of agriculture and in the
rise of the Greek bourgeoisie was great, especially during
the eighteenth century. The story of contraband, sometimes
involving 'feudalized' landlords, is yet to be told. Maurice
Aymard, 'Venise, Raguse et le commerce du blé pendant
la seconde moitié du XVIe siècle' (Paris, SEVPEN, 1966),
for example, minimizes illegal trade. Traian Stoianovich,
however, recognizes its importance. See Land tenure and
related sectors of the Balkan economy, 1600-1800, 'Journal
of Economic History', XIII, 4, 1953, pp. 398-411; Les maïs
dans les Balkans, 'Annales E.S.C.', XXI, 5, 1966, pp.
1026-40; Le maïs arrive dans les Balkans, 'Annles E.S.C.',
XVII, 1, 1962, pp. 84-93.
12 See Halil İnalcik, Capital Formation.
13 The Persian wars can be interpreted as attempts to secure
trade over the silk route, and essential silk supplies (in
the conquest of Tabriz, for example). The wars with the
Karamani aimed at opening trade with Egypt. Balkan cam-
paigns of the fifteenth century aimed at seizing the Belgrade
road (the royal road of the Byzantine Empire) in order to
circumvent Venice. See Radovan Samardžić, Belgrade, centre
économique de la Turquie du Nord au XVIe siècle, 'La Ville
balkanique, XVe-XIXe siècles', 'Studia Balkanica' (Sofia,
1970), pp. 33-44.
14 For the trade of Bursa see Halil İnalcik, Bursa and the
commerce of the Levant, 'Journal of the Economic and Social
History of the Orient', III, 2, 1960, pp. 131-47. For the
Black Sea trade see Halil İnalcik, The Ottoman Empire.
15 For Ragusan trade, see J. Tadic, Le commerce en Dalmatie et
à Raguse et le décadence économique de Venise au XVIIIe
siècle, in 'Aspetti e cause della decadenza economica
veneziana nel secolo XVIII' (Atti del convegno, 27 giugno-2
luglio 1957, Venice, Isola di San Giogio Maggiore), 'Civiltà
Veneziana Studi', 9 (Venice Instituto per la collaborazione
culturale, 1961), pp. 237-74. For Venice, see R. Woolf,

Venice and the Terraferma: problems of the change from commercial to landed activities, in B.S. Pullan (ed.), 'Crisis and Change in the Venetian Economy in the Sixteenth and Seventeenth Centuries' (London, Methuen, 1968); F. Braudel, 'The Mediterranean and the Mediterranean World in the Age of Philip II', I (New York, Harper & Row, 1972); W. Heyd 'Histoire du Commerce du Levant au Moyen Age' (Amsterdam, A.M. Hakkert, 1959); and Aymard, op. cit.

16   The major article on the Mediterranean spice trade is Frederic C. Lane, The Mediterranean spice trade: further evidence of its revival in the 16th century, 'American Historical Review', XLV, 3, 1940, pp. 571–90. See also R. Mantran, L'empire Ottoman et le commerce asiatique aux 16e et 17e siècles, in D.S. Richards (ed.), 'Islam and the Trade of Asia' (University of Pennsylvania Press, 1970), pp. 169–79.

17   See Niels Steensgaard, 'The Asian Trade Revolution of the Seventeenth Century: The East India Companies and the Decline of the Caravan Trade' (University of Chicago Press, 1974).

18   For the silk trade, see Ralph Davis, English imports from the Middle East, 1580–1780, in M.A. Cook (ed.), 'Studies in the Economic History of the Middle East from the Rise of Islam to the Present Day' (Oxford University Press, 1970), pp. 193–206. For English trade in the Levant, see T.S. Willan, Some apsects of English trade with the Levant in the sixteenth century, 'English Historical Review', 70, no. 276, July 1955, pp. 399–410; and A.C. Wood, 'History of the Levant Company' (New York, Barnes & Noble, 1964).

19   For French trade, see P. Massom, 'Histoire du commerce français dans le Levant au XVIIIe siècle' (New York, Burt Franklin, Research and Source Works Series, 1967). See also Robert Paris, 'L'Histoire du commerce de Marseille', V: '1660–1789. Le Levant' (Paris, Plon, 1957).

20   For Selanik, see N. Svoronos, op. cit.; for the overland trade, see Traian Stoianovich, The conquering Balkan orthodox merchant, 'Journal of Economic History', XX, 2, June 1960, pp. 234–313; and Virginia Paskaleva, Osmanlı Balkan Eyaletlerinin Avrupalı Devletlerle Ticaretleri Tarihine Katkı (1700–1850) (A contribution to the history of commerce between the Balkan Provinces of the Ottoman Empire and the European States), 'İstanbul Üniversitesi İktisat Fakültesi Mecmuası', XXVII, 1–2, 1967–8, pp. 265–92.

21   The peasant was 'free' in the sense that he was not a serf. He was, however, restricted in movement. He had to pay a tax if he wanted to leave his village permanently, and his land could be confiscated if left uncultivated for three years in succession. See Ö.L. Barkan, XV. ve XVI. asırlarda Osmanli İmparatorluğunda, and Ö.L. Barkan, Türkiye'de servaj var mi idi? (Was there serfdom in Turkey?), 'Türk Tarih Kurumu Belleten', XX, 1–3, 1956, pp. 237–46.

22   Mehmet II, for instance, confiscated waqf lands of the ulema

faction and expanded timar areas. His son, however,
supported the ulema faction who were instrumental in his
succession to the throne. In the nineteenth century, the
ulema and the civil bureaucrats struggled to control the central
administration.
23 Especially of guilds. Janissaries also became tax-farmers
and engaged in contraband trade.
24 The most famous example was Mehmet Ali in Egypt. Tepen-
delenli Ali in the Balkans, Çapanoğlu, and Karaosmanoğlu
in Anatolia were others. See Kemal H. Karpat, Structural
change, historical stages of modernization, and the role of
the social groups in Turkish politics, in K. Karpat et al.,
'Social Change and Politics in Turkey: A Structural-Historical
Analysis' (Leiden, Brill, 1973), pp. 11-92.
25 Charles and Richard Tilly, Agenda for European economic
history in the 1970s, 'Journal of Economic History', XXXI,
1, March 1971, pp. 184-98.
26 Ö.L. Barkan, Tarahi Demografi Araştırmaları ve Osmanlı
Tarihi (Studies in Historical demography and Ottoman history),
'Türkiyat Mecmuas', X, 1951-3, pp. 1-27, and Ö.L. Barkan,
XVI. Asrin ikinici yarisinda Türkiye'de fiyat hareketleri
(Price movements in Turkey in the second half of the 16th
century), 'Türk Tarih Kurumu Belleten', XXXIV, 136, 1970,
pp. 557-607. M.A. Cook, 'Population Pressure in Rural
Anatolia, 1450-1600' (Oxford University Press, 1972), and
Leila Erder and Suraiya Faroqhi, Population rise and fall in
Anatolia, 1550-1620, 'Middle East Studies' (forthcoming)
demonstrate the increase in Ottoman population, confirming
Braudel's hypothesis of a general upsurge of population in
the Mediterranean during the sixteenth century. The relative
increase in urban population is shown by Ronald C. Jennings,
Urban population in Anatolia in the sixteenth century: a
study of Kayseri, Karaman, Amasya, Trabzon, and Erzurum,
'International Journal of Middle East Studies', VII, 1, 1976,
pp. 21-57.
27 The same argument can be found in Emmanuel Le Roy Ladurie,
'The Peasants of Languedoc' (University of Illinois Press,
1974), deriving from Braudel, op. cit.
28 Avariz, which began as extraordinary taxes, were soon converted into regular taxes due to the endemic revenue crisis.
29 On the levends (or sailors), see Mustafa Cezar, 'Osmanh
Tarihinde Levendler' (Levends in Ottoman history), I;
'Güzel Sanatlar Akademisi Yayinlari', Istanbul, 1965, and
Halil Inalcik, The origin of Ayans, in T. Naff and R. Owen,
(eds), 'Studies in Eighteenth Century Islamic History:
Decline or Change', forthcoming.
30 The Celali uprisings is the collective name given to unrest
and brigandry in Anatolia from the 1590s to the mid-
seventeenth century. Population pressure and the price inflation were the probable causes of this movement which
included smaller sipahis, medrese students, and irregular

soldiers in the retinue of provincial administrators. The central authority, exceptionally, armed the peasantry to combat the bandits, which in turn created more rebels. From a devastated countryside, peasants fled to fortified towns or to unreachable mountainous areas (see Xavier de Planhol, 'Les Fondements géographiques de l'histoire de l'Islam' (Paris, Flammarion, 1968). The literature on the uprisings includes Mustafa Akdağ, 'Celali İsyanlari 1550-1603' (Celali uprisings, 1550-1603) (Ankara Üniversitesi Basimevi, 1963); and William Griswold, Political Unrest and Rebellion in Anatolia, 1605-1609, unpublished PhD. thesis, UCLA 1966. After the emergence of ayans, the bandits may have been absorbed into their retinues. See Neşet Cağatay, Osmanlı İmparatorluğu arazi ve reaya kanunnamelerinde ilhak edilen memleketlerin âdet ve kanunları ve istilahlarin izleri (Traditions and laws of annexed countries in Ottoman land and peasant jurisdiction), Report no. III, Türk Tarih Kongresi, Ankara, 1948.

31 For the effects of the price inflation in the Ottoman Empire, see Barkan, Research on the Ottoman Fiscal Surveys, and Ö.L. Barkan, The price revolution of the sixteenth century: a turning point in the economic history of the Near East, 'International Journal of Middle East Studies', VI, pt 1, 1975, pp. 3-28. See also Mustafa Akdağ, Osmanlı İmparatorluğunun kuruluş ve inkişaf devrinde Türkiye'nin iktisadi vaziyeti (The economic situation in Turkey during the foundation and rise of the Ottoman Empire), 'Türk Tarih Kurumu Belleten', XII, 51, 1949-50, pp. 498-564; and Halil Sahillioğlu, XVII. asrın ilk yarısında Istanbul'da Tedavüldeki Sikkelerin Rāyi (The exchange rate of sikke in circulation in Istanbul in the first half of the XVIIth century), 'Belgeler Dergisi', I, 2, 1964, pp. 223-8.

32 For a description of this assessment and an evaluation of the censuses as sources of data, see Ö.L. Barkan, Research on the Ottoman Fiscal Surveys.

33 They exacted not only salaries, but also 'gifts' from each new Sultan, a significant drain on the treasury. Every thirty-three years, when the lunar calendar fell exactly one year behind the solar calendar, the janissary had to be paid an extra three months' salary. These sıvış years coincided with the worst crises of the treasury and resulted in janissary rebellions. See Halil Sahillioğlu, Sıvış year crises in the Ottoman Empire, in M.A. Cook (ed.), op. cit., pp. 230-52.

34 For the decline of the timar system, see Mustafa Akdağ, Timar rejiminin bozuluşu (Dissolution of the timar system), 'Ankara Üniversitesi Dil ve Tarih Coğrafya Dergisi', IV, 1945, pp. 419-31, and Bistra Cvetkova, L'évolution du régime féodal turc de la fin du XVI jusqu'au milieu du XVIII siècle, in Dimitre Kossav et al. (eds), 'Etudes historiques à l'occasion du XIe congrès international des sciences historiques, Stockholm, août 1960' (Sofia, 1960), pp. 171-206.

35 On this accumulation, see Mehmet Genç, Osmanli Maliyesinde
    Malikane Sistemi (The Malikane System in Ottoman financial
    administration), 'Türk Iktisat Tarihi Semineri', Ankara,
    1975.
36 For the malikane system, see Mehmet Genç, A comparative
    study, and Halil İnalcik, The origin of Ayans.
37 Usury in agriculture is, of course, not well documented.
    See, however, Jennings, op. cit.; also Suraiya Faroqhi,
    Agricultural activities in a Baktashi center: the Tekke of
    Kizil Deli, 1750-1830, 'Südost-Forschungen', 1977, and Halil
    Inalcik, Suleiman the lawgiver and Ottoman law, 'Archivum
    Ottomanicum', I, 1969, pp. 105-38.
38 See Karl Marx, 'Capital', II, ch. 36.
39 See ibid., III, ch. 4.
40 For a discussion of the origins of ayans, see H. İnalcik,
    The origin of Ayans; also on ayans, see Karpat, op. cit.
41 Our primary concern here is to establish the connection
    between tax-farming and the rise of local potentates. We
    are not interested in how their local power bases originated,
    nor do we attempt to assess the significance of this group
    within the context of Ottoman history. It should be men-
    tioned, however, that the development of ayans remains
    one of the least-explored issues in Ottoman historiography,
    and ours is an attempt to place them within the context of
    'feudalization'.
42 For commercialized agriculture in the Balkans, see Christo
    Gandev, L'apparition des rapports capitalistes dan l'économie
    rurale de la Bulgarie du nord-ouest au cours du XVIIIe
    siècle, in 'Etudes historiques à l'occasion du XIe congrès',
    pp. 207-20. See also Traian Stoianovich's three articles
    cited in footnote 11.
43 For the genesis of çitfliks, see H. İnalcik, Adaletnâmeler;
    also on the çitfliks in the Balkans, see T. Stoianovich,
    Land tenure and related sectors.
44 For a discussion of the change in the status of the peasantry
    under the çitflik system, see Suraiya Faroqhi, Rural society
    in Anatolia and the Balkans during the sixteenth century,
    'Tucica: Revue d'études turques', 1977.
45 See E.R.J. Owen, Cotton production and the development of
    the cotton economy in nineteenth century Egypt, in C. Issawi
    (ed.), 'The Economic History of the Middle East, 1800-1914'
    (University of Chicago Press, 1966), pp. 417-29, and Helen
    Rivlin, 'The Agricultural Policy of Muhammed Ali in Egypt'
    (Harvard University Press, 1961).
46 See A.J. Sussnitzki, Zur Gliederung wirtschaftlicher Arbeit
    nach Nationalitäten in der Turkei, in Issawi (ed.), op. cit.
    pp. 115-25.
47 See Dominique Chevallier, Western development and Eastern
    crisis in the mid-nineteenth century: Syria confronted with
    the European economy, in W.R. Polk and R.L. Chambers
    (eds), 'Beginnings of Modernization in the Middle East; the

Nineteenth Century' (University of Chicago Press, 1968), pp. 205-22.

48   See M.A. Ubicini, excerpts from 'Letters on Turkey', in Issawi (ed.), op. cit., pp. 43-5, and Omer C. Sarç, Tanzimat ve Sanayimiz (The Tanzimat and our industry), in Issawi (ed.), op. cit., pp. 48-59.

49   For the text of the Anglo-Turkish Commercial Convention of 1838, see Issawi (ed.), op. cit., pp. 39-40, and for the railway project, see W. von Pressel, Les chemins de fer de Turquie, in ibid., pp. 92-3.

50   See Orhan Kurmuş, The Role of British Capital in the Economic Development of Western Anatolia, 1850-1913, unpublished PhD. thesis, University of London, 1974.

51   See Donald C. Blaisdell, 'European Financial Control in the Ottoman Empire; a Study of the Establishment, Activities, and Significance of the Administration of the Ottoman Public Debts' (New York, AMS Press, 1966).

52   For the revenue problems of the bureaucracy and how these articulated with peripheralization, see Keyder, op. cit.

PRINCIPLES AND CRITIQUE OF THE ASIATIC MODE OF
PRODUCTION*

Lawrence Krader

A. THE ASIATIC MODE OF PRODUCTION, A SYSTEMATIC
OUTLINE

1   The countries of Asia which provided the historical data for
the theory of the Asiatic mode of production were India in the
first place and also China and Persia. These are lands in which
agriculture has been the predominant basis of subsistence over
thousands of years, which has been reported by written history
during the entire period. The great majority of the people of these
countries lived in villages, forming communities in direct relation
with the soil, cultivating it by means which had changed but
little from prehistoric times; but at last the continuity was
broken up by the colonialist incursions, at the time that Marx
engaged in his researches into the subject; the dual means of
the incursions at the hands of the colonialist powers of Europe
were trade and armed force. European capitalism had means
superior to the Asiatic mode of production in both respects.
2   The societies of the Asiatic mode of production were divided
from the earliest known times into a ruling class and a class of
the agricultural producers.
3   The Asiatic mode of production was developed among peoples
who thereby established political societies. These were no longer
primitive peoples, but were ruled by the state and its agencies.
The overarching sovereignty had few connections with the village
communities; the institutional network which interrelates the
state with the life of the modern societies was weakly developed,
being at the beginning of its historical career in the Asiatic
mode of production.
4   The populations of these Asiatic countries were large, but
their archaic agricultural practices left only a small surplus after
the immediate wants of the agricultural families were met. While
the lands were rich in natural bounty, the population was poor.
The tiny surplus production of the individual villages adverted
to the state treasury in the form of taxes; aside from this connec-

tion, the villages and the agencies of the state had few points of contact or interrelation. The taxes were made over to the state in the form of compulsory labor and of compulsory levies of the agricultural product. These taxes were at the same time the respective forms of rent: rent in labor and rent in kind. The village labor was unskilled, when compared to the crafts of the royal court, the compulsory forms of labor were likewise unskilled; the amount produced by each hand in the villages and in the compulsory labor tax made over to the state in the course of a day, was small. The surplus labor and surplus produce were taken from the villages, which were the primary units of tax collection, the mite from each village, by virtue of their great number, was aggregated into a great sum. Hence the societies which were engaged in the Asiatic mode of production gave the impression to the outside observer of having great wealth, but this impression reflected the magnificence of the royal court, not the poverty of the villages. The surplus which the latter provided to the state took the form of the public works of the oriental monarchies. These works were the return to the villages by the agencies of the state in exchange for the involuntary labor and produce extracted from them.

5   The institutional network interrelating the villages was as poorly developed as the network relating the villages to the sovereign power. The relations of exchange between the villages in the Asiatic mode of production existed in a low degree, but they existed; through the exchange between the villages, the commodities made their appearance in history.

6   Production within the villages was conducted primarily for the satisfaction of the immediate wants of the agricultural families and communities. Secondarily, the production of commodities arose within the villages in the Asiatic mode of production, the result of the exchange system.

7   The difference between the villages was not great. Each village produced in the main what the next village produced; the exchange between them took place in the form, and on the basis of, differences in the amounts of agricultural produce and in the natural resources provided by each, which were subject to regional and climatic variations.

8   Each village tended in the main to repeat the kind of production found in the last and the next; thus, although the division of labor among them all was developed, it was not developed in a high degree. The occupation of agricultual labor was found in each; and the non-agricultural employments such as the smith, the potter, the carpenter, in the villages likewise tended to be repeated from one village to the next. In India there have been traditionally twelve such employments in the villages, always the same. The number twelve is a fiction, but it points to the non-variability of these employments. Each employment was engaged in production of whole products or services; there was no further division of labor within the field of enterprise. The smith worked the iron he acquired into plough or knife.

9   There was little circulation of money; gold in the traditional
village communities of India and Persia was buried, or served as
ceremonial prestations, dotations, and ornament.
10   Each village tended to be a self-sustaining unity, with little
dependence on the outside world, having little communication
with each other, or with the sovereign power. They were close
corporations.
11   The self-sustainment and isolation of the villages were founded
on the combination of both agricultural and handicraft production
within them, such as were required to maintain the village life in
its traditional form. Each village formed a community of traditional
kinship structure, common economic interests, and mutual dependency.
12   The opposition between the city and the countryside, and that
between agricultural and manufacture production were developed
but in a modest degree. The low degree of opposition between
city and countryside, and the low degree of opposition between
agricultural and handicraft production was each the determinant
of the other.
13   The labor of the villages was unfree, the laborers were
bound to the soil, in the first place by the positive constraint
of custom: the form of unfreedom in this case was collective,
traditional. In the second place, the labor of the village communities was unfree by virtue of the obligation to provide levies
of compulsory labor and produce to the state.
13a   Although the social classes make their appearance in the
Asiatic mode of production, their express opposition to each
other is not developed. The entire society of the Asiatic mode
of production, in its various classes, is maintained within its
traditional forms and relations, whereby the oppositions, if they
are expressed at all, are brought out in a way that is but dimly
perceptible to the observer coming from capitalist society, and in
a form that is scarcely recognizable to those who are formed in
the European tradition. The folklore is the source for expression
of the opposition.
  The relations of labor in the Asiatic mode of production were
unfree, both in the relation to the soil and to the sovereignty,
the unfreedom was a collective one, bound by the traditions of
the village community, and this relation was the first condition
of the lack of social development, or its development that is other,
and to be recognized by different means than those of capitalist
society. The second condition of the low or other development in
Asian villages lay in the institution of the village headman and the
method of tax collection. The village community served as the
unit of tax collection, the headman was the agency for the trans-
mission of the taxes to the collector and to the treasury. The
headman faced two ways, he was a member of the community, not
always its wealthiest or most influential member, at the same time
he was the extension of the tax collection agency of the state. The
double visage of the tax collection in this way, no less than the
double visage of the tax as rent, reduced and redirected the class

oppositions. The collective unfreedom of labor in the Asiatic mode of production was reinforced by the collective obligation of the village communities for taxation. The internal tradition and the external obligation served to hold each other reciprocally in place.

13b   The villages were bound to the village community, as they were to the soil, by custom. This is an internal and positive factor of the bondage. The obligations to the state in the form of rent and tax are a further positive constraint or bond, but an external one. A negative and external factor that constrained the villagers to their traditional life was the lack of alternatives in other parts of the society, as well as in other parts of the country, for employment. Both the positive constraints, internal and external, worked upon the village as a whole; as such they were collective and tended to maintain the village unity. The negative constraint worked indirectly to the same end; the field of its operation was the individual cultivator and cultivating family. The agency of this latter operation was primarily that of the money-lender.

14   Capital was formed sporadically and in a non-systematic way in the villages and societies of Asia. Within this category were to be found the dealings of the grain- and money-lenders, the village usurers. These relations may be identified in part as capital relations, but there was no system of capital formation, circulation, concentration, and re-investment. These usurers and lenders cannot be identified therefore as capitalists, except in prospect. Speculation in land and grain is a peculiarity of the capitalist system; the hoarding activities, burial of gold in the ground is by no means the same as the miserliness of Western capitalism. The misers depicted by Molière, Balzac, and Dostoyevski do not find their analogue in this way in the villages and cities of Asia; the economic conditions of capitalist society differ from those of the Asiatic mode of production, and the consequences in the psychology of character formation differ, however much the external forms may resemble one another.

15   In the Asiatic mode of production, the sovereignty was subject to the same limitations of tradition and customary right as the villages. The power of the sovereignty was absolute within these limitations, but since the communication with the villages, aside from the levies of taxes in labor and in kind was at a low level, the absolute, despotic rulership had little contact with the villages, and little effect upon them. The sovereignty being despotic in form, and its power absolute, it was exerted upon the courtiers, retainers, clients, the personal following of the monarch, but the villages fell without its circumscription, once their obligations had been met.

16   The agriculture in the Asiatic mode of production was dependent on the storage and conducting, retention, damming, coffering or deflection of the water courses, generally of their management and control. Much of India, Southeast Asia, and China lies within the zone of the monsoon rains, which are

seasonal downpourings, providing great quantities of water in
one season of the year and dearth in another. The water must be
collected in the season of plenty and released in the season of
want. For this reason, a great economic and social value was
placed on the development of the sciences for the predictions of
the water supplies and their control, and the technology necessary for the storage and the distribution of the water. The
early agriculture was developed in the valleys on the Nile,
Tigris, Euphrates, Ganges, Yellow and Yangtze Rivers, valleys
which were inundated by the overflows of the river banks
during the seasons of abundance of water, and provided the
riches for the fertilization of the soil. The flooding of the rivers
was the subject of the predictions by the priestly astronomers
in the ancient days.
17   In India the Brahmins who made the predictions of the
seasonal supplies of water were the residents of the villages.
On the other hand, in ancient Egypt, the science and technology
relative to the control of the water lay in the hands of the priestly
astronomers who were direct agencies of the state. The labor of
ditch-digging, and dam-building for the irrigation systems, was
provided by the agricultural producers. Superficial differences
in the theory of the Asiatic mode of production have been introduced by the consideration of the different models based upon
the different lands which are included in the Asiatic mode of
production as a category, whether India, China, Egypt, Ceylon,
Persia, Burma, or other countries mentioned in this connection.
The centralization of the management of water control is by no
means a feature common to all; it is not a specific and determining
characteristic of the Asiatic mode of production.

    There is a widely held opinion that the Asiatic mode of produc-
tion and the oriental despotism are connected social phenomena.
The despotism in the Orient is sometimes accounted for by the
fiction that the sovereign was the owner of all the land in his
realm, sometimes by the fiction that the state was the sole,
centralized power whereby the water was controlled and managed,
the water control being a necessary condition for agriculture in
the Asiatic mode of production. That the control of the water
supply is indeed a necessary condition for the agriculture in the
Asiatic mode of production may be taken as given, but the agency
of the state in this connection does not take the form that has
been attributed to it by the proponents of this branch of the
theory. The role of the state through its agencies of water control
is attested in some societies associated with the category of the
Asiatic mode of production, but not in others. From this it
follows that while despotism is found in the various societies in
this category, it is not founded upon its centralization of the
control over the water supply. The coincidence of political
despotism and water control is an extrinsic, and not an inherent
characteristic of the relations of the Asiatic mode of production.
Such despotic rule as has been noticed in the traditional history
of India was not founded in principle on the centralization and
management of the water supply.

18   The foundation of the despotism is to be sought elsewhere:
just as there were no alternatives to village employment, which
made the traditional occupations stable, so there were no alternatives to the sovereign power, which made that power absolute.
The dynastic struggles merely altered the succession without
altering the mode of rulership.
19   By raising the question of the processes and relations of the
Asiatic mode of production we raise at the same time the question
of the social evolution of mankind.
20   The village community of the Asiatic mode of production,
according to the theory, was the location of the transition of mankind, the bearer of the transition, from the undivided society to
the society divided into classes and opposed within itself. The
category of the village community in this connection points to
societies of the Asiatic type as the earliest form of class-divided
societies, or the prototype for all such societies, whether the
earliest chronologically or not. On the other hand, the theory of
the gens was advanced, according to which another social institution and another type of society was made into the bearer of the
transition, and served as its model. The prototype, the location,
again setting aside the question of priority in time, was now
transplanted to ancient Greece and Rome. But the gens and the
village community differ not only in the geographic location of
the transition; each points in an opposed direction, and a different problem of social evolution is posed by each. The village
community develops into its given historical form in the process
of the formation of opposed social classes and the state; the
gens by its collapse and dismemberment gives way to the new
form of society. The village community arches over the transition from primitive to civilized society, and new relations of
production as well as new social forms corresponding to them
are thereby released. The gens points backward, and only by
its having been overcome, by its elimination and sublation,
are the new production relations and the new social forms
developed. The two categories stand in a dialectical opposition
to each other.
21   The village community within the Asiatic mode of production
contained and set in motion new productive forces, which transformed the history of mankind. In its internal evolution it
developed commodity exchange between the communities, and
thereby the production of commodities was generated; it created
surplus production, surplus value extracted by the agencies of
the state in the form of involuntary drafts of labor. Out of the
village community was generated the social division of labor thereby. In the evolution of the village community, its productive
forces were transformed, the division of the social classes and
the formation of the political society established. We draw the
attention of the reader to this point of evolution, which is the
indicator of the changeover; the generator of the change was
the release of the productive forces by the increase in the
density of the interrelations of the villages, and in particular,

of commodity exchange. The transformation of the society of the
Asiatic mode of production first took place on the basis of the
exchange relations between communities, and later of the relations
of production, just as, at the end of its history, the transformation of the Asiatic mode of production was effected by the introduction into the social whole of European commerce. The change
was effected in both cases by relations to the outside, in the
former case on the scale of the community, in the latter on the
scale of the society. In the latter case it was also imposed, in
part, from without. The increased density of the exchange relations in the early period of the history of the Asiatic mode of
production had as its effect the production of a surplus which
was extracted by the agencies of the state. This was not a
balanced relation. The generative force lay in the exchange
relations, and not in the production relations. The agency of the
state did not increase the productivity of agriculture.

Much has been written about the stagnation of the Asiatic
mode of production; to be sure, these societies showed a backward form to the outside observer. Nevertheless, the village
communities in the Asiatic mode of production carried out the
transformation of human society and stood to the primitive production in the same way that the capitalist mode of production
stands to the Asiatic. Each constitutes an evolutionary movement,
opening out productive forces and social moments relative to its
forerunners.

22   Within the Asiatic mode of production, both private property
and capital formation were weakly developed; the observations
of the Europeans who wrote of the weakness of private property
were just. But they feared that the weak development in the
form that it had taken in Europe at that time was the determinant
of the despotic power in the Orient; the fears of these seventeenth-
and eighteenth-century men were without foundation. The
oriental despotism did not result from the weakness of private
property; to attribute the unchecked and arbitrary power of the
sovereignty in the Orient to the weakness of private property
is an argument of economic determinism, but of a crude sort,
for it reduces the economic relations to their formal side. We
have seen that both private property and capital formation were
weakly developed in the Orient at that time; the mercantilists,
physiocrats were alerted to the observations about property,
forms of ownership, but not about the relations of labor and of
capital. Their explanations were relative to the categories of
their day. But that the same category of private property should
be the center of the explanation at present is untenable, a
century after the publication of 'Capital', it is likewise untenable
that the discussion of the Asiatic mode of production should be
centered around the problem of despotism even today. Despotism
is a form that emerges out of conditions of political society in
turn bound to relations of political economy. The discussion and
criteria of despotism belong to the superstructure of society,
they withdraw attention from the motor of history, the economic

relations; when the latter are taken up they are reduced to
their formal side.

In the capitalist mode of production, ownership of the means
of production is concentrated in private hands, and ownership
of the same by institutions of the state or by the community
is a residual category, being found where private property has
been given up, or where it has not been moved in. In the
Asiatic mode of production, these relations are turned around:
ownership of the means of production, of the land in the first
place, here rested with the community, or with the overarching
community, the state; in this case, private ownership of the
means of production becomes the residual category. The two
developments, capitalism and the Asiatic mode of production,
have different kinds of property relations. To say that private
property was weakly developed in the Asiatic mode of production
implies a comparison between weaker and stronger, whereby the
standards and criteria of capitalism are used to measure the
traditional Asiatic practices. This is fallacious. The negative
indication of this fallacy is that the application of the standards
from the one to the other is an ethnocentrism or a histori-
cocentrism. The positive indication of the fallacy is that it is
a defective dialectic.

The relation between the Asiatic and the capitalist modes of
production is indeed a dialectic. The capitalist stands to the
Asiatic mode of production as a development out of it and as its
antagonist. The capitalist practices were imposed by force in
the lands of the Asiatic mode of production, which became their
colonies. In Asia during the seventeenth-nineteenth centuries,
capitalism arose, at least in a modest degree, but the forces
that led to this were external to a major extent, and internal
to a minor extent. It is this dialectic that is muddied and obscured by comparisons of weak and strong property relations
and forms.

Further: the relations of the Asiatic and capitalist modes of
production have a bearing on particular historical courses, the
conquests in Asia by European powers in the capitalist period.
They have at the same time a bearing on development of political
economy and society on a world-scale. The Asiatic mode of production both precedes and gives rise to the capitalist in a
general way. The relations of capital and labor, as well as forms
of property, are related as developments and antagonisms
between the two, they are related directly and mediately, in
theory and in practice. Directly and practically they are related
by colonial conquest. On the other hand, they are related
mediately; here we take into account the other modes of production, such as the classical and feudal. Capitalism, from this
point of view, is a realization of potentialities in political economy
and society, in relations of labor to the means of production, of
capital formation, of commodity exchange and production, class
relations, surplus production and extraction, and the evolution
of the state. These potentialities already exist in the Asiatic

mode of production, and are realized in the extent that we
have seen. The capitalist and Asiatic modes of production are
related in theory, this relation has a bearing on the theory of
the social evolution of mankind.

The application of criteria of private property and despotism
to Asia by mercantilists and others, even down to the present
day, decked the societies that they found in borrowed clothes.
Moreover, the relation drawn between private property and
despotism is a superficial one. The despotism did not rest on
the property form. On the contrary, both are the result of
other relations, within the communities, between the communities,
and between the communities and the agencies of the state. Thus,
the strength of the age-old communities was an important factor
in determining the forms of property ownership. The weak
development of commodity relations between these communities
was related in turn to the self-sustaining strength of the village
communities. Thus, that very strength was a condition of the
weakness, or at any rate, the slow development, of the Asiatic
mode of production.

23   Community and society. Human beings are social creatures,
living only in society; the human being is a network of social
relations. The primitive social life of mankind, however, is life
in a community. The break away from the community fully
begins only with the development of political society. The
society of the Asiatic mode of production is near the beginning
of this development; it is a transitional one in the sense that
the political society has been developed, but, at least in the
villages, the social and community relations do not yet diverge,
the community life is the social life. The division of social labor
in these communities is the village division of labor.

These community relations are in part kinship relations, but
to reduce the entirety of the community relations, hence of the
social relations, to those of kinship, whether by descent or by
marriage, is a gross reduction and a simplification. L.H. Morgan
reduced the question of the community to that of the consan-
guineal unity, the gentes, bound as they are, in his conception
to ties of descent; this is likewise the error of H.S. Maine, who
thought that the ancient Indian village community was in its
origin the Joint Family. The social bonds of the ancient community were not only those of kinship; these communities, whose
traces are found in ancient Roman as well as ancient oriental
societies in the historical periods, were at once social groups
deriving their livelihood from their land. They worked upon the
land because they were related as kin, as neighbors, communally
and hence socially; at the same time, because they worked upon
it their relations as kin were socially acknowledged; because
they worked the land, its passage from one generation to the
next in the same descent line was accepted by the social unity
as a whole, in this case, the community. Social labor was
divided in the ancient Indian community, or the labor in the
ancient Indian community was labor in society, which was divided.

24   The great period of social evolutionary theory was the
nineteenth century in Europe, when the progress of mankind
was evidently proved. The weakness of the theory at that time
was its simplism, its naive progressism, its advocacy of a grand
teleology which, it was proposed, aimed at the establishment
of the contemporary social state of the European model of one
sort or another. The theory was uncritical, save for vague
averments by L.H. Morgan. It has had few developments as
such in the twentieth century, whether in critical or uncritical
form, because of its many weaknesses, while writers on the
subject have not coupled their theorizings with social criticism.
On the contrary, the theory of the Asiatic mode of production
contains the theory of transition from the primitive to the
civilized condition on the one hand, and the critique of the
latter on the other. It is the civilized and their 'mission civil-
isatrice' who by colonialist practice imposed European power in
many parts of the world, and brought the Asiatic mode of production to an end. The theory of the Asiatic mode of production
contains the theory of social evolution, in its important phase,
the transition from primitive life; it is at once critical of the
evolutionary process, and judges its outcome. The end of the
Asiatic mode of production is the product of colonialism; the
aim of its theory is the critique of the latter.

NOTE

*   From 'The Asiatic Mode of Production', Assen, Van Gorcum,
     1975, pp. 286-96.

SELECTED FURTHER READINGS: PART IV

P. Anderson, 1974, 'Lineages of the Absolutist State', London, New Left Books.

S.H. Baron, 1975, Marx's 'Grundrisse' and the Asiatic mode of production, 'Survey', 21, pp. 128–47.

C.E.R.M., 1969, 'Sur le Mode de production asiatique', Paris, Editions Sociales.

L.V. Danilova, 1971, Controversial problems of the theory of precapitalist societies, 'Soviet Anthropology and Archeology', IX, pp. 269–328.

J. Friedman, 1974, Marxism, structuralism and vulgar materialism, 'Man', 8, pp. 444–69.

J. Friedman, 1976, Marxist theory and systems of total reproduction, 'Critique of Anthropology', 7, pp. 3–16.

M. Godelier, 1973, 'Perspectives in Marxist Anthropology', Cambridge University Press, 1977.

I. Habib, 1973, Problems of Marxist historical analysis in India, in S.A. Shah (ed.), 'Towards National Liberation', Montreal, Black Rose.

B. Hindess and P. Hirst, 1975, 'Pre-capitalist Modes of Production', London, Routledge & Kegan Paul.

B. Hindess and P. Hirst, 1977, 'Mode of Production and Social Formation', London, Routledge & Kegan Paul.

M. Molnar, 1975, 'Marx, Engels et la politique internationale', Paris, Gallimard.

P. Skalnik and I. Pokora, 1966, Beginning of the discussion about the Asiatic mode of production in the USSR and the People's Republic of China, 'Eirene', V, pp. 179–87.

G. Sofri, 1969, 'Il Modo di produzione asiatico', Turin, Einaudi, 1974.

BIBLIOGRAPHY

N.B. Most of the articles and excerpts in this reader have self-contained bibliographies.
This bibliography contains both the works referred to in the texts written by the editors and additional sources on the AMP. When more than two dates are given for a bibliographical entry, the first date is that of the original date of publication (either in the original language or in the English translation) and the final date refers to the edition from which quotations have been drawn (original dates are given throughout the text).

Abrahamian, E., 1974, Oriental despotism: the case of qajar Iran, 'International Journal of Middle East Studies', 5, pp. 3-31.

Abrahamian, E., 1975, European feudalism and Middle Eastern despotisms, 'Science and Society', XXXIX, pp. 129-56.

Adams, R. Mc., 1960, Early civilizations, subsistence and environment, in Kraeling, C.H. and Adams, R.M. (eds), 'City Invincible', Chicago University Press.

Adams, R.M., 1965, 'Land Behind Baghdad: A History of Settlement of the Diyala Plains', Chicago University Press.

Adams, R.M., 1966, 'The Evolution of Urban Society: Early Mesopotamia and Prehispanic Mexico', Chicago, Aldine.

Adams, R.M., 1972, Some hypotheses on the development of early civilizations, in Leone, M.P. (ed.), 'Contemporary Archeology', Southern Illinois Press (written 1955), pp. 359-64.

Althusser, L., 1958, Despot et monarque chez Montesquieu, 'Esprit', 26, no. 2, pp. 595-614.

Althusser, L., 1959, 'Montesquieu, la politique et l'histoire', Paris, PUF.

Althusser, L., 1965, 'Pour Marx', Paris, Maspéro.

Althusser, L., 1965a (et al.), 'Lire "Le Capital"', 2 vols, Paris, Maspéro (2nd edn, 1969).

Althusser, L., 1969, 'For Marx', New York, Vintage Books, 1970.

Althusser, L., and Balibar, E., 1970, 'Reading Capital', London, New Left Books.

Amer, I., 1958, 'Al-Ard wa'l-Fallah, al Mas'ala al Ziraiyya bi Misr' (The land and the fellah: the agrarian question in Egypt), Cairo.

Amin, S., 1973, 'Le Développement inégal', Paris, Editions de Minuit.

Anderson, P., 1974, 'Lineages of the Absolutist State', London,
New Left Books.
Andrianov, B., 1976, Kontseptsia K. Vitfogelia gidravlicheskoe
obshchestvo i novye materialy po istorii irrigatsii, in
I.V. Bromlei (ed.), 'Kontseptsii zarubezhnoi etnologii:
Kriticheskie etuidy', Moscow, Nauka.
Antoniadis-Bibicou, M., 1966, Byzance et le MPA, 'La Pensée',
129, pp. 47-72.
Armillas, P., 1948, A sequence of Cultural Development in Meso-
america, in Bennett, W.C. (ed.), 'A Reappraisal of Peruvian
Archeology', Menasha, Society for American Anthropology
Memoir no. 4, pp. 105-111.
Armillas, P., 1951, Tecnologia, formaciones socioeconómicas y
religión en Mesoamerica, in Tax, S. (ed.), 'The Civilizations
of Ancient America' (XXIX International Congress of
Americanists).
Asad, T. (ed.), 1973, 'Anthropology and the Colonial Encounter',
London, Ithaca Press.
'Asiaticus', 1963, Il modo di produzione asiatico, 'Rinascita',
5 October, pp. 14-15.
Avineri, S. (ed.), 1969, 'Karl Marx on Colonialism and Modern-
ization', New York, Anchor Books.
Bachelard, G., 1951, 'L'Activité rationaliste de la physique
contemporaine', Paris, PUF.
Bailey, A., 1974, 'On the Specificity of the Asiatic Mode of
Production', unpublished M.Phil. thesis in Social Anthro-
pology, University of London.
Bailey, A. and Llobera, J.R., 1974, The Asiatic mode of pro-
duction: an annotated bibliography, part I, 'Critique of
Anthropology', no. 3, pp. 95-103.
Bailey, A. and Llobera, J.R., 1975, The Asiatic mode of produc-
tion: an annotated bibliography, part II, 'Critique of
Anthropology', nos 4-5, pp. 165-76.
Banu, I., 1967, La formation sociale 'asiatique' dans la perspec-
tive de la philosophie orientale antique, 'La Pensée', 132,
pp. 53-70.
Barceló, M., 1974, Ensayo introductorio, in Amin, S. (ed.),
'Sobre el desarrollo desigual de las formaciones sociales',
Barcelona, Cuadernos Anagrama.
Baron, S.H., 1958, Plekhanov's Russia: the impact of the West
upon an 'oriental' society, 'Journal of the History of Ideas',
XIX, pp. 388-404.
Baron, S.H., 1963, 'Plekhanov: The Father of Russian Marxism',
London, Routledge & Kegan Paul.
Baron, S.H., 1974, Plekhanov, Trotsky and the development of
Soviet historiography, 'Soviet Studies', XXVI, pp. 380-95.
Baron, S.H., 1975, Marx's 'Grundrisse' and the Asiatic mode of
production, 'Survey', vol. 21, pp. 128-47.
Bartra, R., 1965, Ascenso y caida de Teotihuacan, 'El Gallo
Illustrado', 124.
Bartra, R., 1969, (ed.) 'El modo de produccion asiatico',

Mexico, Editions Era (1974, 2nd ed.; 1975, 3rd ed.).
Beals, R., 1955, Discussion; symposium on irrigation civiliza-
tions, in Steward, J. et al., 'Irrigation Civilizations: a
Comparative Study', Washington, D.C., Pan-American Union,
Social Science Monograph.
Bedeker, D.K., 1951, Marxism and Ancient India, 'India Today',
vol. 3.
Beloff, M., 1958, Review of 'Oriential Despotism', 'Pacific
Affairs', vol. 31, no. 2, pp. 186-7.
Boersner, D., 1957, 'The Bolsheviks and the National and
Colonial Question', Geneva, Droz.
Boiteau, P., 1964, Les droits sur le terre dans la société
malgache précoloniale (contribution à l'étude du MPA),
'La Pensée', 117, pp. 43-69.
Bukharin, N., 1921, 'Historical Materialism: a System of
Sociology', University of Michigan Press, 1969.
Buttner, T., 1967, Das präkoloniale Afrika und die Diskus-
sionen zur asiatischer Produktionsweise, 'Jahrbuch für
Wirtschaftgeschichte', vol. IV, pp. 287-312.
Butzer, K.W., 1976, 'Early Hydraulic Civilization in Egypt',
University of Chicago Press.
Canguilhem, G., 1967, Mort de l'homme ou épuisement du cogito,
'Critique', 24, no. 242, July, pp. 599-618.
Canguilhem, G., 1968, 'Etudes d'histoire et de philosophie des
sciences', Paris, Vrin, 1970.
Carneiro, R.L., 1970, A theory of the origin of the state,
'Science', 169, pp. 733-8.
Carrasco, P., 1959, 'Land and Polity in Tibet', American Ethno-
logical Society, Seattle, Wash.
Carrére-d'Encausse, H. and Schram, S., 1969, 'Marxism and
Asia 1853-1964', London, Allen Lane.
Cassirer, E., 1951, 'The Philosophy of the Enlightenment',
Princeton University Press.
C.E.R.M., 1969, 'Sur le "Mode de production asiatique"',
Paris, Editions Sociales.
C.E.R.M., 1970, 'Sur les Sociétés précapitalistes: Textes de
Marx, Engels, Lenine', Paris, Editions Sociales.
Chabod, F., 1964, 'Storia dell'Idea d'Europa', Bari.
Chesneaux, J., 1964, Le mode de production asiatique: une
nouvelle étape de la discussion, 'Eirene', IV, pp. 131-46.
Chesneaux, J., 1964a, Le mpa: quelques perspective de
recherche, 'La Pensée', 114, pp. 33-54.
Chesneaux, J., 1965, Où en est la discussion sur le mpa? 'La
Pensée', 122, pp. 40-59.
Chesneaux, J., 1966, Où en est la discussion sur le mpa? 'La
Pensée', 129, pp. 33-46.
Chesneaux, J., 1968, Où en est la discussion sur le mpa? 'La
Pensée', 143, pp. 47-55.
Chesneaux, J., 1969, Sur le 'Mode de production asiatique',
C.E.R.M., pp. 13-45.
Claessen, H.J.M., 1973, Despotism and irrigation, 'Bijdragen

Tot de Taal- und Volkenkunde', 129, pp. 70-85.
Cohen, Y., 1968, Macroethnology: large-scale comparative
studies, in Clifton, J.A. (ed.), 'Introduction to Cultural
Anthropology', New York, Houghton Mifflin, pp. 402-48.
Coquery-Vidrovitch, C., 1969, Recherches sur un mode de
production africain, 'La Pensée', 144, pp. 61-78.
Cutler, A., Hindess, B., Hirst, P. and Hussain, A., 1977,
'Marx's Capital and Capitalism Today', London, Routledge &
Kegan Paul.
Danilova, L.V., 1968, Controversial problems of the theory of
precapitalist societies, in 'Problemy istorii dokapitalistideskikh
obstesty', vol. 1, Moscow, Nauka, pp. 27-66, in 'Soviet
Anthropology and Archeology', IX, 1971, pp. 269-328.
D.A.S.P., 1931, 'Diskussia ob Aziatskom sposobe proizvodstva
po dokladu M. Godesa' (Discussions on the Asiatic mode of
production according to the report of M. Godes), Moscow-
Leningrad.
d'Hondt, J., 1974 (ed.), 'Hegel et le siècle des Lumières',
Paris, PUF.
Dhoquois, G., 1966, Le mode de production asiatique, 'Cahiers
Internationaux de Sociologie', XLI, pp. 83-91.
Dhoquois, G., 1970, Deux ouvrages de reference sur les
sociétés pré-capitalistes, 'La Pensée', 154, pp. 110-18.
Dhoquois, G., 1971, 'Pour l'Histoire', Paris, Editions Anthropos.
Dhoquois, G., Texter, J., Herzog, P., Gallissot, R. and
Glucksmann, C., 1971, Sur la categorie de 'formation écon-
omique et sociale', 'La Pensée', 159.
Divitcioğlu, S., 1967, 'Asya Tipi Uretim Tarzi ve Az-Gelismis,
Ulkeler' (abstract in Premiéres Sociétés de Classes,
'Recherehes Internationales', 57-8).
Divitcioğlu, S., 1969, Modèle économique de la société ottomane
(XIV et XV siècles), 'La Pensée', 144, pp. 41-60.
Djilas, M., 1957, 'The New Class', New York, Praeger.
Dobb, M., 1966, Marx on precapitalist formations, 'Science and
Society', 30, pp. 319-25.
Draper, H., 1977, 'Karl Marx's Theory of Revolution': vol. I,
'State and Bureaucracy', London and New York, Monthly
Review Press.
Dumont, L., 1966, The village community from Munro to Maine,
'Contributions to Indian Sociology', no. IX, pp. 671-89.
Dumont, L., 1977, 'Homo aequalis', Paris, Gallimard.
Eberhard, W., 1965, 'Conquerors and Rulers', Leiden, Brill.
Eisenstadt, S.N., 1957-8, Review of 'Oriental Despotism',
'Journal of Asian Studies', 17, pp. 435-46.
Elliot, D., 1976, Review of L. Krader, 'The Asiatic Mode of
Production', 'Journal of Contemporary Asia', 6, 4, pp. 489-91.
Engels, F., 1878, 'Anti-Dühring', London, Lawrence & Wishart,
1969.
Engels, F., 1884, 'The Origin of the Family, Private Property
and the State', London, Lawrence & Wishart, 1972.
Études de Marxologie, 1959, 'Cahiers de l'I.S.E.A.', Paris.

Flannery, K., 1972, The cultural evolution of civilizations, in 'Annual Review of Ecology and Systematics', pp. 399-426.

Fletcher, F.T.H., 1939, 'Montesquieu and English Politics (1750-1800)', London, Arnold.

Foucault, M., 1970, 'The Order of Things', London, Tavistock.

Fried, M., 1959 (ed.), 'Readings in Anthropology', vo. 2, New York, Thomas Y. Crowell.

Fried, M., 1967, 'The Evolution of Political Society', New York, Random House.

Friedman, J., 1974, Marxism, structuralism and vulgar materialism, 'Man', 8, pp. 444-69.

Friedman, J., 1975, Tribes, states and transformations, in Bloch, M. (ed.), 'Marxist Analyses and Social Anthropology', London, Malaby.

Friedman, J., 1976, Marxist theories and systems of total reproduction, 'Critique of Anthropology', 7, pp. 3-16.

Friedman, J. and Rowlands, M., 1978, Notes towards an epigenetic model of the evolution of civilization, in Friedman and Rowlands (eds), 'The Evolution of Social Systems', London, Duckworth, pp. 201-76.

Friedrich, C., 1958, Review of K. Wittfogel's 'Oriental Despotism', 'American Slavic and East European Review', 17, pp. 351-2.

Gallissot, R., 1970, Marx et l'Algérie, 'Le Mouvement social', 71, pp. 39-63.

Gallissot, R., 1976, 'Marxisme et Algérie', Textes de Marx / Engels, Paris, 10/18.

Garaudy, R., 1967, 'Le Problème chinois', Paris, Seghers.

Garushiantz, Iu. M., 1966, Ob Aziatskom Sposobe Proizvodstia, 'Voprosy Istorii', no. 2 (extract published in 'Recherches Internationales', 1967, 57-8, pp. 118-40, in Carrère d'Encausse, H. and Schram, S. (1965), 1969, pp. 351-3).

Gerratana, V., 1972, Formazione sociale e società di transizione, 'Critica Marxista', 10, 1, pp. 44-80.

Glick, T.F., 1970, 'Irrigation and Society in Medieval Valencia', Harvard University Press (Belknap).

Godelier, M., 1963, Economie politique et philosophie, 'La Pensée', October 1963.

Godelier, M., 1964, La notion de mode de production asiatique et les schémas marxiste d'évolution des sociétés, in C.E.R.M., 1969.

Godelier, M., 1968, 'Rationality and Irrationality in Economics', London, New Left Books, 1972.

Godelier, M., 1970, Preface to Sur les Sociétés précapitalistes', Paris, Editions Sociales, pp. 13-142.

Godelier, M., 1971, Qu'est-ce que définir une 'formation économique et sociale'? L'exemple des Incas, 'La Pensée', 159, Sept-Oct, pp. 99-106.

Godelier, M., 1971a, On the definition of a social formation: The example of the Incas, 'Critique of Anthropology', 1974, 1, pp. 63-73.

Godelier, M., 1973, 'Horizon, trajets marxistes en anthropologie', Paris, Maspéro.
Godelier, M., 1973a, 'Perspectives in Marxist Anthropology', Cambridge University Press, 1977.
Habib, I., 1973, Problems of Marxist historical analysis in India, in Shah, S.A. (ed.), 'Towards National Liberation', Montreal, Black Rose.
Harris, M., 1968, 'The Rise of Anthropological Theory', London, Routledge & Kegan Paul.
Harris, M., 1969, Monistic determinism: anti-service, 'Southwestern Journal of Anthropology', 25, pp. 1-11.
Harris, M., 1971, 'Culture, Man and Nature', New York, Crowell.
Haupt, G. and Reberioux, M., 1967, 'La Deuxième Internationale et l'Orient', Paris, Cujas.
Hecker, J.F., 1916, 'Russian Sociology: a Contribution to the History of Social Thought and Theory', New York, Columbia University Studies in History, Economics and Public Law, vol. 67.
Hegel, G.W.F., 1830, 'Vorlesungen über die Philosophie des Geschichte', 'Werke', 12, Frankfurt, Suhrkamp, 1970.
Hegel, G.W.F., 1830a, 'The Philosophy of History', New York, Dover, 1956.
Hindess, B., and Hirst, P., 1975, 'Pre-capitalist Modes of Production', London, Routledge & Kegan Paul.
Hindess, B., and Hirst, P., 1977, 'Mode of Production and Social Formation', London, Macmillan.
Hirschman, A.O., 1977, 'The Passions and the Interests: Political Arguments for Capitalism before its Triumph', Princeton Universtiy Press.
Hobsbawm, E.J., 1964, Introduction to 'Pre-capitalist Economic Formations' (K. Marx), ed. E.J. Hobsbawm, London, Lawrence & Wishart.
Hodgen, M.T., 1964, 'Early Anthropology in the Sixteenth and Seventeenth Centuries', University of Philadelphia Press.
Honigman, J.J., 1973, 'Handbook of Social and Cultural Anthropology', Chicago, Rand McNally.
Hunt, C. and Hunt, E., 1976, Canal irrigation and local social organization, 'Current Anthropology', 17, 3, pp. 389-411.
Iolk, E., 1931, K voprosu ob aziatskom sposobe proizvodstva (On the problem of the AMP), in 'Pod znemenem marksizma', Moscow.
Jones, S.B., 1958, Review of 'Oriental Despotism', 'Geographical Review', 48, pp. 306-8.
Kautsky, K., 1887, Die moderne nationalität, 'Die Neue Zeit', V, pp. 442-51.
Kautsky, K., 1927, 'Die materialistiche Geschichtsauffassung', 2 vols, Berlin, Dietz.
Kedourie, E., 1957, Review of 'Oriental Despotism', 'Spectator', 199, 27 September.
Kiernan, V.G., 1967, Marx and India, 'The Socialist Register', pp. 159-89.

Kim, G.F. et al., 1966, 'Obshchee i Osobennoe v Istoricheses-
kom Razvitii Stran Vostoke', Moscow, Nauka.
Kim, G.F. et al., 1971, 'Problemy Dokapitalisticheskikh
Obshchestv v Stranakh Vostoka', Moscow, Nauka.
Kippenberg, H.C., 1977 (ed.), 'Seminar: Die Entstehung der
antiken Klassengesellschaft', Frankfurt, Suhrkamp.
Koebner, R., 1951, Despot and despotism: vicissitudes of a
political term, 'Journal of the Warburg and Courtauld
Institutes', XIV, pp. 275-302.
Kosambi, N.D., 1951, Marxism and Ancient India, 'India Today',
vol. I, no. 2.
Krader, L., 1972 (ed.), 'The Ethnological Notebooks of Karl
Marx', Assen, Van Gorcum.
Krader, L., 1973, The works of Marx and Engels in ethnology
compared, 'International Review of Social History', 18,
pp. 223-73.
Krader, L., 1973a, Karl Marx as Ethnologist, 'Transactions
of the N.Y. Academy of Sciences', Series II, vol. 35, no. 4,
April, pp. 304-13.
Krader, L., 1976, 'Dialectic of Civil Society', Assen, Van
Gorkum.
Kuhn, T.S., 1962, 'The Structure of Scientific Revolutions',
Chicago University Press.
Kuhn, T.S., 1968, The history of science, in 'International
Encyclopedia of the Social Sciences', New York, Collier-
Macmillan, vol. 14, pp. 74-83.
La Grassa, G., 1972, Modo di produzione, rapporti di pro-
duzione e formazione economico-sociale, 'Critica marxista',
10, 4, pp. 84-108.
Laqueur, W.Z., 1957, Review of 'Oriental Despotism', 'Encounter',
9, pp. 83-4.
Leclerc, G., 1972, 'Anthropologie et colonialisme', Paris,
Fayard.
Lecourt, D., 1975, 'Marxism and Epistemology', London, New
Left Books.
Lenin, V.I., 1959, 'Conspectus of Correspondence of K. Marx
and F. Engels 1844-1883' (Russian), Gospolitizdat.
Lenin, V.I., 1960-70, 'Collected Works', 45 vols, London,
Lawrence & Wishart.
Levitt, C., 1978, L. Krader's research on the Asiatic mode of
production, 'Critique of Anthropology', 11, pp. 39-56.
Levy, M.J., 1958, Review of 'Oriental Despotism', 'World
Politics', 10, no. 3, pp. 462-71.
Lewin, G., 1967, The problem of social formations in Chinese
history, 'Marxism Today', 1, pp. 20-5.
Lewin, G., 1969, The Marxist theory of social formations,
'Marxism Today', 13, June, pp. 182-92.
Lewin, G., 1973, 'Die ersten fünfzig Jahre der Song-Dynastie
in China', Berlin, Akademie Verlag.
Lichtheim, G., 1963, Marx and the 'Asiatic mode of production',
'St Anthony's Papers', Oxford/ London, retitled 'Oriental

Despotism', in 'Concept of Ideology and other Essays', New York, Random Books, 1967.

Llobera, J.R., 1978, 'An Epistemological History of the Concept of Mode of Production', unpublished Ph.D. thesis, University of London.

Lowe, D.M., 1966, 'The Function of "China" in Marx, Lenin and Mao', University of California Press.

Luporini, C., 1972, Marx secondo Marx, 'Critica Marxista', 10, nos 2-3, pp. 48-118.

Luxemburg, R., 1913, 'Die Akkumulation des Kapitals', Berlin, 1923 ('Gesammelte Werke', vol. 6).

Luxemburg, R., 1913a, 'The Accumulation of Capital', London, Routledge & Kegan Paul, 1971.

Luxemburg, R., 1925, 'Einführung in die National Ökonomie', Hamburg, Rowohlt, 1972.

Luxemburg, R., 1925a, 'Introduction à l'économie politique', Paris, 10/18, 1971.

Ly, B., 1969, Les classes sociales dans le Sénégal pré-colonial, C.E.R.M., Paris, Editions Sociales, pp. 229-55.

MacNeish, R., 1967, Mesoamerican archeology, in S.J. Siegel and A. Beals (eds), 'Biennial Review of Anthropology', Stanford University Press, pp. 306-31.

MacRae, D.G., 1959, Review of 'Oriental Despotism', 'Man', 59, pp. 103-4.

Mad'iar, L., 1928, 'Ekonomika selskogo khozia'istva v kitaie' (A study of the rural economy in China), Moscow.

Mandel, E., 1971, The AMP and the historical pre-conditions for the rise of capital, in 'The Formation of the Economic Thought of Karl Marx', London, New Left Books, pp. 116-39.

Manivanna, K., 1968, Aspects socioéconomiques de Laos médiéval, 'La Pensée', 138, pp. 56-70.

Marx, K., 1849, 'Wage-Labour and Capital', Moscow, Progress Publishers, 1970.

Marx, K., 1857-8, 'Die Grundrisse der Kritik der Politischen Oekonomie' (1838-41), 1953, Berlin, Dietz Verlag.

Marx, K., 1857-8a, 'Grundrisse: Foundations of the Critique of Political Economy', Harmondsworth, Penguin, 1973.

Marx, K., 1859, 'A Contribution to the Critique of Political Economy', London, Lawrence & Wishart, 1971.

Marx, K., 1867, 'Capital I', London, Lawrence & Wishart, 1970 and 1974.

Marx, K., 1885, 'Capital II', London, Lawrence & Wishart, 1970.

Marx, K., 1894, 'Capital III', London, Lawrence & Wishart, 1972.

Marx, K., 1905, 'Theories of Surplus Value', Part I, London, Lawrence & Wishart, 1969.

Marx, K., 1905, 'Theories of Surplus Value, Parts II and III, London, Lawrence & Wishart, 1972-10.

Marx, K., 1972, 'The Ethnological Notebooks of Karl Marx' (ed. L. Krader), Assen, Van Gorcum.

Marx, K., 1975, 'Articles on Britain', Moscow, Progress Publishers.

Marx, K. and Engels, F., 1845-6, 'Die Deutsche Ideologie', Berlin, Dietz Verlag (1953) 1960.

Marx, K. and Engels, F., 1951-2, 'Ausgewahlte Schriften in Zwei Bände', Berlin, Dietz Verlag, 1974.

Marx, K. and Engels, F., 1953, 'Ausgewahlte Briefe', Berlin, Dietz Verlag.

Marx, K. and Engels, F., 1956-68, 'Werke', 39 vols, Berlin, Dietz Verlag (MEW).

Marx, K. and Engels, F., 1961, 'Werke', vol. 13 ('Zur Kritik der Politischen Oekonomie'), Berlin, Dietz Verlag (MEW).

Marx, K. and Engels, F., 1965, 'Selected Correspondence', London and Moscow, Progress Publishers.

Marx, K. and Engels, F., 1965a, 'Werke', vols 26.1, 26.2, 26.3 ('Theorien über den Mehrwert'), Berlin, Dietz Verlag (MEW).

Marx, K. and Engels, F., 1969, 'Selected Works', 3 vols, Moscow, Progress Publishers.

Marx, K. and Engels F., 1973, 'Werke Erganzungsband Ersten Teil' (Schriften bis 1844), Berlin, Dietz Verlag (MEW).

Marx, K. and Engels, F., 1975, 'Collected Works', London, Lawrence & Wishart (MECW).

Marx, K. and Engels, F., 1976, 'Collected Works', vol. V (1845-6), London, Lawrence & Wishart (MECW).

Masubuchi, T., 1966, Wittfogel's theory of oriental society (or hydraulic society): the studies of Chinese social and economic history in Japan, 'The Developing Economies', IV, pp. 316-33.

McLellan, D., 1973, 'Karl Marx, His Life and Thought', London, Macmillan.

Medvedev, E., 1969, 'Le Régime socio-économique de l'Inde ancienne', Cahiers du C.E.R.M., no. 71.

Meek, R.L., 1962, 'The Economics of Physiocracy', London, Allen & Unwin.

Meek, R.L., 1976, 'Social Science and the Ignoble Savage', Cambridge University Press.

Meillassoux, C., 1964, 'Anthropologie économique des Gouro du Côte d'Ivoire', Paris, Mouton.

Meisner, M., 1963, The despotism of concepts: Wittfogel and Marx on China, 'China Quarterly', 16, pp. 99-111.

Melekechvili, G.A., 1967, Esclavage, féodalisme et MPA dans l'Orient ancien, 'La Pensée', 132, pp. 31-47.

Melekechvili, G.A., 1972, Kharakter sosial 'no-ekonomicheskogo stroia na drevnem vostoke', 'Narodny Azii i Afriki', 4, pp. 53-64.

Melotti, U., 1972, 'Marx e il terzo mondo', Milan, Il Saggiatore.

Melotti, U., 1977, 'Marx and the Third World', London, Macmillan.

Millon, René, 1962, Variations in social responses to the practice of irrigation culture, in Woodbury, R. (ed.), 'Civilizations in Desert Lands', University of Utah Anthropological Papers, no. 62, pp. 56-88.

Millon, R., 1967, Teotihuacán, in 'Avenues to Antiquity'
(readings from the 'Scientific American'), San Fransisco,
Freeman, pp. 221-31.
Mitchell, W.P., 1973, The hydraulic hypothesis: a reappraisal,
'Current Anthropology', 14, pp. 532-7.
Mitchell, W.P., 1976, Irrigation and community in the Central
Peruvian Highlands, 'American Anthropologist', 78, pp. 25-44.
Molnar, M., 1975, 'Marx, Engels et la politique internationale',
Paris, Gallimard.
Molnar, M. and Witzig, C., 1974, L'influence de la mentalité
colonialiste britannique sur le concept asiatique de Marx,
'Relations Internationales', 2, pp. 37-45.
Moore, Barrington, 1958, 'Political Power and Social Theory',
Harvard University Press.
Morris, C., 1972, State settlements in Tawantinsuyu: a
strategy of compulsory urbanism, in Leone, M.P. (ed.),
'Contemporary Archeology', Southern Illinois University
Press, pp. 393-401.
Murphy, R.F., 1967, Cultural change, in Siegel, B.J. and
Beals, R. (eds), 'Biennial Review of Anthropology', Stan-
ford University Press, pp. 1-45.
Murra, J., 1956, 'The Economic Organization of the Inca State',
unpublished Ph.D. thesis, University of Chicago.
Murra, J., 1958, Inca political structure, in Ray, R.F. (ed.),
'Systems of Political Control and Bureaucracy in Human
Societies', Washington, American Ethnological Society,
pp. 30-41.
Nair, S. and Lowy, M., 1973, 'Lucien Goldmann', Paris,
Seghers.
Nambooridiripad, E.M., 1952, 'The National Question in Kerala',
Bombay.
Needham, J., 1959, Review of 'Oriental Despotism', 'Science
and Society', 23, pp. 58-65.
Neimeyer, G., 1958, Review of 'Oriental Despotism', 'Review
of Politics', 20, pp. 264-70.
Nicolaevsky, B.I., 1958, Marx and Lenin on oriental despotism,
'Sotsialisticheski Vestnik', February- March.
Nicolaus, M., 1973, Foreword to 'Grundrisse', by Marx, K.,
London, Allen Lane, pp. 7-63.
Nikiforov, V.N., 1966, 'Zakliuchitel' noe Slovo po Doklady in
Kim, G.F. et al., 'Obshchee i Osobennoe v Istoricheseskom
Razvitii Stran Vostoka', Moscow, Nauka.
Nikiforov, V.N., 1971, 'K. Marks i F. Engels ob Aziatskom
Sposobe Proizvodstva', in Kim, et al., 'Problemy Dokaptial-
istischeskikh Obshchestv v. Stranakh Vostoka', Moscow.
Olmeda, M., 1967, Sur les sociétés azteque et maya, Premières
Sociétés de Classes, 'Recherches Internationales', 57-8,
pp. 254-65.
Palerm, A., 1955, The agricultural bases of urban civilization
in Mesoamerica, in 'Irrigation Civilizations: a Comparative
Study', Washington, D.C., Pan-American Union, pp. 28-42.

Palerm, A., 1970, Mesoamérica y la teoria de la sociedad oriental, 'Ensayos y Conferencias I Cuadernos de Antropologia y Etnologia', Madrid, 1, no. 1, pp. 33-104.

Parain, C., 1966, Protohistoire méditerranéene et la MPA, 'La Pensée ', 127, pp. 24-43.

Pavlova-Silvanskaia, M.P., 1971, Zarubeshnye kritiki o knige Vitfogelia 'Vostochnyi despotizm', 'Narody Azii i Afrika', 2.

Pavlovkaia, A.I., 1965, O Konteseptsii gidravlicheskogo obshchestva K, Vitfogelia, 'Voprosy drevnei istorii', 3, pp. 83-90.

Pecirka, J., 1964, Die sowjetischen Diskussionen über die asiatische Produktionsweise und über die Sklavenhalterformation, 'Eirene', III.

Pecirka, J., 1967, Von der Asiatischen Produktionisweise zu einer marxistischen analyse der frühen Klassengesellschaften, 'Eirene', VI, pp. 141-74.

Pelletier, A. and Goblot, J.J., 1973, 'Matérialisme historique et histoire des civilisations', Paris, Editions Sociales.

Plekhanov, G.V., 1908, 'Fundamental Problems of Marxism', London, Lawrence & Wishart (1929), 1937.

Plekhanov, G.V., 1914-17, 'History of Social Thought in Russia' (in Russian), 3 vols.

Plekahnov, G.V., 1923-7, 'Collected Works' (in Russian), ed. D. Riazanov, 24 vols, Moscow.

Plekahnov, G.V., 1926, 'Introduction à l'histoire sociale de la Russie Paris, Bossard.

Plekahnov, G.V., 1961, 'Selected Philosophical Works', vol. I, London, Lawrence & Wishart.

Pokora, T., 1963, Existierte in China eine Sklavenhaltergesellschaft?, 'Archivnyi Oreintalnyi', XXXI, pp. 353-5.

Pokora, T., 1964, Gab es in der Geschichte Chinas eine durch Sklaverei bestimmte Produktionsweise und Gesellschaftsformation? 'Neue Beitrage zur Geschichte der alten Welt', vol. 1, Berlin, Akademie Verlag, pp. 123-35.

Poliakov, L., 1971, Les idées anthropologiques du siècle des lumières, 'Revue francaise d'histoire d'outre-mer', 3.

Prestipino, G., 1972, Concetto logico e concetto storico di 'formazione-sociale', 'Critica Marxista', X, no. 4, pp. 54-83.

Pulleybank, E.G., 1957-8, Review of 'Oriental Despotism', 'Journal of the Economy and Social History of the Orient', I, pp. 351-3.

Rathje, W.L., 1972, Praise the gods and pass the metates: a hypothesis of the development of lowland rainforest civilizations in Mesoamerica, in Leone, M.P. (ed.), 'Contemporary Archeology', Southern Illinois University Press, pp. 365-92.

Reisman, D., 1976, 'Adam Smith's Sociological Economics', London, Croom Helm.

Rey, P.P., 1972, 'Colonialisme, néocolonialisme et transition au capitalisme', Paris, Maspéro.

Rey, P.P., 1973, 'Les Alliances de classe', Paris, Maspéro.

Riazanov, D., 1925, Karl Marx et la Chine, 'La Correspondance

Internationale', no. 68, 8 July 1925, pp. 563-4.
Ribeiro, D., 1968, 'The Civilizational Process', Washington, D.C., Smithsonian Institution.
Robinson, M., 1966, 'Islam et capitalisme' (English edition 1974).
Rodinson, M., 1966, What happened in history, 'New Left Review', 35, 97-8.
Rosdolsky, R., 1968, 'Zur Enstehungsgeschichte des Marxschen Kapital: der Rohentwurf des Kapital 1857-8', 2 vols, Frankfurt and Vienna.
Rowe, J., 1963, Urban settlements in Ancient Peru, 'Naupa Pacha', 1, pp. 1-27.
Rubel, M., 1956, 'Bibliographie des oeuvres de K. Marx', Paris, Rivière (Supplement: 1950).
Rubel, M., 1975, 'Marx without Myth', Oxford, Blackwell.
Ruben W., 1953-4, Karl Marx über Indien (1853) und die Indienliteratur vor ihm, 'Wissenschaftliche Zeitschrift der Humboldt Universität zu Berlin. Gesellschafts- und sprachwissenschaftliche Reihe', no. 2, Jahrgang 3.
Saad, A.S., 1975, 'L'Egypte pharaonique (autor du mpa)', Cahiers du C.E.R.M., no. 122.
Saad, A.S., 1976, Le MPA et les problèmes de la formation sociale egyptienne, 'La Pensée', 189, pp. 20-36.
Sachs, I., 1967, Nowa faza dyskusji o formacjach, 'Nowe Drogi' (abstract in Premières sociétés de classes, 'Recherches Internationales', 57-8).
Sahlims, M. and Service, E.R., 1960, 'Evolution and Culture', University of Michigan Press.
Sanders, W. and Marino, J., 1970, 'New World Prehistory: Archeology of the American Indian', Englewood Cliffs, Prentice-Hall.
Sanders, W. and Price, B., 1968, 'Mesoamerica: the Evolution of a Civilization', New York, Random House.
Santis, S. de, 1965, Les communautés de village chez les Incas, les Aztéques, et les Mayas, 'La Pensée', 122, pp. 79-95.
Sawer, M., 1974, The sources of Marx's concept of oriental despotism in British political economy, 'Etudes de Marxologie', Série S, 17, pp. 1507-23.
Sawer, M., 1977, 'Marxism and the Question of the Asiatic Mode of Production', The Hague, Martinus Nijhoff.
Sedov, I., 1968, La société angkorienne et le problème du MPA, 'La Pensée', pp. 71-84.
Semenov, Y., 1965, Problema sotsialno-ekonomitcheskogo stroia drevnego vostoka, 'Narody Azii i Afriki', 4, pp. 69-90.
Sereni, E., 1970, Da Marx a Lenin: la categoria di 'formazione economico-sociale', 'Critica Marxista-Quaderni', 4, pp. 29-79.
Sertel, Y., n.d., Les Interprétations de l'évolution et de la structure de la société ottomane.
Service, E.R., 1961, 'Primitive Social Organization', New York, Random House, 1971.
Service, E.R., 1975, 'Origins of the State and Civilization', New York, Norton.

Shah, S.A., 1973 (ed.), 'Towards National Liberation: Essays on the Political Economy of India', Montreal, Black Rose.

Shapiro, M., 1962, Stages of social development, 'Marxism Today', 6, pp. 282-84.

Shiozawa, K., 1958, The theory of the AMP and the ancient Japanese state, 'Rekishinshigaku-Kenkyu', 225, November 1958.

Shiozawa, K., 1965, Les historiens japonais et le 'mpa', 'La Pensée', 122, pp. 63-78.

Shiozawa, K., 1966, Marx's view of Asian society and his AMP, 'The Developing Economies', IV, 3, pp. 299-315.

Simon, J., 1962, Stages in social development, 'Marxism Today', no. 6, pp. 183-8.

Skalnik, P., 1971-2, Engels über die vorkapitalistischen Gesellschaften und die Ergebnisse der modernen Ethnologie, 'Zbornik filozofickej Fakulty univerzity komenskeho Philosophica', XI-XII, pp. 405-14.

Skalnik, P., 1975, Review of M.A. Vitkin, 'Vostok v filososko-istorischesckoi koncepcii K. Marksa i F. Engel'sa', 'Political Anthropology', 1, no. 1.

Skalnik, P. and Pokora, I., 1966, Beginning of the discussion about the Asiatic mode of production in the USSR and the People's Republic of China, 'Eirene', V, pp. 179-87.

Smith, P.E.L. et al., 1972, The evolution of early agriculture and culture in Greater Mesopotamia: a trial model, in B. Spooner (ed.), 'Population Growth', MIT Press.

Sofri, G., 1969, 'Il Modo di produzione asiatico', Turin, Einaudi, 1974.

Spate, O.H., 1959, Review of 'Oriental Despotism', 'Annals of the Association of American Geographers, 59, no. 1, pp. 90-5.

Sprott, W.U.H., 1957, Review of 'Oriental Despotism', 'The Listener', 3 October, p. 527.

Stalin, J., 1938, Dialectical and historical materialism, in Franklin, B. (ed.), 'The Essential Stalin', London, Croom Helm, 1973, pp. 300-33.

Stelling-Michaud, S., 1960-1, Le mythe du despotisme oriental, 'Schweizer betrage zur Allgemeinen Geschichte', 18-19, pp. 328-46.

Steward, J., 1949, Cultural causality and law: a trial formulation of the development of early civilizations, 'American Anthropologist', 51, 1, pp. 1-27.

Steward, J., 1955, 'Theory of Culture Change', University of Illinois Press, 1972.

Steward, J., 1955a, Introduction and Some implications of the symposium, in Steward, J. et al., 'Irrigation Civilizations: A Comparative Study', Washington, D.C., Pan-American Union, Social Science Monograph.

Steward, J., 1970, Cultural evolution in South America, in W. Goldschmidt and J. Hoijer (eds), 'The Social Anthropology of Latin America', University of California Press, pp. 199-223.

Steward, J., 1977, 'Evolution and Ecology', University of
Illinois Press.
Stocking, G.W., 1968, 'Race, Culture and Evolution', New
York, Free Press, 1971.
Struever, S., 1971 (ed.), 'Prehistoric Agriculture', New York,
Natural History Press.
Struve, V., 1965, The concept of the Asian mode of production,
'Soviet Anthropology and Archeology', 4, no. 2, pp. 41-6.
Stuchevski, I. and Vasilev, L., 1966, Three models for the
origin and evolution of precapitalist societies, 'Voprosy
istorii', 5, pp. 77-90 (published in 'Soviet Review', 1967,
6, pp. 26-39).
Suret-Canale, J., 1966, The traditional societies of tropical
Africa (extracts from 'La Pensée', 117), 'Marxism Today',
Feb. 1966, pp. 49-57 (original: 'La Pensée', 117, September-
October 1964, pp. 21-42).
Suret-Canale, J., 1968, A propos du MPA, 'La Pensée', 142,
pp. 89-91.
Ter-Akopyan, N.B., 1965, Razvitie vzgliadov K. Marxa i
F. Engelsa na aziatski sposobo proizvodstva i zemledeltches-
kouyou obchtchinou (The development of K. Marx's and
F. Engels's opinions on the Asiatic mode of production and
the village community), 'Narody Azii i Afriki', 2, pp. 74-88;
3, pp. 70-85.
Terray, E., 1969, 'Le Marxisme devant les sociétés "primitives"',
Paris, Maspéro.
Terray, E., 1969, 'Marxism and "Primitive" Societies', New
York, Monthly Review Press, 1972.
Terray, E., 1975, Class and class consciousness in the Abron
Kingdom of Gyamon, in Bloch, M. (ed.), 'Marxist Analyses
and Local Anthropology', London, Malaby Press.
Thechkov, M.A., 1969, La classe dirigeante du Vietnam pré-
coloniale, 'La Pensée', 144, pp. 28-40.
Thorner, D., 1966, Marx on India and the Asiatic mode of
production, 'Contributions to Indian Sociology', IX, pp. 33-66.
Timasheff, N.S., 1948, The sociological theories of Maxim
M. Kovalevsky, in Barnes, H.E. (ed.), 'An Introduction
to the History of Sociology', University of Chicago Press,
pp. 441-57.
Tökei, F., 1964, Le MPA dans l'oeuvre de K. Marx et F. Engels,
'La Pensée', 114, pp. 7-32.
Tökei, F., 1958, Les conditions de la propriété foncière dans
la Chine de l'époque Tchou, 'Acta Antiqua Acad. Sci. Hung.',
6, nos 3-4, pp. 245-99.
Tökei, F., 1972, Contribution à la nouvelle discussion sur la
mode de production asiatique, 'Nouvelles Études Hongroises',
pp. 80-95.
Tökei, F., 1975, 'Az ázsiai termelèsi mód kérdésehez',
Budapest, Kossuth.
Topfer, B., 1967, Zur Problematik der vorkapitalistischen
Klassengesellschaften, 'Jahrbuch für Wirtschaftsgeschichte',
IV, pp. 259-86.

Tosi, M., 1976, The dialectics of state formation in Mesopotamia,
Iran and Central Asia, 'Dialectical Anthropology', 1, pp.
173-80.

Trotsky, L., 1909, '1905', London, Allen Lane, 1972.

Trotsky, L., 1929, 'Problems of the Chinese Revolution',
University of Michigan Press, 1967.

Trotsky, L., 1932, 'History of the Russian Revolution',
3 vols (1930), London, Gollancz.

Turner, B.S., 1979, 'Marx and the End of Orientalism',
London, Allen & Unwin.

Ulmen, G.L., 1975, Wittfogel's science of society, 'Telos',
summer 1975, 24, pp. 81-114.

Ulmen, G.L., 1979, 'The Science of Society: Toward an Understanding of the Life and Work of Karl August Wittfogel',
The Hague, Mouton.

Varga, E., 1925, La situation économique en Chine, 'La
Correspondance Internationale', 122, 16 December, pp.
1035-7.

Varga, E., 1928, Les problèmes foundamentaux de la révolution chinoise, 'La Correspondance Internationale', 561,
16 June.

Varga, E., 1964, 'Problems of the Political Economy of Capitalism', Moscow.

Venturi, F., 1963, Oriental despotism, 'Journal of the History
of Ideas', 27, pp. 133-42.

Venturi, F., 1966, 'Roots of Revolution: a History of the
Populist and Socialist Movements in Nineteenth-Century
Russia', New York, Grosset & Dunlap.

Vidal-Naquet, P., 1964, Review of 'Oriental Despotism',
'Annales', 19, pp. 531-49.

Veit, Tran Thanh, 1974, Sur quelques obstacles épisté-
mologiques à l'approche du MPA, 'L'Homme et la Société',
33-4, pp. 125-31.

Vitkin, M.A., 1972, 'Vostik v filosfsko - istoricheska Koncepii
K. Marksa i F. Engel'sa', Moscow, Nauka.

Walicki, A., 1969, 'The Controversy over Capitalism',
Clarendon Press.

Wallerstein, I., 1974, 'The Modern World System: Capitalist
Agriculture and the Origins of the European World-Economy
in the Sixteenth Century', New York, Academic Press.

Welskopf, E., 1967, Du rôle des rapports de production dans
l'évolution historique, 'Recherches Internationales', 57-8,
pp. 324-36.

Wendel, H., 1925, Review of K. Wittfogel's 'Geschichte der
bürgerlichen Gesellschaft, Die Gesellschaft', 2, nos 1-6,
pp. 172-7.

Wielenga, B., 1976, 'Marxist Views on India in Historical
Perspective', Studies in Indian Marxism, no. 2, Madras,
Christian Literature Society.

Winzeler, R.L., 1976, Ecology, culture, social organization
and state formation in Southeast Asia, 'Current Anthropology',

17, 4, pp. 623-40 (including comments by other scholars).
Wittfogel, K.A., 1922, 'Vom Urkommunismus bis zur proletarischen Revolution', Part I: 'Urkommunismus und Feudalismus', Berlin.
Wittfogel, K.A., 1922a, 'Die Wissenschaft der bürgerlichen Gesellschaft: Eine Marxistische Untersuchung', Kleine Revolutionäre Bibliothek, vol. 8, Berlin, Malik.
Wittfogel, K.A., 1924, Geschichte der bürgerlichen Gesellschaft', Vienna.
Wittfogel, K.A., 1926, 'Das erwachende China: ein Abriss der Geschichte der gegenwärtigen Probleme Chinas', Vienna.
Wittfogel, K.A., 1927, Probleme der chinesischen Wirtschaftgeschichte, 'Archiv für Sozialwissenschaft und Sozialpolitik', LVIII, 2, pp. 289-335.
Wittfogel, K.A., 1929, Geopolitik, geographischer Materialismus und Marxismus, 'Unter dem Banner des Marxismus', Part I in vol. 3, 1, pp. 17-51; Part II in vol. 3, 4, pp. 485-522; Part III in vol. 3, 5, pp. 698-735.
Wittfogel, K.A., 1929a, Voraussetzungen und Grundelemente der chinesischen Landwirtschaft, 'Archiv für Sozialwissenschaft und Sozialpolitik', LXI, pp. 566-607.
Wittfogel, K.A., 1930, 'Die Ökonomische Bedeutung der agrikolen und industriellen Produktivkräfte Chinas'.
Wittfogel, K.A., 1931, 'Wirtschaft und Gesellschaft Chinas', Part I: 'Produktivkräfte, Produktions und Zirkulationsprozess', Leipzig.
Wittfogel, K.A., 1931a, Hegel über China, 'Unter dem Banner des Marxismus', vol. 5, 3, pp. 346-62.
Wittfogel, K.A., 1932, Die natürlischen Ursachen der Wirtschaftsgeschichte, 'Archiv für Sozialwissenschaft und Sozialpolitik', 67, 4, pp. 466-91; 67, 5, pp. 579-609; 67, 6, pp. 711-31.
Wittfogel, K.A, 1936, Wirtschaftsgeschichtliche Grundlagen der Entwicklung der Familienautorität, 'Studien über Autorität und Familie, Schriften des Instituts für Sozialforschung', Paris.
Wittfogel, K.A., 1938, Die Theorie der orientalischen Gesellschaft, 'Zeitschrift für Sozialforschung', Paris, 7, nos 1-2.
Wittfogel, K.A, 1938a, 'New Light on Chinese Society: an Investigation of China's Socio-economic Structure', Institute of Pacific Relations.
Wittfogel, K.A, 1939-40, The society of prehistoric China, 'Studies in Philosophy and Social Science', Institute of Social Research, 8, pp. 138-86.
Wittfogel, K.A., 1940a, Metereological records from the divination inscriptions of Shang, 'Geographical Review', XXX, pp. 110-33.
Wittfogel, K.A., 1946, Chinese society and the dynasties of conquest, in MacNair, H.F. (ed.), 'China', University of California Press, pp. 112-26.
Wittfogel, K.A., 1947, Public office in the Liao dynasty and the Chinese examination system, 'Harvard Journal of Asiatic Studies', X, pp. 13-40.

Wittfogel, K.A., 1950, Russia and Asia, 'World Politics', 2, 4, pp. 445-62.
Wittfogel, K.A, 1951, The influence of Leninism-Stalinism on China, 'Annals of the American Academy of Political Science'.
Wittfogel, K.A., 1953, Oriental despotism, 'Sociologus', 3, 4.
Wittfogel, K.A., 1953a, The ruling bureaucracy of oriental despotism; a phenomenon that paralysed Marx, 'Review of Politics', 15, pp. 350-9.
Wittfogel, K.A., 1955, Oriental society in transition, 'Far Eastern Quarterly', XIV, 4.
Wittfogel, K.A., 1955a, Developmental aspects of hydraulic societies, in Steward, J.H. (ed.), 'Irrigation Civilizations: a Comparative Study', Social Science Monographs, 1, Pan-American Union, Washington D.C.
Wittfogel, K.A., 1955b, 'Mao Tse-Tung Liberator or Destroyer of the Chinese Peasants?', New York, Free Trade Union Committee, A.F. of L.
Wittfogel, K.A., 1956, The hydraulic civilizations, in Thomas, William L. (ed.), 'Man's Role in Changing the Face of the Earth', Chicago, Wenner-gren Foundation, pp. 152-64.
Wittfogel, K.A., 1957, 'Oriental Despotism: a Comparative Study of Total Power', Yale University Press, 1964.
Wittfogel, K.A., 1957a, Chinese society; an historical survey, 'Journal of Asian Studies', XVI, 3.
Wittfogel, K.A., 1958, Class structure and total power in Oriental despotism, 'Contemporary China'.
Wittfogel, K.A., 1960, A stronger oriental despotism, 'China Quarterly', 1, pp. 29-34.
Wittfogel, K.A., 1960a, The Marxist view of Russian society and revolution, 'World Politics', 12, pp. 487-508.
Wittfogel, K.A., 1961, The Russian and Chinese revolutions: a socio-historical comparison, 'Year Book of World Affairs', 15.
Wittfogel, K.A., 1962, The Marxist view of China, 'China Quarterly', 11 (July-Sept.), pp. 1-20; 12 (October-December), pp. 154-69.
Wittfogel, K.A., 1962a, Agrarian problems and the Moscow-Peking axis, 'Slavic Review', XXI, 4, pp. 678-98.
Wittfogel, K.A., 1963, Russia and the East: a comparison and contrast, 'Slavic Review', XXII, 4.
Wittfogel, K.A., 1963a, Some remarks on Mao's handling of concepts and problems of dialectics, 'Studies in Soviet Thought', III, 4, pp. 251-69.
Wittfogel, K.A., 1967, Review of R.N. Adams's 'The Evolution of Urban society', 'American Anthropologist', 69, pp. 90-2.
Wittfogel, K.A., 1969, Results and problems of the study of oriental despotism, 'Journal of Asian Studies', 28, pp. 257-65.
Wittfogel, K.A., 1970, 'Agriculture: The Key to the Under-standing of Chinese Society Past and Present', Australian National University Press.
Wittfogel, K.A., 1971, Irrigation policy in Spain and Spanish America, in Desai, A.R. (ed.), 'Essays on the Modernization

of Underdeveloped Societies', 2 vols, University of Bombay.
Wittfogel, K.A., 1972, The hydraulic approach to Pre-Spanish Mesoamerica, in Johnson, F. (ed.), 'Chronology and Irrigation', vol. 4, of 'The Prehistory of the Tehuacan Valley', University of Texas Press.
Wittfogel, K., 1977, Nouvelle Préface, 'Le Despotisme oriental', Paris, Editions de Minuit.
Wittfogel, K. and Feng Chia-Sheng, 1949, 'History of Chinese Society', Liao Transactions, American Philosophical Society, Philadelphia, vol. XXXVI.
Wittfogel, K. and Goldfrank, E., 1943, Some aspects of pueblo mythology and society, 'Journal of American Folklore', January-March.
Wittfogel, K. and Schwartz, B., 1960, The Legend of 'Maoism', 'China Quarterly', 1 (January-March), pp. 72-86; 2 (April-June), pp. 16-42.
Wolf, B.D., 1957, Review of 'Oriental Despotism', 'Saturday Review', 8 June, pp. 19-20.
Wolf, E.R. and Palerm, A., 1955, Irrigation in the Old Acolhua Domain, Mexico, 'Southwestern Journal of Anthropology', 11, pp. 265-81.
Woodbury, R.B., 1961, A reappraisal of Hokokam irrigation, 'American Anthropologist', 63, pp. 550-60.
Yaranda Valderrama, A., n.d. Formation précapitaliste dans la civilisation andine, MS.
Yerasimos, S., n.d. Le mode de production asiatique et la société ottomane, MS.
Zakavkazskoe Otdelenie Obschestva Istorikov Marksistov Prikomakademi CIK SSR, 1930, 'Ob aziatskom sposobe proizvodstva' (On the Asiatic mode of production), Zakkniga Publishers.

INDEX